Nils Ellebrecht, Tino Plümecke, Isabelle Bartram, Veronika Lipphardt, Jenny Reardon, Andrea zur Nieden (eds.)
The Order of People

AF839242

Science Studies

Nils Ellebrecht works as a sociologist at the University of Freiburg and the University Hospital Freiburg in Germany. His work focuses on the sociology of organization, technology, security, and medicine.

Tino Plümecke is a sociologist at the University of Freiburg, Germany. His work centers on science and technology studies, theories of discrimination, and critical race studies with a particular emphasis on the history and contemporary developments in genetics and (post)genomics.

Isabelle Bartram is a molecular biologist and works at the Institute of Sociology at the University of Freiburg, Germany. She is interested in interdisciplinary perspectives on biomedical research and questions surrounding responsible use and governance of human genetics.

Veronika Lipphardt is a professor of science and technology studies at the University College Freiburg, Germany. She has worked on the history of the life sciences in the 20th century, particularly on the history of physical anthropology and human population genetics in their political, social, and cultural contexts. Recently, her research has focused on forensic DNA analysis and population genetic studies of vulnerable populations.

Jenny Reardon is a professor of sociology and the founding director of the Science and Justice Research Center at the University of California, Santa Cruz. Her research draws into focus questions about identity, justice, and democracy that are often silently embedded in scientific ideas and practices, particularly in modern genomic research. She has been the recipient of fellowships and awards from, among others, the National Science Foundation, the Max Planck Institute, the Humboldt Foundation, and the Freiburg Institute for Advanced Studies.

Andrea zur Nieden is a sociologist at the University of Freiburg, Germany. Her research operates in the field of science and technology studies resp. sociology of technology, medicine and science, with a recent focus on human diversity in the new life sciences as part of the research group SoSciBio. She has also worked on the sociology and history of psychiatry.

Nils Ellebrecht, Tino Plümecke, Isabelle Bartram, Veronika Lipphardt,
Jenny Reardon, Andrea zur Nieden (eds.)

The Order of People

Contesting Bio-Scientific Human Classifications

[transcript]

The research group 'Human Diversity in the New Life Sciences: Social and Scientific Effects of Biological Differentiations (SoSciBio)' and this publication were funded by the German Federal Ministry of Education and Research (BMBF) under grant 01GP1790.

Federal Ministry
of Education
and Research

Bibliographic information published by the Deutsche Nationalbibliothek
The Deutsche Nationalbibliothek lists this publication in the Deutsche Nationalbibliografie; detailed bibliographic data are available online at https://dnb.dnb.de

This work is licensed under the Creative Commons License BY 4.0. For the full license terms, please visit the URL https://creativecommons.org/licenses/by/4.0/.
Creative Commons license terms for re-use do not apply to any content (such as graphs, figures, photos, excerpts, etc.) not original to the Open Access publication and further permission may be required from the rights holder. The obligation to research and clear permission lies solely with the party re-using the material.

2026 © Nils Ellebrecht, Tino Plümecke, Isabelle Bartram, Veronika Lipphardt, Jenny Reardon, Andrea zur Nieden (eds.)

transcript Verlag | Hermannstraße 26 | D-33602 Bielefeld | live@transcript-verlag.de

Cover design: Kordula Röckenhaus
Printing: Elanders Waiblingen GmbH, Waiblingen
https://doi.org/10.14361/9783839472378
Print-ISBN: 978-3-8376-7237-4 | PDF-ISBN: 978-3-8394-7237-8
ISSN of series: 2703-1543 | eISSN of series: 2703-1551

Printed on permanent acid-free text paper.

Contents

Appendix

Acknowledgments

SO SCI BIO
Research Group

This volume is the result of a fruitful collaboration between the SoSciBio research group (short for Human Diversity in the New Life Sciences: Social and Scientific Effects of Biological Differentiations) and Jenny Reardon (University of California, Santa Cruz), as well as Veronika Lipphardt (University College Freiburg). Comprising Isabelle Bartram, Nils Ellebrecht, Andrea zur Nieden, Tino Plümecke, Laura Schnieder, Isabel Schön, and the research assistants Felix Fink, Lukas Geisthardt, Levin Nikolai, Christoph Panzer, Hannah Schönberger, Charlotte Schulze-Marmeling, Paula Steinbrenner, Annika Troitzsch, Dominik Weber, and Jan Weber, SoSciBio was funded by the German Federal Ministry of Education and Research (BMBF, grant number 01GP1790) at the Institute of Sociology at the University of Freiburg, Germany. The scientific exchanges and networks that emerged through international workshops and panels culminated in this selection of articles.

We extend our heartfelt gratitude to the distinguished reviewers of the chapters in this volume, whose critical insights, thoughtful feedback, and analytically engaging commentaries have significantly enhanced the quality of the contributions. We are especially grateful to Mita Banerjee, Susanne Bauer, Catherine Bliss, Mathias Bös, Cornelia Brink, Elise Burton, Simon Cole, Inga Diederich, Joan Fujimura, Michael Guggenheim, Sibille Merz, Marion Müller, Aaron Panofsky, Ramya Rajagopalan, Katharina Schramm, Andrew Smart, Victor Toom, Sandra Widmer, Matthias Wienroth, and Henri-Michel Yéré for their careful readings, constructive critiques, and valuable suggestions, which have enriched each chapter and greatly contributed to the coherence of the volume as a whole. Special thanks go to Michael Thomas Taylor, who not only edited all the texts for clarity and correctness but also made many suggestions for improving the argumentation.

The Order of People and the Promise of Post-Racial Classifications

Tino Plümecke, Jenny Reardon, Nils Ellebrecht, Andrea zur Nieden, Veronika Lipphardt, Isabelle Bartram

The Nature of Human Classifications in Socio-Political Entanglements

To classify is a fundamental human practice. Systems of classification organize perceived similarities and differences in order to reduce complexity, identify patterns, and recognize relationships. While many who construct these systems argue that they are neutral tools of knowledge and governance, ample literature in the social sciences demonstrates that they are sites of power (Harding 1991; Bowker and Star 2000; Hacking 2002). Perhaps nowhere are these entanglements of systems of classification with matters of power more starkly revealed than in the life sciences. The life sciences constitute authoritative knowledge regimes whose classification schemes routinely exceed the boundaries of laboratory practice. These systems differentiate people using terms such as "race," "ethnicity," or "ancestry"—powerful concepts that shape public discourse, inform healthcare policy, and structure modes of governance (Roberts 2011; Nelson 2016; Reardon 2017). Through their adoption by commercial genetic ancestry companies, law enforcement agencies, and other institutional actors, these systems gain considerable economic and political weight. They shape understandings of how human beings are positioned within various orders, orders that determine who should have access to resources, rights, and recognition (Lipphardt et al. 2021; Bartram, Plümecke, Schultz 2022; M'charek and van Oorschot 2023).

Because of their entanglement with these consequential issues, these life science systems of classification have long been contested and criticized. Rather than treating them as stable or objective ordering devices, many have sought to expose and describe how they emerge through historically situated processes of negotiation, positioning, and epistemic struggle. Charles Darwin, for example, denounced their precarious and constructed quality in *The Descent of Man* (1874), arguing specifically against the excessive proliferation of race classifications among his contemporaries: "Man has been studied more carefully than any other animal, and yet there is the greatest possible diversity amongst capable judges whether he should be classed as a single species or race, or as two (Virey), as three (Jacquinot), as four (Kant), five (Blumenbach), six (Buffon), seven (Hunter), eight (Agassiz), eleven (Pickering), fifteen (Bory St. Vincent), sixteen (Desmoulins), twenty-two (Morton), sixty (Crawfurd), or as sixty-three, according to Burke" (Darwin 1874, 174). For Darwin, this taxonomic chaos did not signal a lack of careful work, but rather the inherent instability of attempts to fix human difference through classificatory reason.

Social theorists in the twentieth and twenty-first centuries increasingly took up the question of whether there are underlying social logics that generate and sustain this instability. One of the most influential interventions came from the French theorist Michel Foucault, who—beginning in the late 1960s—sought to systematically unearth the rules and logics that shape classification schemes in what he called the *sciences humaines*. In *The Order of Things*, he famously asked: When "we say that a cat and a dog resemble each other less than two greyhounds do, what is the ground on which we are able to establish the validity of this classification with complete certainty?" What is the "grid of identities, similitudes, analogies" that enables the creation and ordering of things? (Foucault 2005/1966, xxi).

For Foucault, such grids are not neutral schemes of representation of an underlying reality, but instead are the product of implicit and largely invisible grids of rules, norms, and assumptions that determine what can be thought and said, and what counts as knowledge in a particular historical era. He labeled these grids *epistemes*. According to Foucault, epistemic structures are unique to eras. When one episteme gives way to another, this moment of discontinuity is characterized by an epistemological rupture in which the foundational grids, roles, and norms are restructured. As a result, conceptual vocabularies, modes of inquiry, and even entire research areas and disciplines may become obsolete, while others emerge and gain legitimacy. For Foucault, the shift to the modern episteme laid the groundwork for the emergence of new disciplines such as biology and economics that rendered "the human" an object of study. The result was the epistemologization of "the human" (Foucault 1972).

Foucault's work thus repositioned the human sciences, unsettling them from their position as linear extensions of progress and cumulative knowledge. Rather than revealing pregiven, objective truths, he showed that their classification systems act as instruments of ordering that acquire legitimacy through the very regimes of knowledge they help to constitute.

The consequences of such classificatory regimes—particularly in the life sciences—have been far from benign. Rooted in Enlightenment ideals of systematization, order, and universal reason, early efforts to classify human beings were shaped by the assumption that social and moral differences could be grounded in natural law. Philosophers such as Montesquieu and Kant contributed to this transformation by linking human difference to environmental and physiological explanations, thereby laying the conceptual groundwork for biologically anchored forms of classification and differentiation. The success of this naturalizing classification system lay in its ability to draw upon purported biological knowledge of human differences to justify the social order. As theorists Michael Omi and Howard Winant have pointed out, this system created a conceptual order "that signifies and symbolizes social conflicts and interests by referring to different types of human bodies" (Omi & Winant 2015/1986, 110). These classifications based on "nature" also helped to consolidate further practices of stratification, such as those based on gender, sexuality, class, or illness/disability. All these newly developed scientific differentiations sought not only to bring order to the "nature of society," but also to justify social inequalities created by colonial conquests and the unequal distribution of social goods generated by the gendered division of labor and capitalist relations of production.

These differentiating taxonomies were further elaborated and institutionalized by naturalists, physicians, ethnologists, and anthropologists such as Blumenbach, Linné, Bernier, Meiners, which helped give "race" the status of a core epistemic and structural principle for organizing societies (Mosse 2020). By the first half of the twentieth century, state authorities in multiple nations deployed these taxonomies that ranked humans into inferior and superior races to rationalize exclusionary policies (Proctor 1988; Stern 2016).

In response to the atrocities committed by the Nazis under the banner of racial science, from the mid-1930s a growing number of voices called for the removal of "race" as a concept within the life sciences (Barkan 1992). While this argument is now associated with social scientists and humanists, the argument that race should not be used in biology was put forward by biologists as early as 1935. In that year, Julian Huxley and AC Haddon published *We Europeans: a Survey of "Racial" Problems*, in which they argued that the concept of "race" had been corrupted by its use in society and should be replaced by "ethnic group" (Huxley and Haddon 1935). Social scientists and humanists drew upon these arguments by biologists to argue that race was not a natural reality, but a social construct (Gilroy 2000; Smadley and Smadley 2005).

The Limits of Social Constructivism

Today, this view that race and its derivatives are social constructs has become a central pillar of a powerful epistemic framework that shapes how many people—particularly in the Euro-American West—interpret and make sense of the world. For many theorists, this shift is the result of the kind of epistemological shock that Foucault theorized (Barkan 1992). Yet, whatever its radical and disruptive effect was at the time of its first emergence, we argue that the subsequent uptake of the idea that race is socially constructed has led to the loss of the critical edge of Foucault's genealogical analysis. Too often today the argument is accompanied by a rote acceptance of a flattened, depoliticized assertion of constructivism (Hacking 1999). Natural and social scientists have all too easily and happily acceded to the argument that race is social-political construct and should not be used anymore as a bioscientific category. Perhaps the ease of their acceptance was the result of the productivity of the argument. For natural scientists, arguments about social constructionism have facilitated a distancing of their science from the legacy of scientific racism. Scientific racism, they often argue, was characterized by deterministic typological thought, and not the dynamic perspectives they see in their population-based approaches (Haraway 1989; Gannett 2001). For social scientists, this distancing has buttressed the legitimacy of their work, with the argument that their studies have deconstructed a concept responsible for the loss of millions of lives.

Both camps converged in celebrating the rejection of "race" as a scientific category. The historian of science Elazar Barkan (1992) went so far as to argue in his influential book, *The Retreat of Scientific Racism*, that this rejection of race represented a revolutionary shift ushering in a new scientific paradigm, one in which a dynamic paradigm based on the notion of populations replaced static tradition of thought based on race. Over the course of the 1990s, a growing number of population geneticists and cultural theorists credited this Kuhnian scientific revolution to the triumph of truth over ideology and the demonstration by natural and social scientists alike that race did not map onto reality (Gilroy 2000, Cavalli-Sforza 2001).

Yet this powerful narrative of the social construction of race created a critical lacuna which this volume seeks to address. As research over the past two decades has begun to elucidate, life scientists have continued to mobilize racializing concepts, at times even using the vocabulary of the old systems of racial classification—even when they reject race as a biologically meaningful term (Gannett 2001; Duster 2015). The continued reliance on such concepts despite the widespread adoption of stances that reject race suggests that we may be approaching the limits of constructivist critique. Though social and life scientists reject race as a concept in theory and in recurring proclamations and statements, it remains alive in mundane practices, routines, research designs, and interpretations. The once radical gesture of revealing the socially constructed nature of race and the employment of deconstructivism today risks losing its critical force.

This observation, we argue, should prompt a reconsideration of established analytical strategies. Instead of further refining critiques within the same epistemological framework, what we need is a more profound shift—a new epistemological rupture. Such a rupture would not stop at deconstruction, but would instead direct attention toward the durability, materiality, and institutional embeddedness of racial classifications. It would shift the analytical gaze toward the mechanisms through which these constructs continue to shape lived experiences, reproduce structural inequalities, and persist—even under the banner of emancipatory, egalitarian, or ostensibly postracial commitments.

Postracial Racialization

This volume invites this shift. In so doing, it builds on the work of scholars inspired by Stuart Hall's call to attend to the material, historical reinventions of race and racism (Hall 1986), and especially those that grow out of the very denial of the relevance of race and racialization (Goldberg 2009; Morning and Maneri 2022; Hawthorne 2024).

The articles in this volume pay particular attention to the powerful role that technoscientific realms continue to play in these reinventions of race and the new forms of racism they animate. Authors provide examples of how to critically engage with the continued production and reproduction of race and racialization in these realms—despite the prevailing narrative that scientists moved beyond race at the mid-point of the twentieth century. In so doing, they shed light on what we refer to as the phenomenon of *postracial racialization*.[1] Despite criticisms of the concept of racialization, particularly the tendency to depoliticize the term and obscure the intertwined processes and diverse,

1 This formulation is inspired by the "Theorizing Race after Race" group organized by Jenny Reardon and Camilla Hawthorne at the University of California, Santa Cruz. See https://scijust.ucsc.edu/2018/11/27/theorizing-race-after-race/.

overlapping dynamics that shape race and racism (e.g., Goldberg and Essed 2002), we maintain its analytical value for this volume as it highlights the situated, processual construction of human classifications without reifying race itself (Rattansi 2007; M'charek et al. 2014). This perspective enables nuanced engagement with the practices of sorting, of ascribing similarity and difference, of boundary work, and of defining human groups in contemporary life science research. It is from this vantage point that this volume takes its point of departure.

Specifically, we employ the term racialization not to signal a return of the discredited concept of "race," but to denote a problem space: one in which contested practices, categories, and epistemologies *bio-logicalize* social differences in a manner that historically adheres to the problem of race. Although new terminologies and classificatory tactics have emerged to articulate human classification in the context of what is often described as a "postracial" era, the core issues remain unresolved. Despite the promise made by many genomicists to move beyond race (Fujimura and Rajagopalan 2011)—and indeed all human group categories—over the past two decades, its use of classification systems has intensified. Instead of moving away from the notion of distinct human groupings, scientists in various fields continue to rely on such frameworks to make sense of genetic variation. In archaeogenetics, for example, researchers frequently organize genomic data according to broad continental or national labels in efforts to reconstruct historical patterns of human migration (Reich 2018). Similarly, forensic scientists commonly categorize DNA samples using typologies based on geographically, racially, or ethnically defined reference populations (M'charek, Toom, and Jong 2020; Bartram, Plümecke, and Schultz 2022). In such instances, scientific procedures structure genetic diversity into predefined population groupings. The central question is what kind of tactics and classificatory practices enable these systems that are deployed to reject race, even as they draw on some of the very same epistemic and structural foundations that structure racial thinking.

Unearthing Infrastructural Logics

Postracial racialization is not a universal phenomenon that manifests and operates in the same way around the globe. Rather, it takes shape through locally and historically specific configurations that vary across national, institutional, and disciplinary contexts. This volume is the first to provide a *global transdisciplinary analysis* of how postracial racialization prevails across the life sciences. Highlighting cases from diverse countries around the globe and across multiple scientific disciplines, we illuminate the tactics and techniques through which human biological differentiations are maintained, even when racial logics are being denied. Building on our work as a research group based in Germany (see acknowledgments) that investigates the social and scientific effects of biological differentiations, we have broadened our scope in this volume, to countries outside the dominant research focus on the United States.

Unlike much existing work in the social sciences, which focuses primarily on race in society or policy, this volume turns the spotlight back on the sciences themselves. Crucially, we do not treat "sciences" as a monolith. Instead, we take the life sciences to comprise a heterogeneous set of practices, actors, and institutions, each with its own epistemic cultures, methodological conventions, and political entanglements. Accordingly, the contributions in this volume examine how postracial racialization manifests differently, and sometimes similarly, across disciplinary, national, and institutional sites. The contributions in this volume, therefore, consider human classifications conceptually and analytically as practices that are deeply embedded in broader symbolic, organizational, and political frameworks, in everyday life, in science, and in state infrastructures. These classificatory practices raise several pressing analytical questions: What is the specific utility of classifications of people into a manageable set of groups? What criteria are used to sort people into groups? How are classifications technically constructed, stabilized, and legitimized in scientific infrastructures? And to what extent do they not merely represent, but actively produce and govern classifications of human diversity?

While previous scholarship has critically examined the persistence of racial logics within structures of neoliberal governance and narratives of postracial ideology (e.g., Goldberg 2015; Hawthorne 2022), this volume turns its attention to the situated practices through which postracial racializations are enacted in the life sciences. It examines how forensic scientists attempt to identify suspects, how population geneticists reconstruct ancestral histories, and how biomedical researchers assign group labels to develop targeted medical interventions. Across these fields, postracial racialized classifications operate not as a claim about an explicit biological essence but as an infrastructural logic: they are embedded in datasets, encoded in algorithms, and mobilized through technoscientific reasoning. What unfolds is not the simple return of race, but its transformation—muted and reframed, and yet no less impactful in its social and political effects.

Following the traces of postracial racialization across diverse global contexts also reveals how the discursive and political framings of classifications of human diversity have changed in recent decades. Whereas nineteenth and early twentieth century racial and ethnic classifications often served to legitimate exclusion, oppression, and the legitimization of social hierarchies, today they are increasingly framed as tools for antidiscrimination, inclusion, and social justice initiatives (Epstein 2007; Benjamin 2009; Bliss 2012; Reardon 2012). This transformation, however, is neither linear nor universally embraced. In countries such as Germany, official discourse increasingly seeks to eliminate the use of the term "race" altogether. France, the Netherlands, and Poland refrain from including racial or ethnic categories in censuses, population registers, and official surveys. In countries such as the United States and Brazil, by contrast, racial and ethnic classifications remain central to census-taking, public policy, and biomedical research, and are framed as essential instruments for monitoring inequalities and implementing equity-driven interventions. These divergent strategies reflect not only distinct historical trajectories of ideas and practices of race and racism, but also competing epistemologies regarding how to recognize, measure, and ultimately address social justice.

Contemporary Modalities of Continuing Human Classifications

The contributions assembled in this volume investigate how postracial racializations are enacted, under what conditions they take shape, and through which tactics and logics they are sustained across this wide range of scientific and sociopolitical domains. While each chapter engages with a distinct empirical field, they all share a concern with the continuities, ruptures, and emerging reconfigurations that characterize contemporary practices of human classification. A particular focus lies on the modalities through which race is formally displaced—substituted by adjacent terms such as ethnicity, ancestry, or population—without necessarily abandoning the classificatory impulses and structures associated with it.

Although the chapters approach the dynamics of postracial racialization from diverse disciplinary and methodological vantage points, four interrelated dimensions of human classification in the life sciences can be identified across the volume. These dimensions, outlined below, illuminate how actors in the life sciences navigate the enduring problems of race and respond to critiques of racialization, while simultaneously continuing to register, differentiate, and classify human biological variation.

Beyond Categories and Back Again

New methods—particularly those employed in genetics—seek to move beyond the problem of categorization. In contrast to research approaches that rely on sharply delineated categories, these approaches aim to capture human variation exclusively in terms of gradients (clines), proportions (admixture rates), or clusters (as in genome-wide association studies). Because this approach to variability largely avoids the use of racial typologies, it has come to be associated with the promise of a scientifically grounded postracial order. This promise is further reinforced by the argument that methods which do not rely on predefined categorical distinctions—such as racial classifications—are better aligned with genetic reality, since biological variation does not consist of clearly bounded genetic groups but of continuous frequencies of variation. Advocates of this approach also argue that thinking of differences only in terms of degrees protects bioscientific knowledge from political co-optation. Their hope is that if human differences can be understood fundamentally as transitional, proportional, or clustered, it will be possible to permanently eliminate the idea that genetically defined human groups exist.

Yet this tactic has not succeeded in transcending categories. As the contributions in this section show, the methods employed fail to deliver on their promise to move beyond categorical thinking. Human differentiations, even when initially operationalized through gradients or clusters, remain susceptible to being retranslated into categories. Often it is the scientists themselves who revert to categorization—for reasons of clarity, communication, simplicity, or a connection to what they take to be common sense. Perhaps one of the most prominent examples of this dynamic is the *1000 Genomes Project*, which in its final publications ultimately distinguishes five geographically localized major groups (Auton et al. 2015).

To address this phenomenon, contributions in this section approach classification as a process of meaning-making—one that produces social and epistemic effects. By analyzing the specific decisions, emphases, and omissions within scientific practices, the chapters reveal how certain distinctions are foregrounded and rendered significant, while others are marginalized or deemed irrelevant. This gives rise to fundamental questions: What drives particular forms of human variation to be marked as meaningful differences, while others remain unrecognized? Why are continuous forms of variation repeatedly collapsed into a limited number of discrete categories? And what are the implications of classificatory systems that present themselves as neutral or technical, yet ultimately reproduce familiar patterns of group differentiation?

These questions are taken up concretely in the chapter by Filipa Queirós and Rafaela Granja, who explore how the social function and political charge of classificatory distinctions materialize in the field of forensic genetics. Focusing on the visual representation of suspects, they analyze how extended DNA analysis is used to construct collective images of suspect populations in two distinct cases—an artistic exhibition and a criminal investigation in Canada. Through detailed case studies, they show that while DNA phenotyping is marketed as a cutting-edge technique capable of extracting fine-grained individual traits, in practice it often yields generic, racialized representations. Rather than producing individualized likenesses, the technology tends to reinforce typological, group-based visual cues. Their analysis reveals how such visualizations not only raise concerns of scientific validity but also carry serious social consequences, particularly by reifying stereotypes and contributing to the criminalization of entire communities.

A further in-depth insight into the practices of contemporary DNA analysis is provided by the chapter by Thiago Pinto Barbosa. Focusing on the creation and dissemination of the ancestry categories "Ancestral North Indian" and "Ancestral South Indian" in population genetics, Barbosa analyzes how these terms emerged as seemingly neutral replacements for the racially charged Aryan/Dravidian binary. Situated within the broader context of nationalist politics and caste hierarchies in India, the article traces how bioscientific classifications are crafted through rhetorical and methodological maneuvers that respond to, and simultaneously shape, political sensitivities. While appearing to avoid the overt racialization of older classificatory schemes, the new terminology effectively reconfigures older categories of difference within a nationalized idiom that aligns with dominant Hindu nationalist narratives of unity and indigeneity. The article demonstrates that the act of "making up" new categories is an unstable, contested, and historically layered instance of classification in contemporary genomics.

As the next chapter shows, such classificatory practices do not remain confined to their original domains but travel across scientific disciplines, becoming embedded in new epistemic frameworks and adapted to categories such as "population" or "ancestry." In these new contexts, they are reinterpreted, infused with new meanings, and often give rise to alternative—but similar reductive—narratives. Robert Meunier's chapter examines how typologies of modes of subsistence, originating from hunter-gatherer studies in fields such as archaeology and anthropology, are adopted in microbiome research. He analyses how the transfer of this classificatory model into evolutionary and biomedical contexts entails significant political, ethical, and epistemological implications. In particular, he shows how its application can reinforce essentialized narratives of difference, often along racialized lines. His analysis points to the risks of importing typologies into new contexts without critical scrutiny and demonstrates how classificatory practices can reinforce hierarchies even when cloaked in the language of scientific objectivity.

Taken together, the contributions in this section demonstrate that efforts to move beyond racial categories through postcategorical genetic methodologies often end up reproducing the very logic they seek to overcome. Despite claims of neutrality and precision, such approaches continue to rely on categorical group-based distinctions. As the following section will show, these constraints not only sustain older categories but also stimulate their transformation and rearticulation.

New Categroies but Old Meanings

The second section of this volume turns to a further tactic of postracial classification: the attempt to replace "race" with other terms, which in turn entails certain classification practices. In many societies—particularly outside the United States—the term "race" has been officially abandoned, often due to historical, normative, or political sensitivities. Rather than transcending classification altogether, efforts have emerged to replace "race" with ostensibly more neutral terms such as "ethnicity, migration background, origin, or ancestry." These reformulations are embedded in a broader global trend in which policymakers, institutions, and scholars increasingly aim not only to recognize human variation, but also to make it the object of political action—frequently with the aim of enhancing the visibility of marginalized groups and promoting more equitable access to rights and resources.

The chapters in this section critically examine whether such classificatory practices constitute a genuine break from racial logics—or whether they rearticulate, adapt, or even reinforce them under new labels. The contributions trace how categories are (re)shaped within the intersection of historical knowledge regimes, political strategies, and bureaucratic procedures. Often, these newly legitimized categories remain entangled with older racial distinctions and can also give rise to novel forms of inequality.

Crucially, these analyses challenge the assumption that classifications simply report on preexisting human differences. Instead, the chapters highlight how classificatory schemes are locally negotiated and socially produced—emerging from dynamic interactions among science, state power, and civil society. To understand their effects, classifications must therefore be analyzed not as fixed descriptors but as elements within an epistemic economy in which knowledge-making, political governance, and identity formation are deeply interwoven.

This kind of analysis is taken up in the chapter by Andrea zur Nieden, Laura Schnieder, Isabelle Bartram, Nils Ellebrecht, and Tino Plümecke which examines how the concept of "migration background" became a locally specific substitute for "race" in German epidemiological research. While the term "race" is largely avoided in Germany, as it tends to be identified with biologistic thinking, "migration background" is used to capture social differences in health outcomes—echoing inclusion politics like those in the United States. Drawing on a systematic literature review and qualitative analysis of scientific publications, the authors trace how the concept of migration background operates as an administrative technology of difference that often reintroduces ethnic or racial meanings. They trace how it becomes conflated with ethnicity, nationality, religion, and culture, thereby reinscribing essentialized notions of difference into public health research. By situating these findings within broader debates on public health, diversity, and inclusion, the chapter calls for a more reflexive, critically informed approach in health research and policy.

The chapter by Isabelle Bartram and Tino Plümecke expands the analysis of classificatory practices in the German life sciences by examining the widespread yet conceptually unstable use of the concept "ethnicity." Their contribution reveals how the term functions in highly variable ways, with significant ambiguities and challenges in the assignment, definition, and interpretation of ethnic classifications, drawing on social, biological, and genetic dimensions. Rather than reflecting a coherent or consistently applied concept, "ethnicity" emerges in the German context as a contested and fluid construct, shaped by national political agendas, historical legacies, societal expectations, and disciplinary conventions. The analysis highlights the epistemic and practical tensions that arise when ethnicity is mobilized as a classificatory tool in scientific research, revealing its role in both reproducing and obscuring racializing logics.

A further example of the reinvention and persistence of racialized logics is explored in Jaehwan Hyun's analysis of genetic and genomic research in South Korea from 1945 to 2022. His chapter examines how locally embedded terms such as *injong* and *minjok* have functioned as central grouping concepts and sustained racialized frameworks in scientific discourse, despite efforts to distance the field from explicitly racist terminology. Hyun's chapter not only sheds light on the interplay between global and local scientific practices and shows how translations and interpretations of grouping categories are shaped; it also underscores how translations themselves perpetuate essentialized frameworks.

The chapter by Nicole Sommer analyzes a further conceptual reconstruction. It offers an inquiry into the political semantics of the concepts of "descent" and "ancestry," particularly in the context of the emergence of the term "people of African descent" in the work of international institutions such as the United Nations. Drawing on a qualitative content analysis of UN documents and interdisciplinary literature, the article examines how genetic data serve as both personal tools for self-discovery and as political instruments for collective legal claims-making, political mobilization, and collective memory. While "descent" functions as a legal and political category mobilized by international institutions to recognize histories of enslavement and diaspora, "ancestry"—especially in the context of DNA testing—refers to individualized biogeographical narratives. By analyzing the meanings and implications of ancestry and ancestors, the article questions how these terms influence or diverge from each other to shape the meaning and narrative of people of African descent and interrogates how these concepts contribute to the shaping of identity politics.

Ricardo Gomes Moreira offers another study on the category "ancestry" in the context of modern population genetics in Portugal. His contribution examines, in particular, laboratory practices for mapping genetic diversity and exploring human evolutionary history. Based on ethnographic observations in two genetic laboratories, he examines how ancestry is operationalized in the different stages of scientific work—from planning and sample collection to data analysis and dissemination—and reconstructs how the category takes on a dual role as a molecular construct and sociocultural concept of classification. The chapter also criticizes the danger of conflating molecular ancestry and social identity and emphasizes the need to distinguish between genetic classifications and sociopolitical constructs such as ethnicity and race.

Taken together, the contributions in this section demonstrate how the local efforts to replace the language of race with seemingly neutral alternatives rarely mark a clean epistemic break. Instead, they often reconfigure older classificatory logics within new semantic, administrative, or scientific frameworks—thereby reproducing or even expanding the reach of racialized distinctions.

Global Power and the Circulation of Standards

This section examines a third tactic of postracial classification: the negotiation and rearticulation of globally circulating standards within specific national and disciplinary contexts. As classificatory standards circulate transnationally, they undergo semantic shifts and produce new sets of social and political effects. Rather than moving unchanged across borders, standards have been contested, reinterpreted, and locally adapted—often in ways that reveal tensions between global scientific infrastructures and national epistemic, demographic, or political realities.

These tensions and transformations are especially evident in how regulatory and scientific categories of race and ethnicity developed in the United States have been adopted into international clinical trial protocols, regulatory guidelines, data infrastructures, and publication standards. The application of categories historically and culturally specific to the US in other regions of the world necessarily produces friction, as they often do not translate meaningfully across contexts. How should the category "Hispanic" or "Latino" be interpreted, for instance, outside the Americas, particularly in European contexts? And how can minoritized or marginalized communities be represented in other national contexts if they frequently fall outside of the classificatory schema established by this classificatory grid?

For countries that do not collect racial or ethnic statistics—or that deliberately reject such practices due to historical and political legacies—transnational standardization poses particular challenges. In contexts where the use of the term "race" is legally restricted, demographically misaligned, or normatively discredited, these templates create epistemic tensions and political ambivalence. Researchers working within such environments often justify the adoption of racial and ethnic categories by invoking the need for international comparability—especially with reference to the dominance of US-centered scientific discourse and its influence on publication criteria, funding structures, and regulatory expectations. Yet rather than just harmonizing classification systems, the global dissemination of such standards exposes enduring power asymmetries in the organization, validation, and governance of scientific knowledge.

The contributions in this section thus reveal how dominant classifications are locally adapted, resisted, or tacitly maintained. Racial and ethnic categories influence research not only through explicit labels but also through institutional procedures, data requirements, and funding conditions that implicitly assume the existence of biologically meaningful human groups labeled as "race" or "ethnicity." These classificatory systems are not simply imposed from above; rather, they take shape at the intersection of global regimes of standardization and national histories, disciplinary norms, and sociopolitical conditions. Far from being just tools, such standards are contested, political instruments that actively configure how human difference is made legible, governable, and actionable within scientific and institutional practice.

Anna Bredström and Shai Mulinari explore this coconstructive nature of classification in their analysis of the tensions and contradictions in the use of racial and ethnic categories in clinical research in Sweden. Drawing on interviews with actors involved in pharmaceutical research and regulation, they reveal how race remains embedded in everyday research practices, despite being officially delegitimized as scientifically invalid and morally inappropriate. Their chapter illuminates how international racial classifications imported through international regulatory standards produce friction and ambivalence within national contexts, generating silences, ambivalences, and normative tensions while simultaneously reinforcing forms of biological essentialism.

The chapter by Nils Ellebrecht examines the epistemic and practical conditions under which racial classifications are taken up by medical researchers in Germany. In line with Bredströms and Mulinari's analysis, his contribution reveals how the category of race persists as a tacit classificatory logic within research cultures that self-identify as postracial but are shaped by US standards. Ellebrecht illustrates how ethnicity and race often converge, how processes of ethnicization contribute to the biologization of difference, and how imaginaries of scientific neutrality and cultural homogeneity serve to mask these dynamics. He also shows that German research teams without international partners also employ race classifications to enhance the compatibility of their data in global scientific discourse, presenting local samples as representative of a "Caucasian" or "European" population.

In their contribution, Trudi Buck and Yulia Egorova investigate the afterlives of colonial classification systems in biological anthropology in the UK, as reflected in anthropological collections as well as current textbooks and teaching practices. The chapter begins with a notable absence: the lack of South Asian individuals in forensic anthropological databases. While this absence can be explained in many ways, the authors argue that it is primarily due to the classification of Indian skeletons as either "Caucasoid" or "Pacific Islander" according to the racial classification of the time. As a result, contemporary techniques for estimating anatomical ancestry may exclude unrepresented variability and misidentify remains. The authors also show how this legacy generates confusion and discomfort in forensic teaching contexts, especially for students of South Asian descent. Their contribution underscores the productive force of the absences and gaps in classificatory practices by highlighting how what is missing exerts epistemic and affective effects.

In summary, the contributions in this section show that the use of standards as a postracial classificatory tactic is not a straightforward transfer of heuristic models, but a process shaped by national histories, institutional frameworks, and disciplinary conventions. Frictions emerge not only through semantic mismatches, but also through deeper epistemic tensions—particularly when classifications according to race are officially disavowed but continue to structure research practices. These tensions expose the classificatory force of standards: not only in organizing data, but in producing subject positions.

Postracial Performative Identifications and Subjectifications

In this final section, our focus shifts from the emergence and circulation of classificatory systems to their performative effects—how, once implemented, categories take on a life of their own, shaping perceptions, social positions, and modes of governance. Scientific techniques such as DNA analysis, phenotyping, and ancestry estimation no longer serve merely to describe human variation; rather, they actively produce new kinds of subjects. These practices contribute to the redefinition of group identities and the attribution of biological or visual markers as defining traits—operating within frameworks shaped by sociotechnical conditions, political agendas, and historical legacies.

The chapters in this section explore how postracial classificatory logics are enacted, visualized, and made actionable in applied scientific settings. They examine how sampling and labeling strategies, embedded in biotechnological infrastructures, mediate between scientific categories and lived identities—informing institutional procedures as well as public imaginaries. Across these contributions, the effects of classification become tangible: influencing who is seen, how they are understood, and under what terms they are rendered visible, knowable, or vulnerable. What unites the studies is their shared attention to the entanglement of knowledge production and subject formation—to how scientific classifications do not merely represent but help constitute the subjects they claim to describe.

The section begins with Claude-Olivier Doron's exploration of European-identitarian online communities' appropriation of population genetics to produce alternative epistemologies of race, ethnicity, and ancestry. Extensively analyzing forums, blogs, and participatory genome projects, Doron shows how these communities closely engage and selectively adapt genetic studies to strategically assemble "counterknowledge" that legitimates older racial hierarchies and essentialist notions of European identity. Doron's contribution reveals how epistemic ambivalences and new visualization tools for representing genetic data are enrolled to create the self-identification, community-building, and political boundary-making practices of supremacist whiteness. His analysis of the "social life" of European DNA demonstrates that even the most technologically sophisticated technologies remain entangled with racial and racist typifications that hark back to the early days of race science.

How typologies and past stereotyping affect contemporary human genetics research is examined by Veronika Lipphardt and Mihai Surdu in their analysis of methodologies and representational practices, focusing on studies involving Roma people. Their study scrutinizes the construction of genetic distances and genetic isolation in DNA research on Roma, drawing on a huge dataset of several hundred genetic studies spanning the last one hundred years. By interrogating the interconnections between selective sampling strategies, population labeling, and the use of visualizations, they examine how these practices contribute to constructing truth claims and perpetuating the portrayal of Roma as a supposedly genetically homogeneous and distinct group. Their analysis underscores how these techniques obscure the complexity and diversity of Roma and align with long-standing racializing narratives in Europe.

A quite similar dynamic is examined by Aaro Tupasela and Heta Tarkkala in their study of how Finnish people are represented in human genetics research as an isolated and homogeneous population. Their study examines the supposed uniqueness of the Finnish population, the visual representations that support this narrative, and the population history that has been constructed. Through three analytical perspectives, the authors demonstrate how interpretations are profoundly shaped by the selective choice of reference populations and show how the conceptual flexibility of the "population isolate" concept allows researchers to simultaneously assert distinctiveness while situating Finnish genetics in comparative frameworks with other populations.

Another paradigmatic case of the intertwining of specific historical and social contexts is analyzed by Mahendra Shahare in his critical ethnography, documentary, and interview analysis of the GenomeIndia Project. This national initiative, which aims to sequence the genomes of 10,000 individuals and promises advances in precision medicine and genomic sovereignty, operates with an entanglement of aspects of genomics, caste, and postcolonial nationalism. Shahare analyzes how the project draws on colonial classificatory schemes and how it depoliticizes caste and ethnic identities by embedding them in biological frameworks. The chapter demonstrates how the contemporary GenomeIndia project remains deeply entangled with historical imaginaries of difference and nationalist visions of sameness.

Taken together, the contributions to this volume provide a window onto the range of logics and practices that are currently in use around the globe, especially outside the US context, to navigate the problem of race as it continues to haunt biological classifications of human beings long after the purported demise. Its multiple case studies analytically reveal the dynamics of these postracial racializations—that is to say, of classificatory practices that persist or reconfigure racial logics in contexts where the explicit use of the concept of race is rejected. The contributions illustrate that the effects of these classifications extend well beyond semantic substitution; they are materialized in institutional practices, policy instruments, visual technologies, and epistemic infrastructures. Rather than appearing as residual or outdated remnants, human classification systems that use categories of race and ethnicity and their many proxies continue to be central to scientific inquiry and the ordering of societies. They play a key role not only in the reproduction of social inequalities, but also in shaping frameworks for antidiscrimination and inclusion.

This volume neither posits a linear trajectory of scientific progress beyond race, nor does it suggest a seamless continuity of older racial regimes. It also does not assume a unified logic underlying today's classificatory systems. Instead, it highlights the multiplicity of tactics through which postracial classifications are enacted—ranging from the rebranding of established categories and the reinvention of classificatory labels to the global standardization of classificatory schemes and the emergence of new forms of biosocial subjectivity. Each chapter situates classification within its specific historical, institutional, and epistemic coordinates, showing how classificatory logics are adapted, resisted, or reformulated across different contexts. Together, the studies reveal both the resilience and the flexibility of racialized reasoning within contemporary scientific practice.

Rather than resting on the widely accepted—but increasingly depoliticized—claim that race is socially constructed, the volume calls for a more profound analytical shift. It asks how classificatory regimes operate as technologies of power, recognition, and inequality; how they become embedded in infrastructures, circulate through transnational scientific networks, and shape political imaginaries and material consequences. In doing so, the volume invites a rethinking of the epistemological foundations of classification itself—one that not only critiques the historical entanglements of race and science, but that also interrogates the productive, normative, and political effects of classificatory practices in the present. Such a shift is necessary if we are to understand, contest, and ultimately reconfigure the classificatory regimes that continue to structure our biological and social worlds.

References

Auton, Adam, Lisa D. Brooks, Richard M. Durbin, et al. 2015. "A global reference for human genetic variation." *Nature* 526 (7571): 68–74. doi: 10.1038/nature15393.

Barkan, Elazar. 1992. *The Retreat of Scientific Racism. Changing Concepts of Race in Britain and the United States Between the World Wars*. New York: Cambridge University Press.

Bartram, Isabelle, Tino Plümecke, and Susanne Schultz. 2022. "Genetic racial profiling: Extended DNA analyses as technologies of discrimination." *Science & Technology Studies* 35 (3): 44–69.

Benjamin, Ruha. 2009. "A lab of their own: Genomic sovereignty as postcolonial science policy." *Policy and Society* 28 (4): 341–355. doi: 10.1016/j.polsoc.2009.09.007.

Bliss, Catherine. 2012. *Race Decoded. The Genomic Fight for Social Justice*. Stanford, CA: Stanford University Press.

Bowker, Geoffrey C., and Susan Leigh Star. 2000. *Sorting Things Out. Classification and Its Consequences*. Cambridge, MA: MIT Press.

Cavalli-Sforza, Luigi Luca. 2001. *Genes, Peoples and Languages*. London: Penguin.

Darwin, Charles. 1874. *The Descent of Man, and Selection in Relation to Sex*. 2d ed. London: John Murray.

Duster, Troy. 2015. "A post-genomic surprise. The molecular reinscription of race in science, law and medicine." *The British Journal of Sociology* 66 (1): 1–27. doi: 10.1111/1468-4446.12118.

Epstein, Steven. 2007. *Inclusion. The Politics of Difference in Medical Research*. Chicago: Univ. of Chicago Press.

Foucault, Michel. 1972. *Archaeology of Knowledge and The Discourse on Language*. New York: Pantheon Books. Foucault, Michel. 1980. *Power/Knowledge: Selected Interviews and Other Writings* 1972–1977. Edited by Colin Gordon. New York: Pantheon Books.

Foucault, Michel. 2005, fr. 1966. The Order of Things. An Archaeology of the Human Sciences. London: Routledge.

Fujimura, Joan H., and Ramya Rajagopalan. 2011. "Different differences: The use of 'genetic ancestry' versus race in biomedical human genetic research." *Social Studies of Science* 41 (1): 5–30.

Gannett, Lisa. 2001. "Racism and human genome diversity research: The ethical limits of 'population thinking.'" *Philosophy of Science* 68 (3): S479–S492.

Gilroy, Paul. 2000. *Against Race. Imagining Political Culture Beyond the Color Line*. Cambridge, Mass.: Harvard University Press.

Goldberg, David Theo. 2009. *The Threat of Race. Reflections on Racial Neoliberalism*. Malden MA: Wiley-Blackwell.

Goldberg, David Theo. 2015. *Are We all Postracial Yet?* Cambridge: Polity Press.

Goldberg, David Theo, and Philomena Essed. 2002. "Introduction: From racial demarcations to multiple identifications." In *Race Critical Theories. Text and Context*. Edited by Philomena Essed and David Theo Goldberg, Malden, Mass.: Blackwell, 1–14.

Hacking, Ian. 1999. *The Social Construction of What?* Cambridge, Mass: Harvard Univ. Press.

Hacking, Ian. 2002. *Historical Ontology*. Cambridge, Mass: Harvard University Press.

Hall, Stuart. 1986. "Gramsci's relevance for the study of race and ethnicity." *Journal of Communication Inquiry* 10 (2): 5–27.

Haraway, Donna. 1989. *Primate Visions. Gender, Race, and Nature in the World of Modern Science*. New York, NY: Routledge.

Harding, Sandra. 1991. *Whose Science? Whose Knowledge? Thinking from Women's Lives*. Ithaca, NY: Cornell Univ. Press.

Hawthorne, Camilla A. 2022. *Contesting Race and Citizenship. Youth Politics in the Black Mediterranean*. Ithaca New York: Cornell University Press.

Hawthorne, Camilla. 2024. "Toward a relational theorization of racisms." *Ethnic and Racial Studies* 47 (8): 1620–1628. doi: 10.1080/01419870.2023.2227680.

Huxley, Julian, and Alfred C. Haddon. 1935. *We Europeans: A Survey of "Racial" Problems*. London: J. Cape.

Lipphardt, Veronika. 2012. "Isolates and crosses in human population genetics; or, a contextualization of German race science." *Current Anthropology* 53 (S5): S69–S82. doi: 10.1086/662574.

Lipphardt, Veronika, Mihai Surdu, Nils Ellebrecht, Peter Pfaffelhuber, Matthias Wienroth, Gudrun A. Rappold. 2021. "Europe's Roma people are vulnerable to poor practice in genetics." *Nature* 599 (7885): 368–371. doi: 10.1038/d41586-021-03416-3.

M'charek, Amade, Katharina Schramm, and David Skinner. 2014. "Topologies of race: Doing territory, population and identity in Europe." *Science, Technology & Human Values* 39 (4): 468–487.

M'charek, Amade, Victor Toom, and Lisette Jong. 2020. "The trouble with race in forensic identification." Science, Technology and Human Values 27 (1): 804–828. doi: 10.1177/0162243919899467.

M'charek, Amade, and Irene van Oorschot. 2023. "The politics of face and the trouble with race: Exploring relations at the interface between the individual and the collective in forensic practice." *Social Studies of Science*, 53 (6): 813–825. doi: 10.1177/03063127231211817.

Morning, Ann, and Marcello Maneri. 2022. *An Ugly Word. Rethinking Race in Italy and the United States.* Chicago: Russell Sage Foundation.

Mosse, George L. 2020. *Toward the Final Solution. A History of European Racism*. Madison, Wisconsin: The University of Wisconsin Press, orig. 1978.

Nelson, Alondra. 2016. *The Social Life of DNA*. Boston, MA: Beacon.

Omi, Michael, and Howard Winant. 2015. *Racial Formation in the United States. From the 1960s to the 1980s*. 3rd ed., orig. 1986. New York: Routledge.

Plümecke, Tino. 2010. "Die neuen Differenzen der Lebenswissenschaften. 'Rasse', Genetik und die ungenutzten Potentiale der Soziologie." In: *Ethnowissen. Soziologische Bei-*

träge zu ethnischer Differenzierung und Migration, edited by Marion Müller and Darius Zifonun, 423–450. Wiesbaden: VS.

Plümecke, Tino. 2013. *Rasse in der Ära der Genetik. Die Ordnung des Menschen in den Lebenswissenschaften*. Bielefeld: Transcript.

Proctor, Robert N. 1988. *Racial Hygiene. Medicine Under the Nazis.* Cambridge, MA: Harvard University Press.

Rattansi, Ali. 2007. *Racism. A Very Short Introduction*. Oxford: Oxford University Press.

Reardon, Jenny. 2005. *Race to the Finish. Identity and Governance in an Age of Genomics*. Princeton, NJ: Princeton University Press.

Reardon, Jenny. 2012. "The democratic, anti-racist genome? Technoscience at the limits of liberalism." *Science as Culture* 21 (1): 25–47. doi: 10.1080/09505431.2011.565322.

Reardon, Jenny. 2017. The Postgenomic Condition. Ethics, Justice, and Knowledge after the Genome. Chicago: The University of Chicago Press.

Reich, David. 2018. *Who We Are and How We got here. Ancient DNA and the New Science of the Human Past*. New York, NY: Pantheon Books.

Roberts, Dorothy E. 2011. *Fatal Invention. How Science, Politics, and Big Business Re-create Race in the Twenty-First Century*. New York: New Press.

Smedley, Audrey, and Brian D. Smedley. 2005. "Race as biology is fiction, racism as a social problem is real: Anthropological and historical perspectives on the social construction of race." American Psychologist 60 (1): 16–26.

Stepan, Nancy. 1982. *The Idea of Race in Science*. Great Britain, 1800–1960. London: Macmillan.

Stern, Alexandra Minna. 2016. *Eugenic Nation. Faults and Frontiers of Better Breeding in Modern America*. 2nd ed. Oakland, California: University of California Press.

Beyond Categories and Back Again

From One to(o) Many

The Use of Forensic DNA Phenotyping in the Criminal Justice System

Filipa Queirós and Rafaela Granja[1]

Abstract *Forensic DNA phenotyping (FDP) refers to a set of techniques that aim to derive probabilistic information about certain externally visible characteristics of suspects, such as eye, skin, and hair color, from biological samples. In forensic science, this technology is most commonly used to predict the appearance of an unknown suspect in criminal investigations. In this chapter, we examine the artistic exhibition Probably Chelsea and a real criminal case in Canada as platforms for discussing how, despite narratives claiming that FDP enables a fine-grained approach that might lead to the identification of a suspect, in practice this technology creates a coarse-grained context that casts suspicion on an entire population group.*

1. Introduction

In 2017, the exhibition *Probably Chelsea* premiered at the Fridman Gallery in New York City. Upon entering the space, visitors saw *thirty different portraits*, varying in skin and eye color, as well as facial morphology (figure 1). Despite the very subtle resemblances among them, what they had in common was that they were all created based on the analysis of *one person's DNA*, through a technology known as DNA phenotyping. In the words of the artist, Heather Dewey-Hagborg: "when you enter ... the faces will greet you in a way and invite you in, and hopefully invite you into thinking deeply about genetics and your own identity. I hope that you walk up to these faces and identify with one of them. That there will be a face that you connect to and are drawn into, and that leads you to think, 'that could be my DNA and how many different people are there in my DNA, as well?'"[2]

1 Both authors have contributed equally to this chapter.

2 https://www.youtube.com/watch?v=A-EhRw11JiQ (accessed 4 December 2023).

Figure 1: Exhibition Probably Chelsea by Heather Dewey-Hagborg

Source: https://deweyhagborg.com/projects/probably-chelsea.

In this chapter, we focus on the forensic application of DNA phenotyping, the technology used by Heather Dewey-Hagborg to create the exhibition *Probably Chelsea*. DNA phenotyping aims to derive probabilistic information from biological samples about certain externally visible characteristics, such as eye, skin, and hair color. In forensic science, this technology is most commonly used to predict the appearance of an unknown suspect in criminal investigations.

We take inspiration from the concepts of granularity (Bowker and Star 1999) and enchanted determinism (Campolo and Crawford 2020) to address forensic DNA phenotyping (FDP). Along with the artistic intervention *Probably Chelsea*, we analyze and explore a criminal case in Canada as a platform for discussing how, despite narratives claiming that FDP enables a fine-grained approach that might lead to the identification of a suspect, in practice this technology creates a coarse-grained context that casts suspicion on an entire population group.

This chapter therefore contributes to the body of literature that critically examines FDP (Queirós 2021; Hopman 2021; Granja and Machado 2020; Wienroth 2018b; M'charek 2008; 2020) by presenting both real and fictional cases as platforms for problematizing various issues related to the practical application of FDP technology in criminal investigations. The chapter begins with a brief exploration of some of the visual technologies mobilized in the criminal justice system, followed by a description of the scientific and social controversies that frame FDP. We then explain how our analytical framework is able to provide new insights into the use of FDP in the criminal justice system. Next, we explore the artistic collaboration between Heather Dewey-Hagborg and Chelsea Manning, which culminated in the exhibition *Probably Chelsea*. This case outlines how FDP ignores the complexities that might exist between inferring sex from genetics and transposing it to appearance. We then move on to explore the case of the Edmonton Police Service, where the company Parabon Nanolabs produced a DNA-generated portrait that led to a strong community backlash, pointing out the racist implications of this technology. The final section of the chapter summarizes our reflections.

2. Visual Technologies in the Criminal Justice System

Visual technologies have a long history in the criminal justice system. The quest to categorize "criminal types" and/or identify unknown individuals, such as a suspect or a victim, is replete with attempts employing a variety of techniques, ranging from the more established and routine to those that are currently considered state of the art. The most common techniques used to target the facial image of suspects and/or criminalized populations include photographs, composite images (also known as sketches or composite facial drawings), CCTV footage, facial recognition software, and—the focus of this chapter—FDP. In this section, we focus only on photographs and composite images (sketches), because of their direct (dis)connections with FDP, thereby deliberately leaving out the controversies surrounding CCTV and facial recognition (Introna and Wood 2002; Williams 2020).

The increasing professionalization of police work in the 1880s and 1890s was associated with the institution of the photographic archive (Sekula 1986, 17). However, the use of photography in criminal justice has been at the center of many long-standing controversies. The use of photographs is prominently associated with the work of Alphonse Bertillon, who aimed to create a system of criminal identification that brought together anthropometric descriptions and measurements, photographs, and, when appropriate, records of unusual features such as tattoos and scars—the "portrait parlé" (Machado and Prainsack 2012; Burney and Pemberton 2013; Sekula 1986). Such records were intended as technologies of identification, to be kept as police files; such practices continue to this day, albeit in a different form (Miranda and Machado 2018; Nieves Delgado 2020).

Photographs of deviant bodies were also used in studies that considered criminality to be a physical manifestation of the body (Lombroso 1876), with particular links to the eugenic movement, most notably through the work of Sir Francis Galton. In the late 1870s, Galton used photography to compile the "typical" characteristics of "criminals" in order to categorize individuals into "criminal types" (Hopman and M'charek 2020). In particular, Galton layered multiple photographic negatives of portraits to make composite portraits. In his words:

> The photographic process ... enables us to obtain with mechanical precision a generalized picture; one that represents no man in particular, but portrays an imaginary figure possessing the average features of any given group of men. These ideal faces have a surprising air of reality. Nobody who glanced at one of them for the first time, would doubt its being the likeness of a living person, yet, as I have said, it is no such thing; it is the portrait of a type and not of an individual. (Galton 1879, 132–133)

As posed by Allan Sekula, "Bertillon's nominalist system of identification and Galton's essentialist system of typology constitute not only the two poles of positivist attempts to regulate social deviance by means of photography, but also the two poles of these attempts to regulate the semantic traffic in photographs" (Sekula 1986, 55). Despite their significant differences, both systems show how the image of suspected and criminalized populations is a focal point of interest, either for identification or categorization purposes.

In addition to photography, another visual technology commonly used in the criminal justice system is composite images, also known as sketches or composite facial drawings. These are facial representations of a criminal suspect created through an interview process in which a forensic artist combines personal expertise with the descriptions of eyewitnesses, to produce an image of the individual, usually a suspect (Bleumink, Jong, and Plájás 2021; Taylor 2000; Hopman and Bleumink 2023; Plájás 2023). Its first documented use in police investigations dates back to 1881 in the United Kingdom where, as part of a murder investigation, a caricature of a suspect was drawn and distributed to the public in the form of a police wanted poster (Taylor 2000). Whether sketches are by hand or created with the assistance of computer and software-based systems, their use has become an established and routine forensic practice (Bleumink, Jong, and Plájás 2021; Hopman and Bleumink 2023) within law enforcement agencies around the world. This technology is seen as particularly valuable in cases where the police have no significant leads other than the eyewitness's description of the suspect.

Because the creation of sketch images depends in part on human testimony and memory, forensic artists rely on a variety of techniques to produce the final image. The use of descriptive categories, reference images, and standardized photographic databases of different population groups is common practice among forensic artists producing sketches (Bleumink, Jong, and Plájás 2021; Taylor 2000; Nieves Delgado 2020; Plájás 2023). This allows forensic artists not only to narrow down the broader conception of the suspect within a smaller population group, but also to focus on the more specific characteristics and traits of that same population (Bleumink, Jong, and Plájás 2021). As noted by Ildikó Plájás, "For the facial composite to fulfill its purpose, that is, to lead the investigators to an individual, it needs to help narrow down larger population groups into smaller ones" (Plájás 2023, 942). Thus, although the primary goal of a sketch image is to reunite information to identify the suspect through this performative (re)construction of their externally visible features, the final image aims to translate a fine balance between generic and specific representations of their face. This means that when a sketch image is released to the public, it must simultaneously enable the public to recognize and identify the suspect's distinctive features and to interpret the ambiguity that was left in the less detailed ones. As Bleumink et al. note:

> When the composite drawing is too generic, criminal officers, who have to trace every single lead, face the risk of receiving too many leads pointing to a range of different individuals. As such, the composite loses its function. This is where circulating a composite representing a minority population, or an "uncommon" face, becomes more informative than a composite that resembles someone from the majority population in a specific area. (2021, 20–21)

Despite the continuing relevance of photographs and sketch images to the criminal justice system, both techniques have significant drawbacks. In the context of a criminal investigation, photographs allow only for the inclusion/exclusion of certain previously identified and photographed individuals. Sketch images rely heavily on eyewitness and victim testimonies, which are based on memory and recollections of the criminal event, often considered to be fallible, subject to emotional fluctuations and fading memory, and therefore inaccurate. On this basis, the increasing relevance of science within policing has opened the door to the development and use of other visual technologies, anchored in what is seen as a more "robust" science, namely forensic genetics, perceived by several stakeholders as a "truth machine" (Lynch et al. 2008). FDP therefore emerges as another visual technology entering the criminal justice toolkit. However, as noted by Roos Hopman, despite its innovative potential some of its processes are reminiscent of previous techniques: "facial renditions produced through FDP make adjacent practices from different places and times. In the application of contemporary technologies to phenotypic data, nineteenth-century statistical averaging techniques as well as physical anthropological measuring and data-collection practices resurface" (2021, 4).

3. Forensic DNA Phenotyping

The ability to match a DNA profile collected at a crime scene to a specific individual ushered in a new paradigm in forensic science that has been expanding ever since. In the early 2000s, technologies such as forensic DNA phenotyping (FDP) began to be developed, expanding the field's potential to contribute to criminal justice. Forensic DNA phenotyping refers to a set of techniques that aim to derive probabilistic information about certain externally visible characteristics of suspects—such as eye, skin, and hair color—from biological samples. FDP is generally paired with biogeographical ancestry (BGA) inference, which provides probabilistic information about a person's genetic ancestry as being of African American, Southern European, or Northern European geographic-genetic origin. That is, although BGA does not analyze phenotypic traits, it still informs the prediction of externally visible characteristics (Hopman 2021; Koops and Schellekens 2008, 211; M'charek 2020).

In the field of forensic science, FDP is primarily designed to assist in criminal cases where a *one-to-one* logic (comparison of DNA profiles) is not possible (Kayser and de Knijff 2011; Børsting and Morling 2015; Kayser 2015). Aiming to assist police investigations in the decision-making process, FDP uses probabilistic representations that reveal a set of variable biological characteristics shared by certain population groups to which the suspect may belong. It thus operates in a *one-to-many* logic where the population is (dis)aggregated into different configurations of groups of individuals based on certain criteria (a specific eye color, skin color, biogeographical ancestry group, etc.). FDP thereby shifts the locus of police investigation from *one* (the suspect) to *many* (a collective population).

FDP has been presented by its proponents as a technology characterized by scientific and technological progress. Drawing on the scientific legitimacy of DNA (Lynch et al. 2008), FDP is wrapped in an aura of progress, objectivity, immutability, impartiality, and reliability. It has therefore been presented as "the production of a 'biological witness' from crime scene DNA" (Walsh and Kayser 2016, 417). "In essence," Kayser writes, "FDP outcomes can serve as 'biological witness,' and may potentially provide even more accurate information than human eyewitnesses do, who are known to be unreliable" (Kayser 2015, 34).

Despite such overly optimist promises associated with FDP, this technology is still characterized by several controversies and complex politics of legitimation and contestation (Granja and Machado 2020), both within and beyond the scientific community of forensic genetics.[3] Despite ongoing controversies regarding the accuracy of the technology (Sharma et al. 2019; Lipphardt 2018), currently, the application of FDP tends to focus on predicting pigmentation characteristics, such as eye, hair, and skin color and biogeographic ancestry (Kayser 2015; Kayser and Schneider 2009; Kayser et al. 2023). Given the complexity in understanding the genetic-based settings that result in the physical expression of other characteristics—such as the variation of face shapes, body weight/structure, hair loss, baldness and hair morphology—scientific research concerned with these predictions is still underdeveloped (Kayser 2015). Some publications have presented results using genetic kits and computer software for the application of FDP. Although they aim to allow the accuracy of this technology to be assessed, they have proved to be inconsistent and unsatisfactory (Sharma et al. 2019).

On the one hand, social scientists have played a crucial role in debating FDP controversies by questioning the conceptual ambiguity associated with the concept of origin and its interrelations with racial categories. Contributions in this domain criticize how FDP is able to legitimize the notion that race is a category of differentiation with genetic validity (Vailly 2017; Queirós 2021). On the other hand, social sciences also emphasize the potential of FDP to reproduce stigmatization and reinforce criminalization. By increasing the visibility of racial or ethnic differences, FDP holds the potential to affect disproportionately population groups already vulnerable to the action of the criminal justice system, such as racial and ethnic minorities (Toom et al. 2016; Queirós 2021; Hopman 2020; Hopman and M'charek 2020; Duster 2008; Ossorio 2006; Skinner 2018; M'charek 2013; Bartram, Plümecke, and Schultz 2022).

In addition to academic research and police interest in FDP, the development of this technology has been characterized by a fetishization of the new (Brown and Michael 2003) and has therefore attracted considerable interest from for-profit forensic providers. In January 2015, a private company based in the United States, Parabon Nanolabs, launched Snapshot. Snapshot is a tool that allegedly enables the prediction of the appearance, including face morphology, of an unknown individual. Its commercialization is directly targeted at criminal investigations and its development has received directed funding from the Department of Defense.[4] Snapshot differentiates itself from other existing products on the market by providing a DNA-based portrait of an individual, typically a suspect.

3 Within the scope of this article the expression "community of forensic genetics" or "scientific community of forensic genetics" is used to refer to what Simon Cole calls "forensic genetic scientists," that is, individuals working on criminal cases and that are employed by a forensic laboratory, and "research scientists," which represent individuals employed by universities whose primary professional occupation is scientific controlled laboratory research with applications in forensic science (Cole 2013).

4 https://media.defense.gov/2022/Nov/23/2003120778/-1/-1/0/PARABON_STORY.PDF (accessed February 16, 2023).

Owing to its overly optimistic promises, Snapshot has received wide attention from high-impact media outlets, such as *National Geographic* and *The New York Times* (Greenwood 2016; Pollack 2015; Southall 2017) and significant interest from police departments across the world. According to the Parabon website, "Snapshot has been used by hundreds of law enforcement agencies around the world to help generate leads, narrow their suspect pools, and solve human remains cases, in both active and decades-old investigations." This claim is followed by a list of cases allegedly solved using Snapshot.[5]

However, Snapshot has received considerable backlash from several members of the forensic genetics' scientific community (Wienroth 2018a). Forensic geneticists involved in academic research dedicated to the development of the FDP describe the impossibility of assessing the scientific robustness of the services offered, thus raising questions of validation and transparency, and more broadly, of scientific ethics and responsibility (Granja and Machado 2020). Despite the ongoing controversies surrounding Snapshot, Parabon NanoLabs has irrevocably linked the use of FDP technology to the possibility of creating a DNA-based portrait.

4. Analytical Lens

Framing FDP as part of the technolegal worlds of forensic genetics (Toom, Wienroth, and M'charek 2023), in this chapter we draw inspiration from the concepts of granularity (Bowker and Star 1999) and enchanted determinism (Campolo and Crawford 2020). On the one hand, Bowker and Star's seminal work on classifications and classification systems as powerful technologies (1999, 319) helps us to understand how the fine-grained aspect of granularity (referring to specificity) is repeatedly mobilized to refer to the informational potential of DNA, which is seen as the ultimate identifier (Lynch et al. 2008). As described earlier, such a narrative is recurrently adopted by proponents and supporters of FDP who invoke DNA uniqueness to argue for the technology's fine-grained approach, especially when comparing FDP to other visual technologies. On the other hand, the coarser component of granularity captures how FDP ends up categorizing broad groups that share the same particular trait(s) or characteristic(s). In other words, although the technology uses DNA as a fine-grained source of information, in practice its operationalization and results reveal a coarse-grained approach, as there are multiple ways of translating its results. As Bowker and Star also point out, "all information systems are necessarily suffused with ethical and political values [...] These systems are active creators of categories in the world as well as simulators of existing categories" (Bowker and Star 1999, 321). Applied to the case of FDP, this sheds light on how this technology makes use of both old and (re)new(ed) scientific practices and classification systems that are embedded in complex configurations of scientific and cultural frameworks. In practice, it has the power to (re)create dynamics of collectivization of suspicion towards certain population groups (Queirós 2021; Hopman 2021).

The concept of enchanted determinism is proposed by Campolo and Crawford (2020) within the context of artificial intelligence. It refers to "a discourse that presents deep

5 https://snapshot.parabon-nanolabs.com/phenotyping (accessed February 16, 2023).

learning techniques as magical, outside the scope of present scientific knowledge, yet also deterministic, in that deep learning systems can nonetheless detect patterns that give unprecedented access to people's identities, emotions and social character" (2020, 3). A *translation* of the concept (Callon 1986; Latour 1986; 1992) into the context of the FDP allows a particularly fruitful perspective on the role of Parabon. Although Parabon bases its service on some of the FDP's predictions (e.g., eye, hair, and skin color), the scientific community of forensic genetics consider its current promise to be outside the scope of current scientific knowledge, as several doubts remain about the scientific validity of inferring facial morphology from DNA (Granja and Machado 2020; Wienroth 2018a). In addition, even more important for the purposes of this chapter is to understand how the discourse of enchanted determinism has crucial social implications: "[it] values the technical 'accuracy' of predicting social identities over a deeper contextual knowledge of how those predictions are made, their limitations, or the impacts those predictions have in the world (Campolo and Crawford 2020, 12). Considering this, in the remainder of this chapter we explore how the concept of enchanted determinism is particularly useful for exploring how Parabon services, particularly Snapshot, are fraught with political and social risks. We examine, in other words, how Snapshot's claims of high levels of accuracy and objectivity, when applied in real criminal cases, raise complex issues of responsibility and liability with unfolding implications in terms of individual and, more importantly, collective suspicion, discrimination, and criminalization.

5. Case Studies

5.1 Probably Chelsea

In 2010, Chelsea Manning, a former US Army soldier was sentenced to thirty-five years in prison for leaking classified and sensitive military and diplomatic documents to WikiLeaks. While in prison, Manning openly identified as a trans woman. Placed in solitary confinement on several occasions during her imprisonment, Manning was subject to a strict visiting policy and was not allowed to be photographed.

Heather Dewey-Hagborg is an "artist and biohacker interested in art as research and technological critique."[6] One of her best-known projects is *Stranger Visions* (2012–2013), in which she used a technology known as DNA phenotyping to produce 3D-printed facial sculptures from the analysis of genetic material (hair, cigarette butts, chewed gum) collected from public places. In 2015, Heather Dewey-Hagborg received an email from *Paper* magazine asking if she would be interested in making a 3D facial sculpture of Chelsea Manning based on her DNA, as she was not able to receive visits or be photographed at the time. This led to an exchange of letters between Chelsea Manning and Heather Dewey-Hagborg, and eventually to Chelsea Manning sending the artist a letter containing her biological samples (cheek swabs and hair clippings). Based on this data, in 2015, the artist created *Radical Love*, a project featuring two life-size 3D-printed models of Chelsea Manning's face, based on her DNA. The project premiered at the World Economic Forum in

6 https://deweyhagborg.com/bio (accessed February 22, 2023).

January 2016. In light of Manning's transgender identity, one of the 3D DNA-based facial sculptures was intended to represent Manning's face as gender-neutral. The other was rendered as a woman. The project is described by the artist as "an homage to and exploration of gender identity stereotypes in forensic DNA phenotyping."[7] But the collaboration between Chelsea Manning and Heather Dewey-Hagborg did not end there. In 2015, the two of them created a graphic short story entitled "Suppressed Images." And two years later, after Manning's sentence was commuted by President Barack Obama, the artist premiered the exhibition *Probably Chelsea* (introduced above). This collection comprised thirty different possible 3D-printed facial sculptures of Chelsea Manning, each generated through DNA phenotyping technology. At the exhibition, the sculptures were strategically placed at diverse heights to create the illusion of a crowd as visitors entered the space. Sculptures varied in their skin and eye color and morphology, while bearing an almost imperceptible resemblance to each other. According to Dewey-Hagborg, "genomic data can tell a multitude of different stories about who and what you are. Probably Chelsea shows just how many ways your DNA can be interpreted as data, and how subjective the act of reading DNA really is."[8]

Chelsea Manning's case brings into focus a topic that up until this point has remained underexplored in the body of literature exploring FDP. Skin and eye color have received significant attention in the literature on FDP due to their relation to the notion of race (Hopman and M'charek 2020; M'charek 2008; Queirós 2021; Bartram, Plümecke, and Schultz 2022). However, this artistic interpretation of DNA phenotyping technology also foregrounds how gender performance influences individuals' appearance in ways that cannot be disclosed by DNA. Given how bodies are socially situated and contingent upon gender performance (Butler 1993), it is crucial to bring into the discussion the differences between sex and gender identity. That is, despite the fact that sex can be inferred from DNA (and such information can be useful for forensic purposes), the leap from there to creating images of what a suspect might look like involves a set of assumptions and stereotypes that completely ignore gender performativity. Chelsea Manning's case therefore shows how the diversity and plurality of gender performance, which might be mutable, changing not only across time but also space, do not find translation into genetic readings of the human body. Consequently, this poses additional challenges to the application of FDP technology as a means of generating information about the externally visible characteristics of individuals.

FDP technology has been presented by its advocates and supporters as a tool that is capable of providing law enforcement with a fine-grained approach and specific information of intelligence value about the appearance of criminal suspects, particularly in the absence of additional clues or eyewitnesses. The case of *Probably Chelsea* shows that not only do the results of DNA phenotyping not provide definitive answers—thus contradicting the imagery of unity and singularity associated with DNA—but they are also open to different interpretations, thereby outlining a coarse-grained translation outside forensic laboratories (Bowker and Star 1999). As shown by the artist, too many and different faces can be drawn from the same DNA sample analysis.

---

7 https://deweyhagborg.com/_archived/chelsea/ (accessed February 22, 2023).

8 https://deweyhagborg.com/projects/probably-chelsea (accessed February 22, 2023).

5.2 The Edmonton Police Service Case

On 10 March 2019, a Black female in her mid-20s was walking home at dawn in the city of Edmonton in Alberta, Canada. On her way, she passed a bus stop where several people were waiting, including an unknown male. The bus arrived, picked up the people waiting, and left. The unknown man did not enter the bus. He followed the young female as she continued walking home, eventually dragging her to a field and sexually assaulted her. During the attack, the victim lost consciousness. She was left alone, seriously injured, wearing only a t-shirt in 27-degree Celsius weather. When she regained consciousness, she walked towards a street where a resident saw her and called the police. According to the Edmonton Police Service, the victim specifically described the suspect as a 162-cm Black man with an accent, wearing bulky winter clothes and a black toque that covered most of his face. The police collected the DNA of the suspect through a rape kit but it did not match any of the profiles stored in forensic DNA databases. With no eyewitnesses or CCTV footage and after all traditional investigative options had been exhausted, the case went cold.

Three and a half years later, the Edmonton Police Service decided to contract Parabon to use their Snapshot service in this case. Although they had never used this technology, the decision was perceived as a "last resort," as declared by Colleen Maynes, a detective in the Edmonton Police Service Sexual Assault Section, and Enyinnah Okere, chief operations officer for the Community Safety and Well-Being Bureau of Edmonton Police Service (Okere 2022; Edmonton Police Service 2022).

Hoping to generate new leads on the suspect's identity, on 4 October 2022 the Edmonton Police Service released a statement on their website and social media along with the the DNA-generated portrait that Parabon had produced. Confirming the description given by the victim, the DNA-based portrait presented the suspect as a Black man with African ancestry, adding new information about his brown/black eyes and black hair. In the statement issued to accompany the disclosure of the DNA-generated portrait, the police mentioned some of Snapshot's limitations. For example, Parabon uses a default—not DNA-based data—age of twenty-five and body mass index of twenty-two in their predictions (Edmonton Police Service 2022). The Edmonton Police Service thus outlined how "the suspect may be older, [or] may have a different body composition," but also "may have facial hair or a different hairstyle than depicted in the photo" (ibid.). The statement ended with a brief acknowledgment of the potential effects the release of the DNA-based portrait could generate in the Black community of Edmonton:

> Following consultation with community stakeholders, the EPS [Edmonton Police Service] is aware of the impact this release may have on a marginalized community. Due to the severity of the occurrence, the need to advocate for a victim of a violent sexual assault and in consideration of the public safety interest, investigators believe the release of this image based on DNA evidence is required in order to further the investigation. As always, any leads generated from the release of a composite image would require further investigative steps. (Edmonton Police Service 2022)

Notwithstanding the alerts on the limitations of the DNA-generated portrait and the mentioning of the potential discriminatory effects on the Black community, the release of that DNA-based portrait triggered an unprecedented backlash. Shortly after the disclosure of this "new" information, reactions deriving from different groups quickly arrived. Nine Edmonton-based Black community activist groups addressed a letter to the Edmonton Police Commission chairman calling for an end to the police use of FDP technology. They argued that information produced by Parabon and its subsequent release to the public further stigmatized and criminalized racialized groups within the Edmonton community (Issawi 2022).

The release of the portrait also inflamed social media. Haruun Ali, a community activist, while juxtaposing a picture of himself with the one released by the police, stated: "You can take any random Black man that you know, put him next to it and to be frank, it will probably look similar to them. ... This is modern-day racial profiling."[9] Adam Rutherford, a geneticist, was one of those linked to science who had an active voice in this controversy. Reposting the tweet that police had made disseminating the DNA-generated portrait, Rutherford addressed reliability and accuracy issues related to the work of Parabon. In his own words: "You can't make facial profiles or accurate pigmentation predictions from DNA, and this is dangerous snake oil."[10] Finally, Doug King, professor of justice studies at Mount Royal University, also addressed issues related to the idea of accuracy that Snapshot's DNA-based portrait conveys and the risks of racial profiling:

> "We see the composite picture and then we automatically assume it's an accurate composite picture. ... By putting a composite together, we fix it in people's minds that this is what the person is supposed to look like. So, you have to guard against that because then the issue becomes, 'I see someone who looks like that—my next-door neighbor.' And it turns out it has nothing to do with that person" (Gibson 2022).

In the face of myriad accusations of racism from Black community groups, activists, and scientists dismissing Parabon's work as irresponsible science, a statement of apology was issued by the Edmonton Police Service just forty-eight hours after releasing the DNA-generated portrait. This was signed by Chief Okere, who is responsible for overseeing the Sexual Assault Crimes Team and is himself a member of the Black community. The statement announced the immediate removal of the DNA-generated portrait from the police website and social networks:

> We were not and are not oblivious to the legitimate questions raised about the suitability of this type of technology. The potential that a visual profile can provide far too broad a characterization from within a racialized community and in this case, Edmonton's Black community, was not something I adequately considered. There is an important need to balance the potential investigative value of a practice with the all too real risks and unintended consequences to marginalized communities. ... I prioritized the investigation—which in this case involved the pursuit of justice for the victim, herself

9 https://twitter.com/HaruunYEG/status/1583622302528606209 (accessed February 20, 2023).

10 https://twitter.com/adamrutherford/status/1577556029499355136 (accessed February 20, 2023).

> a member of a racialized community, over the potential harm to the Black community. This was not an acceptable trade-off and I apologize for this. (Okere, 2022)

Finally, not ruling out future uses of Parabon's Snapshot service or of FDP, despite protests in the media calling for a ban of this technology, Okere also indicated the police will review their internal and consultative processes to prevent a controversy like this from happening again (ibid.). Up to the date of writing this article, the criminal case remains unresolved.

This case is a clear example of the potential effects that Parabon's use of FDP technology brings. From one DNA sample, Parabon produces *one* portrait image, one face. And while that face, at first glance, conveys the same fine-grained idea of unity and singularity as DNA, that imagery vanishes, as the case so well demonstrates, with the acknowledgment of its breadth within society. The portrait created through DNA has a coarse-grained (Bowker and Star 1999) translation outside the forensic laboratory. In simple words, despite the illusion of an individual genetically based portrait, it produces a broad and stereotyped generic face from which *too many* faces can be drawn.

In addition, the Edmonton Police case also constitutes an exemplary case of the potential for overinvestment in innovative DNA technologies by police forces and, simultaneously, on the effects of disinformation on what type of data the application of specific technologies can bring to the case under investigation. Given that the police already had a description from the victim of the suspect as a Black man, it is unclear what the purpose of the DNA-based portrait was. In this case, promises of accuracy surpassed the possible social and political consequences: "accuracy measures ascribe to them a power of social prediction and categorization without assessing the underlying social and political risks, hidden behind counter-intuitive properties or uninterpretable mechanisms" (Campolo and Crawford 2020, 12).

We therefore argue that the Edmonton police case is an example of the enchanted determinism (Campolo and Crawford 2020) effect of the Parabon representation of FDP. At the heart of this case is the promise, promoted by Parabon, to provide a DNA-based portrait. Attracted by the alleged technical accuracy of predicting social identities, in this case the social identity of the criminal suspect, the police ended up disregarding: (i) the deeper contextual knowledge of how these predictions are made, which in this case refers to the black-box knowledge of Parabon's product, protected by commercial logics of secrecy and patent control (Wienroth 2018a; Granja and Machado 2020); (ii) Parabon's limitations, namely the possibility of producing a DNA-generated portrait, which is highly contested by most members of the forensic genetics community directly working with FDP—something that, as previously noted, was also voiced by geneticist Adam Rutherford when he commented on the case; and (iii) more importantly in this particular case, the impact that these predictions have in the world by focusing on a racialized and highly stigmatized community, as the community backlash so well demonstrates.

In addition to the aforementioned impacts on racialized communities, Snapshot's results affect not only the company's public image and scientific credibility, but also the perceived validity, scientific credibility, and precision associated with FDP in the context of forensic science applications. The controversy surrounding the Edmonton Police Service has sparked movements advocating for a ban on using FDP technology. This ban targets not only Parabon as a forensic services provider or this particular service, Snapshot, but calls for a complete prohibition on the use of this technology.

This case therefore raises complex questions about accountability. As noted by Campolo and Crawford (2020, 1): "Enchantment shields the creators of these systems from accountability while its deterministic, calculative power intensifies social processes of classification and control." Although the Edmonton Police Service issued a statement of apology and announced the immediate removal of Parabon's DNA-based portrait from the police website and social networks, as of the time of writing and to the best of our knowledge, the company responsible for producing such data not only keeps the suspect's DNA-based portrait on their website, which is still linked to (and, thus, identifies) the Edmonton Police Service,[11] but also has not taken a stand, thereby distancing themselves from ethical responsibility and legal liability (see Campolo and Crawford, 2020, 3).

6. Final Remarks

This chapter aims to contribute to the body of literature focusing on forensic genetics and their technolegal worlds (Toom, Wienroth, and M'charek 2023) by fostering new avenues of discussion related to the complexities associated with the use of FDP in the criminal justice system. It focuses on two publicly available cases: an artistic intervention by Heather Dewey-Hagborg and the criminal case of the Edmonton Police Service that made use of Parabon's Snapshot service. The chapter mobilizes such cases as platforms for discussing how FDP: (i) ends up materializing a coarse-grained approach; (ii) fails to capture the multiplicity of social categories, such as gender identity; (iii) holds the potential to reinforce stigmatization and criminalization of certain population groups.

The allure of producing an image of suspects has irrevocably marked the history of police work, as shown, for example, by the use of composite images proposed by Galton and, later, by sketches. In both cases, there is something appealing in supposedly creating a generalized image of the criminal suspect that does not necessarily represent an individual but characteristics shared by a group. As outlined by Galton, the goal is to create an image "that represents no man in particular, but portrays an imaginary figure possessing the average features of any given group of men. Nobody who glanced at [it] for the first time, would doubt its being the likeness of a living person, yet ... it is the portrait of a type and not of an individual" (Galton 1879, 132–133). Such conceptualization shares remarkable similarities with the coarse-grained approach of FDP.

Although the proponents of FDP argue that this technology surpasses limitations of previous visual technologies, it ends up reproducing similarities by relying on broad classification systems to create "a materialized image of a suspect's likeness [that] is never final" (Bleumink, Jong, and Plájás 2021, 29). Therefore, despite claims of enabling a fine-grained approach due to the informational potential and specificity of DNA, the cases presented in this chapter show that FDP ultimately operates within and creates a coarse-grained context. On the one hand, the case of *Probably Chelsea* is particularly useful for showing how FDP ignores the complexities that might exist between inferring sex from

11 https://snapshot.parabon-nanolabs.com/posters (accessed February 27, 2023).

genetics and transposing it to appearance. On the other hand, the Edmonton Police case pinpoints how the production of a genetically based individual portrait as possessing inherent value to criminal investigations is illusory. Such a portrait fails to provide specific individualizing information about the suspect's unique physical appearance. Instead, it casts suspicion within a group of individuals who possess the same set of characteristics (M'charek and Wade 2020). Thus, rather than assisting the process of identifying the suspect, the portrait creates a broad and stereotyped generic face, from which *too many* faces can be derived.

At the same time, and similarly to earlier visual technologies, the DNA-based portrait reinforces a particular image of the suspect in people's minds. It appears to be authentic and conclusive, conveying the idea that "*this* is what the suspect looks like." This happens, as M'charek posits, because "the face—just like DNA—is associated with individuality" (2020, 1). But while the DNA-generated image appears to be authentic and conclusive, as both cases demonstrate, it paradoxically omits all the other *many* possible interpretations of genetic data that are not represented in that *one* portrait.

In addition to reproducing old patterns and contributing to the production and reification of stereotypes with crucial implications for the affected communities, the cases analyzed in this article also raise issues of liability, responsibility and accountability. The Edmonton Police case particularly illustrates how the use of products from for-profit forensic providers can affect the perceived validity, scientific credibility, and accuracy associated with FDP in the context of forensic science applications. Shielded by commercial logics of secrecy and patent control and anchored on accuracy promises, Parabon's service creates liminal spaces of accountability that distance it from ethical responsibility and legal liability (see Campolo and Crawford, 2020, 3). The case of FDP is therefore particularly useful for interrogating the regulatory norms, social values, and competing interests entailed in the technolegal worlds of forensic genetics (Toom, Wienroth, and M'charek 2023), considering its situated and broad legal, cultural, and political implications.

Acknowledgments

This work has received funding from the Fundação para a Ciência e Tecnologia (FCT—Portuguese Foundation for Science and Technology) under the Programme Scientific Employment Stimulus—Individual Call, attributed to Filipa Queirós (2021.00686. CEECIND/CP1698/CT0006) and Institutional Call, attributed to Rafaela Granja (CEECINST/00157/2018). In addition, this work is also supported by the FCT under the project UIDB/00736/2020 (base funding), UIDP/00736/2020 (programmatic funding) and the pluriannual R&D funding (UIDP/50012/2020).

The authors would like to express their gratitude to the editors of this book and to the participants of the workshop "*Mere labels, racial concepts, or scientific methods? The politics, epistemologies and ethics of naming populations in the life sciences and beyond*" for their insightful comments on an earlier version of this chapter. The authors also would like to thank the anonymous reviewers for their very helpful and insightful comments.

7. References

Bartram, Isabelle, Tino Plümecke, and Susanne Schultz. 2022. "Genetic Racial Profiling." *Science & Technology Studies* 35 (3): 44–69. DOI: https://doi.org/10.23987/sts.101384.

Bleumink, Ryanne, Lisette Jong, and Ildikó Zonga Plájás. 2021. "Composite Method." *Science & Technology Studies* 34 (3): 17–37. DOI: https://doi.org/10.23987/STS.80293.

Børsting, Claus, and Niels Morling. 2015. "Next Generation Sequencing and Its Applications in Forensic Genetics." *Forensic Science International: Genetics* 18: 78–89. DOI: https://doi.org/10.1016/j.fsigen.2015.02.002.

Bowker, Geoffrey, and Susan L. Star. 1999. *Sorting Things Out: Classification and Its Consequences*. Cambridge, MA and London: The MIT Press. DOI: https://doi.org/10.1109/MAHC.2000.841148.

Brown, Nik, and Mike Michael. 2003. "A Sociology of Expectations: Retrospecting Prospects and Prospecting Retrospects." *Technology Analysis & Strategic Management* 15 (1): 3–18. DOI: https://doi.org/10.1080/0953732032000046024.

Burney, Ian, and Neil Pemberton. 2013. "Making Space for Criminalistics: Hans Gross and Fin-de-Siècle CSI." *Studies in History and Philosophy of Biological and Biomedical Sciences* 44 (1): 16–25. DOI: https://doi.org/10.1016/j.shpsc.2012.09.002.

Butler, Judith. 1993. *Bodies That Matter: On the Discursive Limits of "Sex."* New York: Routledge.

Callon, Michel. 1986. "Some Elements in the Sociology of Translation: Domestication of the Scallops and the Fishermen of St. Brieuc Bay." In *Power, Action and Belief: A New Sociology of Knowledge*, edited by John Law, 196–223. London: Routledge.

Campolo, Alexander, and Kate Crawford. 2020. "Enchanted Determinism: Power without Responsibility in Artificial Intelligence." *Engaging Science, Technology, and Society* 6: 1–19. DOI: https://doi.org/10.17351/ests2020.277.

Cole, Simon. 2013. "Forensic Culture as Epistemic Culture: The Sociology of Forensic Science." *Studies in History and Philosophy of Biological and Biomedical Sciences* 44 (1): 36–46. DOI: https://doi.org/10.1016/j.shpsc.2012.09.003.

Duster, Troy. 2008. "DNA Dragnets and Race: Larger Social Context, History and Future." *GeneWatch* 21 (3–4): 3–5.

Edmonton Police Service. 2022. "Police Use DNA Phenotyping in Unsolved Sexual Assault." Edmonton Police Service Webpage. 2022. https://www.edmontonpolice.ca/News/MediaReleases/DNAPhenotypeOct4.

Galton, Francis. 1879. "Composite Portraits, Made by Combining Those of Many Different Persons into a Single Resultant Figure." *The Journal of the Anthropological Institute of Great Britain and Ireland* 8: 132. DOI: https://doi.org/10.2307/2841021.

Gibson, Caley. 2022. "Edmonton Police Apologize after Concerns Raised over Release of DNA Phenotyping Composite Sketch." *Global News*, 6 October 2022. https://globalnews.ca/news/9181617/edmonton-police-apology-dna-phenotyping-composite-sketch/.

Granja, Rafaela, and Helena Machado. 2020. "Forensic DNA Phenotyping and Its Politics of Legitimation and Contestation: Views of Forensic Geneticists in Europe." *Social Studies of Science* 53 (6): 1–19. DOI: https://doi.org/10.1177/0306312720945033.

Greenwood, Veronique. 2016. "How Science Is Putting a New Face on Crime Solving." *National Geographic*, July 2016. https://www.nationalgeographic.com/magazine/2016/07/forensic-science-justice-crime-evidence/.

Hopman, Roos. 2020. "Opening up Forensic DNA Phenotyping: The Logics of Accuracy, Commonality and Valuing." *New Genetics and Society* 39 (4): 1–17. DOI: https://doi.org/10.1080/14636778.2020.1755638.

Hopman, Roos. 2021. "The Face as Folded Object: Race and the Problems with 'Progress' in forensic DNA phenotyping." *Social Studies of Science* 53 (6): 1–22. DOI: https://doi.org/10.1177/03063127211035562.

Hopman, Roos, and Ryanne Bleumink. 2023. "Between Pencils and Genetic Markers: Rethinking Innovation in Policing through Forensic Face-Making Technologies." *International Journal of Police Science and Management* 25 (3): 280–296. DOI: https://doi.org/10.1177/14613557231173.

Hopman, Roos, and Amade M'charek. 2020. "Facing the Unknown Suspect: Forensic DNA Phenotyping and the Oscillation between the Individual and the Collective." *BioSocieties* 15: 438–462. DOI: https://doi.org/10.1057/s41292-020-00190-9.

Introna, Lucas, and David Wood. 2002. "Picturing Algorithmic Surveillance: The Politics of Facial Recognition Systems." *Surveillance & Society* 2 (2/3): 177–198. DOI: https://doi.org/10.24908/ss.v2i2/3.3373.

Issawi, Hamdi. 2022. "Africa Centre, Black Community Groups, Call for End to Police DNA Phenotyping." Edmonton Journal, 20 October 2022. https://edmontonjournal.com/news/local-news/africa-centre-black-community-group-call-for-end-to-police-dna-phenotyping?utm_medium=Social&utm_source=Twitter#Echobox=1666272594.

Kayser, Manfred. 2015. "Forensic DNA Phenotyping: Predicting Human Appearance from Crime Scene Material for Investigative Purposes." *Forensic Science International: Genetics* 18: 33–48. DOI: https://doi.org/10.1016/j.fsigen.2015.02.003.

Kayser, Manfred, Wojciech Branicki, Walther Parson, and Christopher Phillips. 2023. "Recent Advances in Forensic DNA Phenotyping of Appearance, Ancestry and Age." *Forensic Science International: Genetics* 65: 102870. DOI: https://doi.org/10.1016/J.FSIGEN.2023.102870.

Kayser, Manfred, and Peter de Knijff. 2011. "Improving Human Forensics through Advances in Genetics, Genomics and Molecular Biology." *Nature Reviews Genetics* 12 (3): 179–192. DOI: https://doi.org/10.1038/nrg2952.

Kayser, Manfred, and Peter Schneider. 2009. "DNA-Based Prediction of Human Externally Visible Characteristics in Forensics: Motivations, Scientific Challenges, and Ethical Considerations." *Forensic Science International: Genetics* 3 (3): 154–161. DOI: https://doi.org/10.1016/j.fsigen.2009.01.012.

Koops, Bert-Jaap, and Maurice Schellekens. 2008. "Forensic DNA Phenotyping: Regulatory Issues." *The Columbia Science and Technology Law Review* IX: 158–202.

Latour, Bruno. 1986. "The Powers of Association." In *Power, Action and Belief: A New Sociology of Knowledge*, edited by John Law, 264–280. London: Routledge.

Latour, Bruno. 1992. "Where Are the Missing Masses? The Sociology of a Few Mundane Artefacts." In *Shaping Technology/Building Society: Studies in Sociotechnical Change*, edited by Wiebe Bijker and John Law, 225–258. Cambridge: MIT Press.

Lipphardt, Veronika. 2018. "Vertane Chancen? Die Aktuelle politische Debatte um erweiterte DNA-Analysen in Ermittlungsverfahren*." *Berichte zur Wissenschaftsgeschichte* 41 (3): 279–301. DOI: https://doi.org/10.1002/bewi.201801900.

Lombroso, Cesare. 1876. L'Uomo Delinquente [O Homem Delinquente]. Milan: Hoepli.

Lynch, Michael, Simon Cole, Ruth McNally, and Kathleen Jordan. 2008. *Truth Machine: The Contentious History of DNA Fingerprinting*. Chicago: University of Chicago Press.

Machado, Helena, and Barbara Prainsack. 2012. *Tracing Technologies: Prisoners' Views in the Era of CSI*. Farnham, UK: Ashgate.

M'charek, Amade. 2008. "Silent Witness, Articulate Collective: DNA Evidence and the Inference of Visible Traits." *Bioethics* 22 (9): 519–528. DOI: https://doi.org/10.1111/j.1467-8519.2008.00699.x.

M'charek, Amade. 2013. "Beyond Fact or Fiction: On the Materiality of Race in Practice." *Cultural Anthropology* 28 (3): 420–442. DOI: https://doi.org/10.1111/cuan.12012.

M'charek, Amade. 2020. "Tentacular Faces: Race and the Return of the Phenotype in Forensic Identification." *American Anthropologist* 122 (2): 369–380. DOI: https://doi.org/10.1111/aman.13385.

M'charek, Amade, and Peter Wade. 2020. "Doing the Individual and the Collective in Forensic Genetics: Governance, Race and Restitution." *BioSocieties* 15 (3): 317–328. DOI: https://doi.org/10.1057/s41292-020-00193-6.

Miranda, Diana, and Helena Machado. 2018. "Photographing Prisoners: The Unworthy, Unpleasant and Unchanging Criminal Body." *Criminology & Criminal Justice* 19 (5): 591–604. DOI: https://doi.org/10.1177/1748895818800747.

Nieves Delgado, Abigail. 2020. "The Face of the Mexican: Race, Nation, and Criminal Identification in Mexico." *American Anthropologist* 122 (2): 356–368. DOI: https://doi.org/10.1111/AMAN.13383.

Okere, Enyinnah. 2022. "EPS Statement on DNA Phenotyping." Edmonton Police Service. 2022. https://www.edmontonpolice.ca/News/MediaReleases/StatementonDNAPhenotyping.

Ossorio, Pilar N. 2006. "About Face: Forensic Genetic Testing for Race and Visible Traits." *Journal of Law, Medicine & Ethics* 34 (2): 277–292. DOI: https://doi.org/10.1111/j.1748-720X.2006.00033.x.

Plájás, Ildikó Zonga. 2023. "InterFaces: On the Relationality of Vision, Face and Race in Practices of Identification. A Multimodal Intervention." *Social Studies of Science* 53 (6): 938–953. DOI: https://doi.org/10.1177/03063127231151237.

Pollack, Andrew. 2015. "Building a Face, and a Case, on DNA." New York Times, 23 February 2015. https://www.nytimes.com/2015/02/24/science/building-face-and-a-case-on-dna.html.

Queirós, Filipa. 2021. "The (Re)Invocation of Race in Forensic Genetics through Forensic DNA Phenotyping Technology." In *Racism and Racial Surveillance*, edited by Sheila Khan, Nazir Can, and Helena Machado, 199–222. London: Routledge. DOI: https://doi.org/10.4324/9781003014300-12.

Sekula, Allan. 1986. "The Body and the Archive." *October* 39: 3–64. DOI: https://doi.org/10.2307/778312.

Sharma, Vishakha, Krupa Jani, Pavan Khosla, Erin Butler, Donald Siegel, and Elisa Wurmbach. 2019. "Evaluation of ForenSeqTM Signature Prep Kit B on Predicting Eye and Hair Coloration as Well as Biogeographical Ancestry by Using Universal Analysis Software (UAS) and Available Web-tools." *Electrophoresis* 40 (9): 1353–1364. DOI: https://doi.org/10.1002/elps.201800344.

Skinner, David. 2018. "Race, Racism and Identification in the Era of Technosecurity." *Science as Culture* 29 (1): 77–99. DOI: https://doi.org/10.1080/09505431.2018.1523887.

Southall, Ashley. 2017. "Using DNA to Sketch What Victims Look like – Some Call It Science Fiction." The New York Times, 19 October. https://www.nytimes.com/2017/10/19/nyregion/dna-phenotyping-new-york-police.html?_r=0.

Taylor, Karen T. 2000. *Forensic Art and Illustration*. Boca Raton: CRC Press.

Toom, Victor, Matthias Wienroth, and Amade M'charek. 2023. *Law, Practice and Politics of Forensic DNA Profiling: Forensic Genetics and Their Technolegal Worlds*. New York: Routledge.

Toom, Victor, Matthias Wienroth, Amade M'charek, et al. 2016. "Approaching Ethical, Legal and Social Issues of Emerging Forensic DNA Phenotyping (FDP) Technologies Comprehensively: Reply to 'Forensic DNA Phenotyping: Predicting Human Appearance from Crime Scene Material for Investigative Purposes' by Manfred Kayser." *Forensic Science International: Genetics* 22: e1–4. DOI: https://doi.org/10.1016/j.fsigen.2016.01.010.

Vailly, Joëlle. 2017. "The Politics of Suspects' Geo-Genetic Origin in France: The Conditions, Expression, and Effects of Problematisation." *BioSocieties* 12 (1): 66–88. DOI: https://doi.org/10.1057/s41292-016-0028-x.

Walsh, Susan, and Manfred Kayser. 2016. "Predicting Human Appearance from DNA for Forensic Investigations." In *Handbook of Forensic Genetics: Biodiversity and Heredity in Civil and Criminal Investigation*, edited by António Amorim and Bruce Budowle, 415–448. New Jersey: World Scientific.

Wienroth, Matthias. 2018a. "Socio-Technical Disagreements as Ethical Fora: Parabon NanoLab's Forensic DNA SnapshotTM Service at the Intersection of Discourses around Robust Science, Technology Validation, and Commerce." *BioSocieties* 15 (1): 28–45. DOI: https://doi.org/10.1057/s41292-018-0138-8.

Wienroth, Matthias. 2018b. "Governing Anticipatory Technology Practices. Forensic DNA Phenotyping and the Forensic Genetics Community in Europe." *New Genetics and Society* 37 (2): 137–152. DOI: https://doi.org/10.1080/14636778.2018.1469975.

Williams, Damien Patrick. 2020. "Fitting the Description: Historical and Sociotechnical Elements of Facial Recognition and Anti-Black Surveillance." *Journal of Responsible Innovation* 7 (sup1): 74–83. DOI: https://doi.org/10.1080/23299460.2020.1831365.

ANI versus ASI, or Aryans versus Dravidians?

Science-Politics Entanglements in the Making of New Categories of Difference in India

Thiago P. Barbosa

Abstract *The article examines the entanglements between population genetics of India and its deep-rooted political implications by analyzing the construction of the categories "Ancestral North Indian" (ANI) and "Ancestral South Indian" (ASI). Beginning with the historical backdrop of the Aryan migration theory and its connection to racial frameworks, I explore how recent (a)DNA research has rekindled debates about ancestral origins, the caste system, and inequality in India. Scrutinizing the crafting of ANI and ASI categories by the US-based geneticist David Reich and his collaborators, I discuss these scientists' attempts to disentangle genetics from politics and show how they, however, inadvertently align with Hindu nationalist narratives. Examining the limitations and implications of this nationalization of ancestral categories, the article reflects on the inescapable politicization of such terms and their exclusionary effects. I argue that the persistent challenges posed by the legacies of race within the sciences of human diversity cannot be resolved solely through categorical rebranding or by efforts to isolate science from politics. The conclusion argues for a more nuanced and reflexive understanding of science-politics entanglements and gestures toward the importance of community involvement to better navigate the ethical and political complexities of population genetics.*

Introduction

Discussions about human diversity in India have long simmered around the search for the origins of human differences in the past. How did differentiations among social groups come about? How did the so-called caste system emerge? Do the different castes in India have different ancestries? In dealing with these questions, many scholars have focused on the "Aryan migration theory" (AMT), earlier mostly known as Aryan invasion theory. Theories about an Aryan *people* or *race* took on a prominent role in Nazi ideology, and, in fact, nineteenth-century German Indologists played a key role in theorizing and popularizing ideas about Aryan ancestry, often in an imagined dichotomy with another racialized group, whether Jews in Germany or so-called Dravidians in South Asia (Figueira 2002). Accounts of AMT encompass the idea that an ancestral group of people known as Aryans migrated from Eurasia down to India a few thousands of years BCE, encountering (or conquering) at least one other ancestral original population, known in some accounts as Dravidians. The many various versions of this model define Aryans and Dravidians in different ways as linguistic, racial, and/or ancestral groups that shaped the diversity of current populations in the Indian subcontinent. Many human groups usually in the North of India are identified with an ancestry/language/race that is Aryan or Indo-European (depending on the model's version), while most groups in the South are identified with the one defined as Dravidian.

AMT became an "obligatory passage point" (Latour 1987, 132) in inquiries about ancestry and the peopling of India. While AMT had been laid to rest in the past, the recent development of new DNA analysis technology reopened this debate. However, this molecular reassessment of old questions about origins has not been free of controversies—both political and scientific. The reassessment of this Indian ancestral history debate through new genomic and ancient DNA technology has stirred up tensions regarding social differences and inequality in the subcontinent. This is also because many versions of AMT had located South Asia's caste differentiations with the incoming Aryans, who are in some accounts regarded not only as the ancestors of today's uppermost caste groups (such as Brahmins) but also as the writers of texts that are considered foundational to Hinduism. Concomitant with the recent rise of Hindu nationalism, this new research refueled and reignited debates about diversity and inequality in India, about the relation of casteism to AMT and the relation between casteism and racism, and, more specifically, about the concepts and categories used to denote ancestral groups (Egorova 2009, 2010; Subramaniam 2019).

In this scenario, the terms "Ancestral North Indian" (ANI) and "Ancestral South Indian" (ASI) have recently emerged in the debate. In this article, I draw from ethnographic research and a close reading of population genetics publications to examine the political and scientific entanglements in which the categories ANI and ASI were crafted.[1] Against the backdrop of the racial legacies in scientific inquiries about human diversity in India, I discuss how key scientists in this field deal with what they see as the "trouble" of the entanglement of their sciences with politics. In the following section, I provide a background on discussions on AMT since the nineteenth century and its relation to racial theories. Then, I analyze how, and with what political reverberations, recent studies on population genetics reignited the flame of racialized understandings of caste and difference in India. The subsequent section examines how the crafting of ANI and ASI relied on inventive—rhetorical, methodological, and practical—efforts by scientists such as David Reich and his collaborators in their attempt to overcome the political (and racialized) weight of other categories of difference but, as a result, ended up being complicit with Hindu nationalistic political framings and other discourses of Indian national integration. Finally, I discuss the limitations and implications of the nationalization of categories for ancestral difference and reflect on the entanglements between science and politics in the field of population genetics of India. I argue that while contemporary scientists might perceive national terms as a suitably "apolitical" replacement for troubling

1 I conducted ethnographic fieldwork between 2017 and 2020 in India, whereby I interviewed several human geneticists and anthropologists and participated at their scientific meetings.

racialized categories, in fact national categories are inevitably politicized and can have important exclusionary consequences. I conclude by contending that the persistent troubles of race in sciences of human diversity will not be solved by simple categorical rebranding or by isolating the realm of science from politics.

Racial Categories in an Old Debate

Inquiries about the migration of so-called Aryans to what is now the territory of India have been replete with ambiguities as much as they have had heightened political affordances. Thapar (2019) explains that defining the entity "Aryan" has been a problem since it was referred to for the first time several centuries ago. "In the public mind," writes Thapar (2008, 1), "the meaning of the term [Aryan] remains a bit vague and arbitrarily conflates race, ethnicity, culture, language, religion and geography." Similar to "race," as much as the term "Aryan" might be ambiguous, it has also been remarkably elastic and therefore enabled different interpretations and rearticulations in public and scientific discourses.

The history of studies on Aryans coming into India goes back to the late eighteenth century, when the British orientalist Sir William Jones talked about an ancient "Aryan invasion" and proposed "the idea of a racial difference existing between northern and southern Indians, and between high and low castes" (Bates 1995, 233). While Jones's theories "were only weakly supported by linguistic and archaeological evidence" (ibid.), they were taken up in the nineteenth-century scholarship on ancient Sanskrit scriptures. German Oxford professor Friedrich Max Müller, for instance, was emblematic of the time in theorizing on the similarities between Sanskrit and European languages to then elaborate that Sanskrit-speaking Aryans moved from somewhere in Eurasia down to India en masse, in approximately 3000 BCE, conquering the Aboriginal Dravidian-speaking peoples of the subcontinent (Thapar 2008). Building upon the widespread perception that South Indians generally have darker skin color tones than North Indians, Müller used the word "race" to describe these two linguistic groups and, based on philological evidence, asserted that the Aryans described the people they encountered as "noseless" and "dark," which he hastily interpreted as racial markers (Trautmann 1997; Channa 2003). Along these lines, Müller suggested a racial distinctiveness and ancestral origin of the speakers of what he called Indo-Aryan languages (the Sanskrit-related modern languages spoken in the northern areas of the Indian subcontinent), which he contrasted to the Dravidian languages mostly spoken in the Indian South (Channa 2003; Sur 2011). The binary racial contrast between Aryans and Dravidians runs parallel to the contrast, also drawn by Müller, between Aryans and Semites; in both binary constructions, the Aryans stood as an idealized ancestral people, over which eugenicist, nationalistic (and masculinist) mythical fantasies would be projected, not only in India but also, notoriously, in Germany (Figueira 2002).

This orientalist work, in turn, inspired the racial classifications of the British colonial administrator and anthropologist Sir Herbert Risley, who was responsible for formulating a methodology for government surveys on "castes" and "tribes" of British India, to which he included a series of anthropometric measurements. Risley believed that anthropometry could distinguish between Aryan and non-Aryan descent among different tribal and caste groups. For Risley, the migration of Aryans was at the root of the preserved diversity of Indian populations: he attributed the institution of caste endogamy to the Aryans' wish to maintain their racial distinctiveness. For him, the idea of a "community of race ... is the real determining principle, the true *causa causans* of the caste system" (Risley 1891, 259). Risley elaborated a taxonomy which prescribed that "the people of India" could be racially classified mainly in two main types: Aryan and Dravidian. Thus, like Müller's, Risley's racial typological reasoning was shaped by a two-race historical model that considered Aryans as the major migrating group to the interior of India, which had previously been inhabited by groups that could be racially classified as Dravidians (Sur 2011).

Risley's ideas were discussed by subsequent generations of Indian scholars.[2] Sociologist-anthropologist G. S. Ghurye argued in his famous book *Caste and Race* (1957 [1932]) that caste endogamy emerged from a racial purity wish among upper-caste Brahmins, who he believed were incoming Aryans, with an ancestral Nordic racial origin. Ghurye based this argument on the interpretation that "varna," the Sanskrit word that describes the four main ranks of castes, also meant "color," which he attributed as a sign that this *varna* differentiation corresponded to skin color differences. In this logic, the four *varnas*—Brahmins, Kshatriyas, Vaishyas, and Shudras—had different skin color tones. In sum, Risley's, Ghurye's, and many other scholars' racial articulations of AMT composed what Trautmann summarizes as a "racial theory of Indian civilization":

> This is the theory that the light-skinned, Sanskrit-speaking Aryans clashed with the dark-skinned indigenous peoples of India, speaking non-Indo-European languages, whom they conquered. The sequel was the creation of the caste system, understood (in this perspective) to be the enduring constitution of Indian civilization, keeping the peoples of India separate from one another in respect of marriage ... (Trautmann 2019, 272).

In turn, intellectuals opposed to casteism criticized some aspects of this racial theory of Indian civilization but not necessarily the Aryan narrative or the racial framework embedded in it. Whereas Ghurye—himself an upper-caste Brahmin—stopped short of questioning caste and the principle of racial purity, anticasteist intellectuals such as Phule theorized on the Brahmanical origin of caste endogamy with the goal of criticizing it. Instead of justifying the institution of caste by Brahmins, Phule denounced it, blaming the *invading* Aryans' Brahmanist eugenicist thinking for the intentioned caste segregation and inequality in India (Figueira 2002). Thereby, Phule and other anticasteist intellectuals still articulated the Aryan migration/invasion theory in their writings, even

---

2 While Risley's ideas resonated widely, they were also contested, including by the German Indologist Müller, who later questioned the use of racial anthropometric methods to study the distinction of modern descendants of Aryans and Dravidians. Although Müller had suggested the equation of racial and language groups in many previous texts (Thapar 2008), in response to Risley Müller advocated for a strict separation between what he called "a phonological race" and a "ethnological race," cautioning against the collation between language and anthropometrically observed physical appearance (Müller 1891, 179). However, "Müller spoke too little and too late" (Figueira 2002, 45).

ascribing upper-caste communities with Aryanness both in the sense of origin and racial traits. By doing so, they argued that invading Aryans oppressed—and, through instituting the rules of caste endogamy and untouchability, segregated themselves from—the original inhabitants of India. This account presented these original inhabitants of India as the ancestors of those who today suffer the most from caste segregation, namely low-castes, Dalits, and Adivasis. Hence while these theories for the origin of the caste system were anticasteist in their political orientation, they nevertheless tended to confirm long-standing racial frameworks in the search for the origins of difference and inequality among these social groups.

Concurrently, most positions within the discourse of Hindu nationalism today oppose such anticasteism-oriented Aryan *invasion* theory articulations and stress that the Aryans' homeland is India. Because the Aryans were believed by many to be the writers and protagonists of scriptures that are considered foundational to Hinduism, many Hindu nationalist narratives claim that the Aryans were the ancestors of the Hindus. Therefore, such narratives associate the birth of Hinduism and by extension the territoriality of India with the homeland and ancestral holy land of the Aryans and, by extension, the Hindus.[3] As Hindu supremacy ideologue Savarkar (1989 [1923], 85) declared: "All Hindus claim to have in their veins the blood of the mighty race incorporated with and descended from the Vedic fathers." This racial and religious association between Hindus and the glorious times of the Aryans of the Vedic scriptures infuses this discourse of Hinduness, also known as Hindutva, with a profound weight of timeless tradition, bonding it with a patrilineal familial-reproductive temporality (today's Hindus being the children of the "Vedic fathers" of the past). The temporal and affective affordances granted by the Aryan figure in association with the roots of Hinduism in India are key to the imaginations of a national belonging in which Hindutva is seen as the country's *Leitkultur*, the leading cultural bearer that grants social cohesion to the nation. Some take this narrative as far as to claim that the heartland of Aryan civilization was in India and that the Aryans migrated from there toward Europe. Some versions of this narrative argue that AMT "is a colonial conspiracy that erases India's glorious precolonial history" (Subramaniam 2019, 150).

Thus, the ongoing aggravation and mainstreaming of Hindu nationalistic discourses since prime-minister Modi's strident Hindu nationalistic government (2014–present) amplifies the political stakes of research about Aryan migration. Because arguments that challenge the sacred territoriality and origins of Aryans (and their descendants) can be read as anti-Hindu(tva) and antinational, scholarly articulations and activist mobilizations of Aryan migration/invasion theories have had a pronounced political potency.

3 One needs to note, nevertheless, that the current Indian homeland–oriented discourse on Hinduism is also a rather recent historical formation, which is tightly enmeshed with the mainstreaming of Hindu nationalism. There were widespread positions in the previous centuries that rather aligned Vedic scriptures with a foreign (often even Nordic) Aryan origin (e.g., Tilak 2011 [1903]; see also Banerjee 2016).

Troublesome Legacies in the Population Genetics of India

As a result of these scientific and political debates on ancestry and origins in India, up to this date Aryan migration stories carry very high political stakes that are inevitably tied to caste-related anxieties. These anxieties also accompany new studies based on genetics research that have, since the 2000s, revisited these stories of ancient migrations to India. Even when genetics narratives on ancestral migrations to India are not articulated with the loaded AMT terminology, population geneticists' euphemizing deployment of continental categorizations did not prevent anxious public repercussions, similar to the old troubles of racializing Aryan migration stories.

An emblematic case in this regard is the controversy that followed the publication of the article "Genetic Evidence on the Origins of Indian Caste Populations," first-authored by Utah-based geneticist Bamshad (et al. 2001). Tracing ancestry in distinction between paternally inherited Y-chromosome and maternally inherited mitochondrial DNA, the paper asserted that "European" or "West Eurasian" *males* migrated to the Indian subcontinent and, once there, either happened to be accepted as belonging to upper-caste groups or formed a caste system hierarchy in which they were at the top. Although the paper deflected from the term "Aryan," its argument confirmed a key assumption in some versions of AMT discussed above, namely that the Aryans were either the direct ancestors of upper castes or that they were responsible for caste differentiations and segregation as known today. Multiple other interpretations of the paper's arguments flourished in its public rearticulations. Associating the continental category "European" used in the paper with long-standing accounts of Aryan invasion, some interpretations picked on the Europeanness or foreignness of the origins of upper-caste individuals and therefore of the caste system itself, while other responses read the evidence of a European Y-chromosome lineage as a sign of this male mass migrations' sexual violence against indigenous women in the subcontinent (Egorova 2009; 2010). This also stirred up caste-based resentments and other historical anxieties concerning sexual violence, which disproportionally affects socially minoritized—often lower caste and Dalit—women (see Rao 2005). Furthermore, anticasteism and Dalit activists mobilized Bamshad's article to sustain their plea for the inclusion of caste-based discrimination as a topic of discussion at the United Nations' 2001 World Conference against Racism in Durban. This in turn sparked a heated public debate in India on the relation between race and caste, and racism and casteism. While Indian scholars were divided on both sides of this debate (pro and against the understanding of casteism as a form of racism), the Indian government—and Hindu nationalists—successfully blocked any mention of caste in the proceedings of the World Conference (Natrajan and Greenough 2009; Egorova 2009; 2010; Benjamin 2018).

In this scenario of heightened political sensibility, some population geneticists started to newly consider how they denominated categories of ancestral populations in their research in India. The terms Ancestral North Indian (ANI) and Ancestral South Indian (ASI) came out for the first time in a 2009 paper that made the cover of *Nature*, titled "Meet the Ancestors: Indian Population History from Gene Screening." The paper was written by collaborating teams led by the Harvard-based David Reich and the Indian biologist Kuramasamy Thangaraj, a scientific director at the Centre for Cellular and Molecular Biology (CCMB) in Hyderabad. Reflecting how Thangaraj proudly displayed a sample of that edition of Nature on his coffee table in my visit to his office in 2019, the 2009 paper is considered a turning point in this scientific field.

Figure 1: Cover of Nature 461, issue 7263, 24 September 2009.

Since 2009, the acronyms ANI and ASI have constituted an integral part of the vocabulary of the population genetics of India (e.g., Moorjani et al. 2013; Haak et al. 2015; Narasimhan et al. 2019). If "any claimed connection between a present population and a hypothesized past population always arises from a specific chain of technological interventions and discursive acts" (Oikkonen 2017, 204), the formation of the hypothetical ANI and ASI populations also relied on rhetorical and practical efforts by these scientists. These efforts were driven by a worry about the political troubles of the debate.

To illuminate the scientific and political aspects of the construction of these categories of difference, in the following I examine Reich's best-selling science book *Who We Are and How We Got Here* (2018) in closer detail. In first-person narration, Reich describes in detail how his cooperation with the CCMB started and how his team, in order to solve an impasse with the Hyderabad-based scientists, came to the idea of forging the terms ANI and ASI.

Figure 2: Reich's graphic representation of the Indian Cline of northern and southern ancestry.

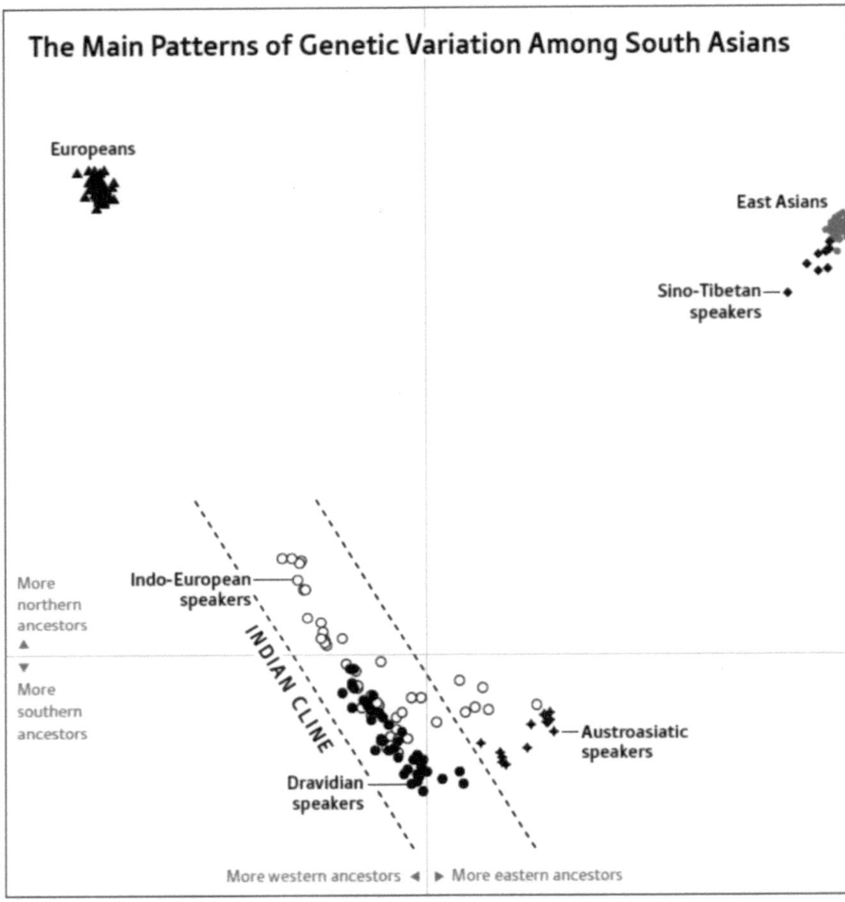

Source: Reich 2018, 131.

Reich introduces the book's chapter about his population genetics research in India by addressing the old central question in the field: AMT. He starts by lamenting that the idea of Aryans coming into India has been "difficult to discuss in an objective way" because it "has been seized on by nationalists in both Europe and India"—"including Nazis"—and blames what he calls the "politicization" of the question for the regrettable fall of the "Aryan *invasion* theory," a term that he would have otherwise preferred (Reich 2018, 125, emphasis mine). Although he acknowledges the lack of archaeological evidence that would support this invasion theory, he contends in response that evidence from genetics can solve the debate on ancient migrations to India, a country which he frames as a "the product of collisions of culture and people" (ibid., 126).

As Reich rearticulates the Aryan invasion theory in his work, his model of ancestry and admixture of Indian populations is informed by the binary of Dravidians versus Aryans, even though this assumption is not clearly laid out in his papers or book. As the India chapter in his book progresses, the argument that comes to the fore is that the binary division of Indian populations simply inductively emerges from the genetic data. Reich argues that it was with use of a new DNA analysis technology—the single nucleotide polymorphism microarray (Reich 2018, 130)—that the Indian DNA samples brought to him personally by Thangaraj could be analyzed in Harvard; the analysis showed a "gradient of variation" in which today's Indian groups (sampled as castes and tribes) seem to present different degrees of admixture between two poles: a West Eurasian–related ancestral population and another very different pole that is more prevalent in the southernmost inhabitants of India, where Dravidian languages are spoken (ibid., 132). He called this gradient "the Indian cline" (see Figure 2). Those groups of Indians outside the "Indian cline," especially Austroasiatic and Sino-Tibetan speakers, were left outside Reich's model.[4]

As Reich (2018) narrates, he did not initially have the categories ANI and ASI in mind; at first, he worked with the category "West-Eurasian" to describe the ancestral population which overlaps with accounts about Aryans and to a large extent genetically matches those who today speak Sanskrit-related languages. But then, during what he calls "the tensest twenty-four hours of [his] scientific career," he was pushed to review the terms articulated in his genetics model for India: after he presented the results of his DNA analysis in a meeting with his collaborators in Hyderabad, Thangaraj and other Indian colleagues "seemed to be threatening to nix the whole project," writes Reich (ibid., 134). According to the Harvard-based scientist, Thangaraj opposed the use of the term "West Eurasians" because it would suggest that West-Eurasian people migrated en masse into India. This was a conclusion for which—as the Indian colleagues "correctly pointed out," Reich admits—his "data provided no direct evidence" (ibid.). In fact, population genetics models usually work in a way that allows the *direction* of ancient migrations to be guessed at best; to construct such migration hypotheses, geneticists usually interpret their data on genetic similarities between populations in different geographies (Oikkonen 2017). Questioning Reich's hypothesis, the Indian colleagues reasoned that, instead, "there could have been a migration in the other direction, of Indians to the Near East and Europe"; they therefore preferred to frame the relation observed in Reich's data as "genetic sharing" instead of precipitately asserting a direction of migration in which a foreign West-Eurasian population is ancestor to current Indian populations (ibid.). Facing this confrontation, Reich writes that he "felt that we were being prevented by *political considerations* from revealing what we had found" (ibid., 135, emphasis mine).

However, it was only after thinking about and around those "political considerations" that Reich came up with the following solution to the clash between the Harvard-based team and the Hyderabad group of scientists: the categories of ancestry needed to be reframed to conform to what Reich called the "cultural resonances" of their research (ibid.). This meant that the new categories should not suggest that present-day Indian groups had a foreign ancestry.

4 This exclusion of Sino-Tibetan and Austroasiatic speakers has been a major point of criticism in Reich and Thangaraj's paper in its discussion within India (Majumder 2018, 973–974; Majumder and Basu 2015).

Making Up New Categories to Avoid the Trouble? Nationalizing "Ancestry"

How to talk about ancient migrations without disturbing a sense of a *national* population? How to avoid any trouble tied to ancestral population categories, especially when they invoke histories of foreign migrations and invasions? These were questions that Reich possibly thought about while he tried to find a way out of the conundrum in which he found himself in Hyderabad after his Indian collaborators questioned his hypothesis about ancestral West-Eurasian migrations into the Indian subcontinent.

It was through a conceptual maneuver that Reich achieved a way out of the impasse with his Indian collaborators: rebranding those ancestry categories as *Indian*. Reading between the lines of Reich's book, we see that the political background to his nationalizing terminological maneuver is present in his narration through the allegory of a national Hindu festivity: the Diwali festival marks the scene in which he was enlightened with a solution to the clash in that first meeting in Hyderabad. Reich narrates:

> That evening, as the fireworks of Diwali, one of the most important holidays of the Hindu year, crackled, and as young boys threw sparklers beneath the wheels of moving trucks outside our compound, [my Harvard colleague] and I holed up in his guest room at ... Thangaraj's scientific institute and tried to understand what was going on. The cultural resonances of our findings gradually became clear to us. So we groped toward a formulation that would be scientifically accurate as well as sensitive to these issues.
>
> The next day, the full group reconvened. We sat together and came up with new names for ancient Indian groups. We wrote that the people of India today are the outcome of mixtures between two highly differentiated populations, "Ancestral North Indians" (ANI) and "Ancestral South Indians" (ASI) .... The ANI are related to Europeans, central Asians, Near Easterners, and people of the Caucasus, but we made no claim about the location of their homeland or any migrations. The ASI descend from a population not related to any present-day populations outside India. We showed that the ANI and ASI had mixed dramatically in India. The result is that everyone in mainland India today is a mix, albeit in different proportions, of ancestry related to West Eurasians, and ancestry more closely related to diverse East Asian and South Asian populations. No group in India can claim genetic purity. (Reich 2018, 135)

It might not be coincidental that Reich's eureka moment of sensibilization toward the cultural and political resonances of his work in India—as well as the inspiration to the subsequent creative category-making—happened within the strident soundscape of Diwali, which is in fact the only Hindu holiday that is celebrated virtually all over India and not only by Hindus but also by different religious and nonreligious groups. By adding the adjective "Indian" to the categories of difference that convey the ancestry of present-day Indians, the collaborating teams of scientists truly Indianized, or nationalized, the reference point of that difference, in addition to deciding, at first, to not pinpoint the geographical "homeland" of that ancestry, which could fall, depending on the temporal reference, outside of current Indian borders. By separating language-based difference categories from ancestry, the researchers put forward a sense of distance between the existing groups of today and their ancestors, while also counterbalancing this distance through the geographical reinforcement that both ancestral groups are *Indian*.

This nationalization of difference is reinforced by the conclusion that Reich draws from that category-making meeting with Thangaraj: he asserts that "no group in India can claim genetic purity"; all Indians are mixed (Reich 2018, 135). Here, one can sense the scientists' carefulness in striving to prevent their research from giving leverage to political claims of genetic distinctiveness within India, unlike Bamshad et al.'s 2001 paper and other works that had used continental categories and ended up being mobilized by difference-based political claims. In this regard, Reich and Thangaraj's argument that "all Indians are mixed" is similar to other genetic imaginations of national unity in mixedness. While such an argument in India resonates with anthropologist Irawati Karve's post-Partition theorizations on how all Indians are "mongrels" (quoted in Barbosa 2022, 160; see also Deshpande and Barbosa 2024; Barbosa 2025; Mukharji 2022), narratives of national mixedness have also been observed internationally, for example in Latin America (Kent et al. 2015), East Asia (Tsai 2010), and the Middle East (Burton 2021). Even in national territories that have been at the "crossroads" of ancient migrations and home of supranational ethnic groups, such as Iran and Turkey, geneticists tend to nationalize categories of difference, conforming human variation to nation-state boundaries and emphasizing national admixture (Burton 2021). While the intentionality of such nationalizing moves is not always straightforward, Reich and his collaborators' insistence on the mixedness of Indians demonstrates some concern with political appropriations of their work within the country.[5]

Thus, while Reich seemed to worry about *certain* political and cultural reverberances of his work about India, he and his collaborators rearticulated knowledge based on population genetics that would achieve *certain* political affordances. Specifically, they hoped to convey genetics categories and conclusions that would conform to the idea of unity in Indian nationality while avoiding providing material for possible political mobilizations based on genetic difference that might challenge national cohesion. At the same time, they continued to investigate the same old questions related to the debate on Aryan ancestry. In publications after that tense meeting in Hyderabad, Reich and his coauthors continued to analyze their genetic data of castes and tribes to, via the new ancestry categories ANI and ASI, discuss old questions related to the origins of human diversity in India (Reich et al. 2009; Moorjani et al. 2013; Haak et al. 2015). For example, in their first joint publication, Reich, Thangaraj, and coauthors aimed to pinpoint the time of the beginning of the caste system (framed by them as a drastic reduction in genetic admixture). And while they avoided the word "Aryan," they nevertheless strove to locate the existence of ANI-related genetic markers in different Indian populations in the Indian North and South (Reich et al. 2009).

In sum, Reich continued to explore the same questions addressed by population geneticists in India at least since the mid-twentieth century, while aiming to shed new light on them through new technologies and denominations for categories of ancestry. An important continuity lies in how Reich's model maintains a dichotomy between two

5 This assertion of mixedness of Indians has also been further reflected by more popular publications in India that reported on Reich et al.'s research (e.g., Joseph 2018).

ancestral populations to explain the peopling of India, which in orientalist scholarship was framed in the binary Aryans and Dravidians but which he now framed as ANI and ASI. Besides being grounded in the Aryan vs. Dravidian binary, Reich's model might have been designed in this way also due to the sake of methodological pragmatism: statistical methods of population geneticists work better in a model that only considers two ancestral populations in the mix (see Rajagopalan and Fujimura 2012). Possibly, it was also due to their avoidance of loaded terms such as "Aryan" or "invasion" and thanks to the Indianization of categories of difference that their paper did not spark a politically explosive debate—especially if compared to the controversy unleashed by Bamshad et al. (2001). Particularly, the use of ANI and ASI and the emphasis on admixture might have muffled the paper's appropriation by political movements, making more difficult this time for minoritized social groups in India, most notably Dalits, Adivasis, and other groups affected by casteism, to claim that dominant upper-caste groups in India had a foreign (West-Eurasian or Aryan) ancestry, as they had done before. Due to the nationalizing affordances of Reich and Thangaraj's terminology and conclusions, the mobilization of such an ancestry-based claim would be semantically and discursively more complicated this time. In this sense, their paper was much less threatening to the political status quo in the current context of the mainstreaming national unity claims that have been tightly enmeshed with Hindu nationalism. In fact, their paper—and the categories of difference they invented—politically afford the maintenance of the Hindu majority status quo.

Discussion and Conclusions

Population genetics has inherited troublesome legacies of race. The tie between racial frameworks and the population genetics of India can be perceived in the use of racialized difference categories (including "caste" or "tribe") or in the employment of theories with under-analyzed assumptions that associate ancestry, heredity, geography, Volk, culture, language, and biological markers. The Aryan versus Dravidian binary model configures one of these theories with racialized categories, one that runs deep in scientific genealogies that aim at explaining the peopling of India as well as the birth of the caste system, which is associated by some accounts with the incoming Aryans. While the Aryan migration/invasion theory has been intensively discussed in India, reracializing readings of it have been reinjected in this discursive field since the advent of new DNA technology in the 2000s mainly through the work of Global North–based geneticists, such as Bamshad and Reich.

This article has shown how the entanglement between science and politics shapes not only the public reception and impact of population genetics research but also the work of these researchers in the first place. We have seen how Harvard-based geneticist Reich took up the debate on AMT but, due to the pressure of what he saw as "politicization," tried to circumvent *certain* possible political effects of his research. He did so through the strategy of a new categorical denomination, coining the now widely accepted terms Ancestral North Indian (ANI) and Ancestral South Indian (ASI). While confined to a binary ancestry model—which overlaps with the Aryan/Dravidian dichotomy—the avoidance of terms that connect this ancestry to foreignness was meant to constrict possible status quo-defying political affordances. Reich and colleagues tried to avoid the repercussion of Bamshad et al.'s paper a few years before, when anticasteism activists mobilized the foreignness associated to a Y-chromosome lineage and certain upper-caste groups with a political goal: to denounce the caste system as a violent, exogenous imposition, and to parallel casteism to racism. Instead, by nationalizing the ancestrality of the genetic lineage of the social groups sampled in India, Reich and his collaborators aimed at muffling any possible political appropriation of their research that would question the endogeneity of certain caste groups or of the caste system itself. Thereby, this nationalization of ancestry silenced the hauntings of caste by bringing the ghosts of Aryans home to India.

At the same time, rebranding categories of ancestral difference as "Indian" conformed to Hindu nationalism. What is at stake in population genetics for this religious-nationalistic discourse is the location of ancestral groups such as Aryans and, by association, the roots of Hinduism within the realm of India and Indians—even if "India" might mean, for some Hindu nationalistic accounts, a much larger territory beyond the (relatively recent and still disputed) Indian northern borders. If orientalist and colonial anthropological attempts to define "India" operate through reductionisms—often via an essentializing objectification of "Hinduness" that nevertheless formed a sense of Indian nationalism (Inden 2000)—population geneticists working with a pan-Indian national framing fall into the same traps of reductionism. In their attempt to formulate ambitious pan-Indian scale conclusions, both colonial scholars and today population geneticists can be blamed for essentializing and exclusionary reductions.

But as reductionism might be inevitable when scientists' ambitions go as far as to cover such a large scale ("India") and the human diversity encompassed in it, we can ponder what kinds of exclusions and inclusions are accentuated in such reductions and with what effects. If the exclusions inherent to Hindu supremacist discourse are more obvious, the reductionisms in this population genetics of India can also have important consequences. These scientists have made laborious efforts to stabilize categories of difference which, especially given the deep temporalities unlocked by new genomic technology, need to be temporally frozen or put in relation with scientifically fabricated ancestral populational categories. Severe reductionisms also take place in sampling strategies: we have seen that social groups that are not primarily associated with ideas of the Indian Hindu nationhood, such as speakers of Tibeto-Burmese and Austroasiatic languages—not to mention Muslims— were excluded from sampling or from the analysis. In a circular logic, these reductionist exclusions confirm these scientists' conclusions regarding the genetic image of India that they imagine and paint.

Therefore, the somewhat inevitable reductionism of both the Hindu nationalism–inclined scholarship and the population genetics of India analyzed here results in the creation of a picture of India that brings certain social groups to the fore, while excluding, invisibilizing, enclosing, or assimilating others. The hardly commensurable social, cultural, and biological diversity of human groups living under the roof of the "Indian national" is thereby met with a coherence seeking effort of *domestication* (Subramaniam 2019, 147; Oikkonen 2017, 163; Burton 2021): this human diversity is flattened or simplified through the national scale of the interpretative grids of this India-framed genetics knowledge. The genetic diversity is framed in a way that it does "not undermine the essential unity of the nation-state" (Burton 2021, 215).

Coupled with the nationalization of ancestral categories, we have also seen the emergence of an idea of an Indian national unity in shared ancestry. However, this "unity in sameness" also reinforces—and is coconstitutive of—Hindu nationalist visions of India. Thus, we can expect that the conflict-domesticating efforts of the "unity in a pan-Indian shared ancestry" rhetoric of prominent geneticists like Reich will limit the transformative power of the mobilization of biohistorical arguments by anticaste activists. In this way, the pan-Indian biological-familial portrait fabricated in these geneticizing comprehensions of India has analogous limitations to that of a biological family. From a queer critique perspective (e.g., Anzaldúa 1987), one could think of the heteropatriarchal biological family as a space where male authority-centered and adultist structures enforce norms, secure privileges, and defuse conflict. In light of the "unity in shared ancestry" biological renderings of India, too, groups that see themselves outside the pan-Indian national family ideal have great difficulty in operationalizing biohistorical claims to differ and challenge the cohesive future of the nation-state. In other words, the linear familial-reproductive temporality articulated in population genetics knowledge that frame "Indians" as having a unified genetic ancestry might have a disciplining effect against unity-defying difference-based political claims. The discursive space of the biological-national family, after all, has not been prone to impulses toward progressive or social justice-oriented sociohistorical transformations—at least not beyond its anticolonial affordance.[6]

Even if now less explosive, the debate on AMT is far from settled. Dalit intellectuals still mobilize Aryan invasion arguments in their critical assessments on the origins of caste (e.g., Yengde 2021). And population geneticists did not give up on the topic: just in 2019, a *Science* paper published by Reich, Thangaraj, and as many as 116 coauthors again touched upon the most contentious aspects of this issue, this time avoiding the word "Aryan" by mobilizing its equivalent ancestral group "Yamnaya Steppe pastoralists," but again resurrecting the argument of linkage between male (i.e., Y-chromosome) "Steppe ancestry" and the priestly caste groups, like Brahmins, that are, in their words, the "traditional custodians of literature composed in early Sanskrit" (Narasimhan et al. 2019, 12). We can, thus, expect further reverberation from this and upcoming genomic research. The multiple connections between humans that can be engendered through different temporal and spatial frames of DNA analysis can lead to flexible but also ambiguous appropriations of group-level genetic belonging (Oikkonen 2017). Therein, precisely, lies the high political appeal of population genetics knowledge.

Finally, which lessons can we learn from the science-politics entanglements in the population genetics of India? Reich's pretension of politically unthreatening—and politically undisturbed—"purely factual" scientific objectivity seems to be, in fact, never accomplishable. This case is another demonstration of the entanglement of politics and science in population genetics: it shows that genetics research that aims at constructing knowledge about a nation will ultimately be important to the politics of its respective national context. This happens not only because genetic sciences operate on national infrastructures (Burton 2021), but also because such research delivers deep biological-historical narratives that feed group-making and political subjectivity processes (Oikkonen

6 See Oikkonen (2017, 139–142) for a queer critique of linear reproductive temporality in population genetics.

2017). In fact, as Burton (2021) argues, population genetics and nationalism have been "thoroughly co-constituted." Thereby, population genetics research deals with very crucial elements related to nationhood and national belonging and exclusions. If "biologically founded pasts" might be used to "open up future prospects" and to "enforce or undermine current privileges" (Sommer 2016, 18–19), any scientific articulation on past or present human diversity might have important implications for political mobilizations about the inequalities that are intertangled with the difference categories that those scientists work with. This is especially so in the case in India, where we see a historical, intrinsic interlocking between difference, social cohesion, and social inequalities.

In sum, we can see that in this field of research, like in all scientific fields, there are several layers of questionable—but left largely unquestioned—assumptions that are built into theories, concepts, and categories, with consequences to the knowledge outcomes that, once taken up, might further shape society and politics. However, not only in Reich's (2018) writings but also among almost all the scientists I interviewed in India (from anthropologists and archaeologists to geneticists), there was a general lament about the "politicization" of scientific inquiries about the ancient past. These scientists complained about how ancestry and ancient migration research in India is often politically taken up by the media and activists, in ways that infect this field of research with politics. In other words, these scientists see the political sphere as a contaminating threat or, at best, a cause for troubling headache on the way of their objective scientific research and output. Furthermore, when they are forced to reflect about possible political impacts of their research, their reflection falls short in considering many possible detriment political implications. We have seen that Reich's effort to forge an ancestral category that would be, in his view, as politically unthreatening as possible vis-à-vis a sense of national (Hindu) religiosity ended up enabling an easier Hindu nationalistic appropriation of the team's research, while making it more difficult to be used to leverage status quo-defying arguments that would address historical caste-based inequalities. Thus, the case of Indian population genetics confirms the general observation that population geneticists—regardless to how they define or denominate categories of difference—cannot "produce politically neutral data about ancestry" (Burton 2021, 248–249).

Therefore, these scientists lack a sustained engagement with the multifaceted ways in which their science and its surrounding social world coconstitute each other. As a result, they can only insufficiently envision, reflect upon, and respond to the political implications of their research. In other words, their positivistic understanding of science severely comprises their well-intentioned efforts (when existent) of reflecting upon the consequences of their work. This is also evident in a more recent paper on global guidelines for ethical ancient DNA research, in which Reich and his coauthors (Alpaslan-Roodenberg et al. 2021) neatly separate "scientific" and "community" concerns and even argued that involving community members in decisions about publication and data sharing would be unethical. Echoing the response by Kowal et al. (2023), I argue instead that one possible orientation to provide a firm ethical ground to these scientists' response-ability vis-à-vis the many possible political troubles of population genetics is, precisely, the enhancement of community partnerships in all processes of research.

The issue of the racial legacy in sciences of human diversity will hardly be solved by relegating the responsibility for the effects of scientific knowledge to the political world as if the realm of science were in isolation from politics (see also Reardon 2005; 2008; Subramaniam 2014). Nor will the hauntings of past racial research be avoided by strategies of categorical rebranding. If the sciences of human diversity are to break free from racialization, researchers are to engage in more reflective ways with the entanglements of their sciences with the political world.

References

Alpaslan-Roodenberg, Songül, David Anthony, Hiba Babiker, et al. 2021. "Ethics of DNA Research on Human Remains: Five Globally Applicable Guidelines." *Nature* 599 (7883): 41–46. DOI: https://doi.org/10.1038/s41586-021-04008-x.

Anzaldúa, Gloria. 1987. *Borderlands: The New Mestiza = La Frontera*. San Francisco: Aunt Lute Books.

Bamshad, Michael, Toomas Kivisild, W. Scott Watkins, et al. 2001. "Genetic Evidence on the Origins of Indian Caste Populations." *Genome Research* 11 (6): 994–1004. DOI: https://doi.org/10.1101/gr.173301.

Banerjee, Subrata Chattopadhyay. 2016. "Brahmo Samaj as an Actor in the Dissemination of Aryan Invasion Theory (AIT) In India." *International Journal of Asian Studies* 13 (1): 19–59. DOI: https://doi.org/10.1017/S1479591415000182.

Barbosa, Thiago P. 2022. "Racializing a New Nation: German Coloniality and Anthropology in Maharashtra, India." *Perspectives on Science* 30 (1): 137–166. DOI: https://doi.org/10.1162/posc_a_00405.

Barbosa, Thiago P. 2025. *Racializing Caste: Anthropology between Germany and India and the Legacy of Irawati Karve (1905–1970)*. Berlin: De Gruyter.

Bates, Crispin. 1995. "Race, Caste, and Tribe in Central India: The Early Origins of Indian Anthropometry." In *The Concept of Race in South Asia*, edited by Peter Robb, 218–257. Delhi: Oxford University Press.

Benjamin, Ruha. 2018. "Recasting Race: Science, Politics, and Group-Making in the Postcolony." In *Reconsidering Race: Social Science Perspectives on Racial Categories in the Age of Genomics*, edited by Kazuko Suzuki and Diego A. von Vacano, 173–189. New York: Oxford University Press.

Burton, Elise K. 2021. *Genetic Crossroads: The Middle East and the Science of Human Heredity*. Stanford: Stanford University Press.

Channa, Subhadra. 2003. "Colonialism, Caste and the Myth of Race: A Historical Perspective on the Interaction of Indian Beliefs and Western Science." In *Understanding People of India: Anthropological Insights*, edited by A. K. Kalla and D. K. Bhattacharaya, 535–553. New Delhi: University of Delhi.

Deshpande, Urmilla, and Thiago P. Barbosa. 2024. *Iru: The Remarkable Life of Irawati Karve*. New Delhi: Speaking Tiger Books.

Egorova, Yulia. 2009. "De/geneticizing Caste: Population Genetic Research in South Asia." *Science as Culture* 18 (4): 417–434. DOI: https://doi.org/10.1080/09505430902806975.

Egorova, Yulia. 2010. "Castes of Genes? Representing Human Genetic Diversity in India." *Life Sciences, Society and Policy* 6 (3). DOI: https://doi.org/10.1186/1746-5354-6-3-32.

Figueira, Dorothy Matilda. 2002. *Aryans, Jews, Brahmins: Theorizing Authority Through Myths of Identity*. Albany: State University of New York Press.

Ghurye, Govind Sadashiv. 1957 [1932]. *Caste and Race in India*. Bombay: Popular Book Depot Bombay.

Haak, Wolfgang, Iosif Lazaridis, Nick Patterson, et al. 2015. "Massive Migration from the Steppe Was a Source for Indo-European Languages in Europe." *Nature* 522 (7555): 207–211. DOI: https://doi.org/10.1038/nature14317.

Inden, Ronald B. 2000. *Imagining India*. Bloomington: Indiana University Press.

Joseph, Tony. 2018. *Early Indians: The Story of Our Ancestors and Where We Came from*. New Delhi: Juggernaut.

Kent, Michael, Vivette García-Deister, Carlos López-Beltrán, Ricardo Ventura Santos, Ernesto Schwartz-Marín, and Peter Wade. 2015. "Building the Genomic Nation: 'Homo Brasilis' and the 'Genoma Mexicano' in Comparative Cultural Perspective." *Social Studies of Science* 45 (6): 839–861. DOI: https://doi.org/10.1177/0306312715611262.

Kowal, Emma, Laura S. Weyrich, Juan Manuel Argüelles, et al. 2023. "Community Partnerships Are Fundamental to Ethical Ancient DNA Research." *HGG Advances* 4 (2): 100161. DOI: https://doi.org/10.1016/j.xhgg.2022.100161.

Latour, Bruno. 1987. Science in Action: How to Follow Scientists and Engineers Through Society. Cambridge, MA: Harvard University Press.

Majumder, Partha P. 2018. "Understanding the Aryan Debate: Population Genetic Concepts and Frameworks." *Current Science* 114 (05): 971. DOI: https://doi.org/10.18520/cs/v114/i05/971-975.

Majumder, Partha P., and Analabha Basu. 2015. "A Genomic View of the Peopling and Population Structure of India." *Cold Spring Harbor Perspectives in Biology* 7 (4): a008540. DOI: https://doi.org/10.1101/cshperspect.a008540.

Moorjani, Priya, Kumarasamy Thangaraj, Nick Patterson, et al. 2013. "Genetic Evidence for Recent Population Mixture in India." *American journal of human genetics* 93 (3): 422–438. DOI: https://doi.org/10.1016/j.ajhg.2013.07.006.

Müller, Friedrich Max. 1891. "Address to the Anthropological Section of the British Association at the Meeting Held at Cardiff in August, 1891." *The Journal of the Anthropological Institute of Great Britain and Ireland* 21: 172–192.

Mukharji, Projit Bihari. 2022. *Brown Skins, White Coats: Race Science in India*, 1920–66. Chicago: The University of Chicago Press.

Narasimhan, Vagheesh M., Nick Patterson, Priya Moorjani, et al. 2019. "The Formation of Human Populations in South and Central Asia." *Science* 365 (6457). DOI: https://doi.org/10.1126/science.aat7487.

Natrajan, Balmurli, and Paul Greenough, eds. 2009. *Against Stigma: Studies in Caste, Race and Justice Since Durban*. New Delhi: Orient BlackSwan.

Oikkonen, Venla. 2017. *Population Genetics and Belonging: A Cultural Analysis of Genetic Ancestry*. Cham, Switzerland: Palgrave Macmillan.

Rajagopalan, Ramya, and Joan H. Fujimura. 2012. "Making History via DNA, Making DNA from History: Deconstructing the Race-Disease Connection in Admixture Mapping." In *Genetics and the Unsettled Past: The Collision of DNA, Race, and History*, edited by Keith Wailoo, Alondra Nelson, and Catherine Lee, 143–163. New Brunswick: Rutgers University Press.

Rao, Anumpama, ed. 2005. Gender and Caste. London and New York: Zed Books.

Reardon, Jenny. 2005. *Race to the Finish: Identity and Governance in an Age of Genomics*. Princeton: Princeton University Press.

Reardon, Jenny. 2008. "Race and Biology: Beyond the Perpetual Return of Crisis." In *Race and Genomics: Old Wine in New Bottles? Documents from a Transdisciplinary Discussion*, edited by Staffan Müller-Wille and Hans-Jörg Rheinberger, special issue, *NTM* 16 (3): 373–377.

Reich, David. 2018. Who We Are and How We Got Here: Ancient DNA and the New Science of the Human Past. Oxford: Oxford University Press.

Reich, David, Kumarasamy Thangaraj, Nick Patterson, Alkes L. Price, and Lalji Singh. 2009. "Reconstructing Indian Population History." *Nature* 461 (7263): 489–494. DOI: https://doi.org/10.1038/nature08365.

Risley, Herbert Hope. 1891. "The Study of Ethnology in India." *The Journal of the Anthropological Institute of Great Britain and Ireland* 20: 235–263.

Savarkar, Vinayak Damodar. 1989 [1923]. *Hindutva: What Is a Hindu?* New Delhi: Bharatiya Sahitya Sadan.

Sommer, Marianne. 2016. *History Within: The Science, Culture, and Politics of Bones, Organisms, and Molecules*. Chicago: University of Chicago Press.

Subramaniam, Banu. 2014. *Ghost Stories for Darwin: The Science of Variation and the Politics of Diversity*. Urbana: University of Illinois Press.

Subramaniam, Banu. 2019. *Holy Science: The Biopolitics of Hindu Nationalism*. Seattle: University of Washington Press.

Sur, Abha. 2011. *Dispersed Radiance: Caste, Gender, and Modern Science in India*. New Delhi: Navayana.

Thapar, Romila. 2008. *The Aryan: Recasting Constructs*. Gurgaon: Three Essays Collective.

Thapar, Romila. 2019. "Foreword: The Search for the 'Aryan.'" In *Which of Us Are Aryans? Rethinking the Concept of Our Origins*, edited by Romila Thapar et al., vii–xvi. New Delhi: Aleph.

Tilak, Bal Gangadhar. 2011 [1903]. *The Arctic Home in the Vedas: Being Also a New Key to the Interpretation of Many Vedic Texts and Legends*. Fourth edition. London: Arktos.

Trautmann, Thomas R. 1997. *Aryans and British India*. Berkeley: University of California Press.

Trautmann, Thomas R. 2019. "The Aryan Question." In *Questioning Paradigms, Constructing Histories: A Festschrift for Romila Thapar*, edited by Kumkum Roy and Naina Dayal, 261–275. New Delhi: Aleph in association with The Book Review Literary Trust.

Tsai, Yu-yueh. 2010. "Geneticizing Ethnicity: A Study on the 'Taiwan Bio-Bank.'" *East Asian Science, Technology and Society* 4 (3): 433–455. DOI: https://doi.org/10.1215/s12280-010-9146-x.

Yengde, Suraj. 2021. "A Modern History of Caste." Public lecture, ICI Berlin Institute for Cultural Inquiry, Berlin, 15 November. Accessed November 10, 2024. https://youtu.be/RUqB2D5ZPYE.

Who Is a Hunter-Gatherer?

Anthropological Concepts and Their Use in Microbiome Research

Robert Meunier

Abstract *In this chapter, I critically assess the use of the category of hunter-gatherers in human microbiome research. Reviewing debates on the category within the interdisciplinary field of hunter-gatherer studies, I reconstruct the resulting complex and critical stances of archaeologists and anthropologists. I show how these often get lost when microbiome researchers assume this category. I use these insights to complement a growing literature on race and related categories in human microbiome research. This literature emphasizes political and ethical issues concerning racism and representational violence, as well as exploitation and bioprospecting. I argue that awareness of debates in hunter-gatherer studies can add more depth and nuance to this critical assessment. Additionally, I show how these debates can improve microbiome research on hunter-gatherers, as they highlight important methodological and epistemological problems concerning generalization and analogical reasoning.*

Introduction

The use of race or related categories in various fields involved with genomic concepts and technologies, from human evolutionary biology to biomedicine to forensics, has been analyzed and criticized by many authors in the humanities and social sciences since the formation of the Human Genome Project.[1] A relatively recent field of research that is enabled by genomic technologies is microbiome research. This chapter addresses human microbiome research and asks how naming populations plays out in this field. One way of naming populations that gained prominence in microbiome research is based on typologies of modes of subsistence (Ellen 1994), which were developed and are used in archaeology and various subdisciplines of anthropology, and consist of categories such as hunter-gatherer, pastoralist, agriculturalist, etc. This transfer of concepts can be problematic when important insights from the context of origin get lost. Here I investigate

1 For the relevant literature, see references in the introduction to this volume. The history of race in biology and medicine and its criticism goes back much further; see, e.g., Ernst and Harris (1999).

to what extent this is the case and what the consequences are for research on hunter-gatherer microbiomes.

Adding to the literature on race in biomedicine and other genomics-driven fields, there is a small, but growing literature from the humanities and social sciences that scrutinizes how race and related categories figure in human microbiome research (Benezra 2020; Chellappoo and Baedke 2023; De Wolfe et al. 2021; Helmreich 2016, Ch. 6; Hobart and Maroney 2019; Hutchison and Núñez Casal 2023; Nieves Delgado and Baedke 2021; Raffaetà 2022; Rawson 2024; Rowland 2020). Some authors thematize microbiome studies on populations with perceived traditional ways of life, including hunter-gatherers, but often discuss them in combination with other studies that compare microbiomes of populations within industrialized societies that are identified in terms of race or ethnicity. The conceptualizations of modes of subsistence and race or related categories are doubtless historically and practically intertwined. It thus makes sense for scholars with a focus on race to cover a continuum of cases where race categories, processes of reification of race, and the reproduction of racial stereotypes and racialized injustice play out. Nonetheless, race and subsistence modes are quite distinct ways to categorize people, and each follows a different logic. Subsistence types are decidedly seen as not pertaining exclusively to groups identified by geography or ancestry. The scheme runs orthogonal to race classification. Hence, while the above-mentioned commentators put the focus on race, where the categorization of populations as hunter-gatherers is but one among other problematic aspects, I complement this work by focusing on the category of hunter-gatherer directly, where its role in racialized discourse in but one problem, among others.

The category of hunter-gatherers or foragers is usually part of a classification of different modes of subsistence, including fishers, pastoralists, or herders, horticulturalists, agriculturalists, and other categories (Ellen 1994). The number of categories recognized, as well as their names and meanings, differ between authors, fields, or traditions, and change over time. However, the category of hunter-gatherer takes a special position. Paleoanthropologists believe that archaic hominins, at least from *homo erectus* evolving almost two million years ago, lived as hunter-gatherers (Robinson 2014). And so did modern humans, *homo sapiens*, which emerged about 300,000 years ago, until ca. 12,000 years ago, when the domestication of plants and animals set in (Ellen 1994). Hence, other modes of subsistence only emerged after the evolution of hunter-gatherer ways of life, and only recently in evolutionary time-scales. For this reason, from science to pop-culture, modern humans are seen as having evolved as hunter-gatherers (Lavi et al. 2024). This fact lends special significance to this category as opposed to other modes of subsistence and is one reason why otherwise methodologically and conceptually very diverse disciplines came together in a field called "hunter-gatherer studies." It also explains the expansive interest in contemporary populations that are identified as hunter-gatherers, despite the small numbers. Knowledge about these populations is often considered to be informative about prehistoric humans. This is also one of the motivations for microbiome researchers to study these populations, in addition to an interest in synchronic, epidemiological differences between populations.

My analysis is based on three separate bodies of literature. I will first reconstruct some debates on the category of hunter-gatherers within hunter-gatherer studies and the resulting complex and critical stances towards the category in the field. Next, I will show how this complexity initially is lost when microbiome researchers use the concept to classify populations from which they collect samples. I will then draw on the critical science studies literature, mentioned above, to highlight the political and ethical consequences of this neglect. I aim to show how learning from the debates in hunter-gatherer studies can add to the critical assessment of research on hunter-gatherer microbiomes. I use the insights from my analysis to articulate further epistemological criticism of using the hunter-gatherer category in an unqualified manner. Eventually, these considerations could help to improve scientific practice (Benezra et al. 2012).

Hunter-Gatherer Studies and Its Debates

Based on travel reports, populations that were later referred to as foragers or hunter-gatherers have been part of European scholarly discourse at least since the sixteenth century.[2] They were subject to empirical fieldwork in archaeology and anthropology since the second half of the nineteenth century. Here, I will focus on the debates that unfolded after the 1960s and shaped the contemporary understanding of the category. The 1966 conference "Man the Hunter" and the subsequent publication of a volume by the same title (Lee and DeVore 1968) are usually perceived as crucial events in the emergence of hunter-gatherer studies as an interdisciplinary field, bringing together anthropology, archaeology, and other disciplines (Jordan and Cummings 2014). The field consolidated through the Conferences on Hunting-and-Gathering Societies (since 1978), the journal *Hunter-Gatherer Research* (since 2002), the *International Society for Hunter-Gatherer Research* (since 2014), and numerous collected volumes and handbooks.[3]

While the category of hunter-gatherer has functioned productively in catalyzing interdisciplinary exchange, it thereby became subject to critical debate. Initial work in the field was strongly influenced by cultural ecology originating in the United States and associated with Julian Steward (1902–1972). This was a materialist, natural science–oriented approach in which aspects of culture and social organization were analyzed as adaptations to environmental circumstances (Barnard 2014; Jordan and Cummings 2014). The view encouraged empirical approaches, but the focus was on those aspects that could be measured and related to environmental factors, such as energy expenditure, division of labor, or demography (Guenther 2007). Hunter-gatherer studies was thus a field in which cultural anthropology interacted with biological anthropology, the study of biological variation in relation to geography and environment. Furthermore, the goal to formulate generalizable socioeconomic, cultural, or ecological patterns that apply, albeit with local variations or exceptions, to many hunter-gatherer societies, strongly supported archaeologists' strategy of analogical inference from contemporary to ancient populations. Not only did anthropology inform archaeology, but archaeologists began to conduct fieldwork themselves, a program referred to as "ethnoarchaeology" (Lane 2014).

The editors of *Man the Hunter*, Irven DeVore (1934–2014) and Richard B. Lee (*1937), attempted to provide a paradigmatic characterization of hunter-gatherer societies, past

2 For histories of the idea of hunter-gatherers, see Barnard (2014) and Pluciennik (2001).

3 https://www.ishgr.org/, last accessed 11 October 2024.

and present, referred to as the "nomadic style" or later "foraging mode of production" (Lee and DeVore 1968, 11–12; see Guenther 2007; Jordan and Cummings 2014). On this view, the key social and economic features were that hunter-gatherers live in small groups, or "bands," which are, however, linguistically and reproductively interconnected; that they move in larger geographical areas, but have a camp as their home base; and that males hunt, while females gather, where in general all food acquired is shared among the community. From this, the authors derived further implications regarding behavioral patterns, including an egalitarian social order with little personal property, a dynamic exchange between groups and reformation of groups with a tendency to keep group size low and for strategies to avoid conflict, as well as immediate food production with little storing of surpluses. While the conference participants did not agree on a definition and while the above characterization was not as consensual as it was presented, it provided a baseline against which research questions could be formulated and nonconforming patterns of behavior could be described, thus helping the interdisciplinary research program of hunter-gatherer studies to coalesce around a shared set of questions and themes. For instance, feminist anthropologists such as Sally Slocum (*1939) questioned the neglect of the contribution of woman to subsistence and social organization (Sterling 2014).

Lee and DeVore had been codirecting the Harvard Kalahari Project since 1963, which focused on the !Kung San.[4] This population thus served as the exemplar in the formulation of the paradigm. However, anthropologists working in other geographical areas, as well as archaeologists, found it difficult to align their findings with the paradigmatic model. Environments vary substantially across geographical regions inhabited by foragers. For instance, in the presence of waters, populations will engage in fishing and according to what is available, some populations predominantly hunt, while others predominantly gather. Accordingly, the diets of some populations involve large amounts of meat, while others are mainly plant-based (Ellen 1994). Differences in environments and diets could also be observed or assumed for ancient populations. For early hominins, the degree of hunting and the role of scavenging is still discussed (Robinson 2014). Additionally, some groups might exhibit more sedentary behavior, larger groups, or more complex organization. Such variation was observed for contemporary as much as for ancient hunter-gatherers (Guenther 2007). Accordingly, some authors suggested distinctions, such as simple vs. complex or immediate vs. delayed return hunter-gatherers. However, even these came to be criticized "for reducing the inherent flexibility and variability of forager behavior, social structure, and ideology into rather simplistic oppositional categories" (Jordan and Cummings 2014, 11). As early as the 1970s, researchers also became more attuned to the increasing external threats to hunter-gatherer societies and the resulting changes. Furthermore, while there were already anthropologists interested in the symbolic or religious aspects of hunter-gatherer societies involved in *Man the Hunter*, most famously Claude Levi-Strauss (1908–2009), to whom the volume was dedicated, towards the end of the twentieth century, there was a marked move "beyond ecology" (Guenther 2007, 378). This not only added to the sense of diversity of hunter-gatherer

4 While they initially used "!Kung Bushmen," !Kung San and Ju/'hoansi, became the most commonly used designations in the literature. Nevertheless, the dynamics of foreign and self-designation of groups and subgroups remain contested.

lifeways, but also indicated that not all aspects, even of the hunting and gathering activities themselves, could be explained in ecological-adaptive terms.

While the nomadic-style paradigm served as a foil to describe diversity, diversity undermined generalizations in the field. In the light of observed variation, it appeared increasingly difficult to identify characteristics that hold for hunter-gatherers *as such*. Some suggested negative definitions: "It is the absence of elaborate mechanisms for regulating the environment which is perhaps the most significant characteristic of a food-collecting way of life" (Ellen 1994, 202). However, it has been suggested that hunter-gatherers manage their resources and through their activities strongly shape the environments which they inhabit (Fowler and Turner 1999). Others then take an even more relativistic stance and define hunter-gatherers as "people whom anthropology has traditionally recognized as hunter-gatherers" (Kelly 2013, 2). Variability then becomes central as that which is to be explained. Patterns of similarity and difference need to be explained with respect to similarities and differences in environments or contact with other populations or agents.

Accordingly, regarding modes of subsistence, there is little coherence in the classification of (sub)types. The literature on typological reasoning in science suggests that typologies are often implicit, their usefulness and methodological soundness is context dependent, and they usually remain subject to revisions (Love 2009). When dealing with biological or social entities, variability is not simply a question of noise in the data that can be filtered out by the appropriate statistical or rhetorical means; instead, it is constitutive of the very domain of phenomena. Nonetheless, the history of biological and social sciences shows that typologies, which turn out to be useful in some contexts, tend to become interpreted in an essentialist manner (Dupré 1993). Their categories are taken to name natural kinds, i.e., a class of things characterized by defining properties and stable across contexts and time.

From the late 1980s onwards, the initial essentializing and universalist assumptions of the field of hunter-gatherer studies came under intense criticism, from external commentators, e.g., in feminist science studies (Haraway 1989), as well as from within the field. In a controversy that came to be known as the "Kalahari debate" or "Revisionist debate," some archaeologists and anthropologists termed "revisionists" described others as "traditionalists" or "isolationists." Traditionalists were accused of treating living hunter-gatherers as representing a type that resulted from human evolution and that was maintained intact as an archaic, but in itself ahistoric entity, isolated from the history of changes that came with agriculture, colonialism, and industrialization. This idea of the primitive and pristine "Bushmen," revisionists argued, was a fiction about the Other of civilization that was created in reaction to anxieties of modernity felt in Europe and North America, especially in light of the threat of nuclear annihilation and a sense of disenchantment (Wilmsen 1989; Wilmsen et al. 2009). Thus, the concept of hunter-gatherers was seen as a result of the specific Cold War and counterculture sensitivities of the 1960s, the period when the field of hunter-gatherer studies took shape (Guenther 2007). Revisionists, instead, pointed to a long history of contact between hunter-gatherers and their pastoralist or agriculturalist neighbors, as well as the more recent history of colonial encounters and interaction with state agencies and NGOs. They gathered evidence that living and past hunter-gatherers typically were engaged in trade, worked for herders and farmers, and even pursued limited horticulture and herding themselves. Many traits of these populations, including some features described as archetypical by traditionalists, were described as the result of diverse and specific histories of interactions (Headland and Reid 1989). Some described hunter-gatherers as "a kind of impoverished rural underclass, which had been pushed into marginal areas through the impacts of wider economic and political developments" (Jordan and Cumming 2014, 14; see Wilmsen 1989). The position has been rejected as postmodern cynicism and the oversimplified framing of traditionalists' views, as well as the evidence for and notion of "rural proletarians" have been criticized (Lee 1992, 42). Nonetheless, the debate focused on the concerns with the central category of hunter-gatherer studies and its essentialist understanding. Most in hunter-gatherer studies today accept the importance of histories of interactions (Cannon 2014).

In response to these debates, a quarter of a century after the publication of *Man the Hunter*, one of the editors diagnosed a "crisis in hunter-gatherer studies" and concluded: "Even if it is agreed that hunters and gatherers exist, almost everything else about them is a matter for contestation" (Lee, 1992, 31). The reframing of hunter-gatherers as having a history "called into question the very empirical and epistemological basis of modern hunter-gatherer studies" (Jordan and Cummings 2014, 14). As Jordan and Cummings observe, "[p]articularly problematic was the implicit choice within hunter-gatherer studies to portray modern forager groups as relatively pristine, isolated, and self-sufficient units, whose internal dynamics were stripped of both history, as well as the effects of participating in wider spheres of interaction" (ibid.). Especially the analogy between living populations described as hunter-gatherers and Paleolithic populations, or what has been termed the "living fossil trope" (Guenther 2007, 373), was questioned by observations of ecological and cultural diversity and the histories of interactions and change. Hence, neither could living hunter-gatherers be seen as pristine representatives of an original humanity, nor could findings about extant groups be straightforwardly extrapolated to ancient populations. The question of which and how inferences can still be made from contemporary to prehistoric populations, then, is highly contested (Lane 2014; Marlowe 2005).

Many in hunter-gatherer studies continue to work in a narrow behavioral-ecology framework. Some still find a generalizing characterization on hunter-gatherers useful (Winterhalder 2001), while others acknowledge diversity (Kelly 2013). In any case, most researchers in the field are aware of the expanding insights on the historical and contemporary interactions and the cultural complexity, as well as the critical debates on previously unquestioned assumptions, and at least take them into account as important caveats when interpreting their data. Many work towards research agendas that forego generalization attempts and synthesize ecological and historical views, or focus on local, historically contingent social, economic, and political situations of what has been termed "postforaging societies," e.g., regarding the effects of tourism (Guenther 2007, 382). Additionally, ethnographic research is increasingly seen as conducted for and with members of the (post-)hunter-gatherer communities, partly as active investigators, partly in the context of their political activism (Guenther 2007). Hunter-gatherer studies (as cultural and physical anthropology and archaeology in general) came to reflect on its colonial past, the exploitative aspects of its research, and its postcolonial positioning, especially also in the context of apartheid South Africa (Barnard 2007; Schramm 2016).

In the following, I will turn to research on the microbiomes of hunter-gatherers and assess this work in the light of these developments and debates in hunter-gatherer studies.

Hunter-Gatherers in Human Microbiome Research

Research on the Hunter-Gatherer Microbiome

The interest, from a medical perspective, in hunter-gatherer lifeways is not new. In Darwinian medicine, researchers speak of an environment of evolutionary adaptedness or EEA (see Méthot 2015). It is often said that the species *homo sapiens* was characterized by a hunter-gatherer way of life for 99 percent of its existence.[5] Hence, the argument goes, the environment encountered by Paleolithic hunter-gatherers constitutes the environment to which modern humans are adapted. On this view, due to changes in lifeways, in particular the emergence of agriculture, urbanization, and industrialization, the environments in which most humans live today are quite different from the EEA. Under the assumption that genetic evolution happens at a slower pace than these changes, a mismatch between human physiology and the conditions under which most people live can occur. While some aspects of this maladaptedness have been mitigated by technology, it is nonetheless seen as having various health consequences. In particular, low-fiber, high-fat diet and sedentary lifestyle have been recognized as leading to health problems, such as type 2 diabetes or cardiovascular diseases, prevalent in industrialized societies. In this context, hunter-gatherers are seen as a baseline or model for estimating the divergence of populations, especially Western ones, from the diet and behavior to which humans are thought to be adapted and the related physiological processes, e.g., regarding energy expenditure (Pontzer et al. 2018).

Likewise, research on the human microbiome is interested in the evolutionary dimension of the ecological relation of human hosts and their microbiota.[6] Microorganisms not only fulfill specific metabolic functions, such as fiber degradation, but are also thought to influence and regulate the immune system, especially during development. The hygiene revolution, beginning in the nineteenth century, came to be seen as not only protecting humans from dangerous infectious diseases, but also as limiting the exposure to infections in general and to specific microorganisms that were part of the EEA

5 This phrase, often cited in microbiome and Darwinian medicine research, goes back to the volume *Man the Hunter* (Lee and DeVore 1968, 3). For a critique of this number, see Lavi et al. (2024, n33).

6 The sum of microbes present in an ecological setting of interest, e.g., the human gut, is typically referred to as "microbiota." The term "metagenome" is used to designate the sum of the genomic material collectively harbored by the microbiota (sometimes together with the host). "Microbiome" is used to refer to the microbiota as communities and as represented through their metagenome, i.e., as studied through genomics and other "omics" technologies (Berg et al. 2020). While these terms can include archaea, protists, fungi, and viruses inhabiting the ecosystem in question, they are often used to refer only to bacteria.

of humans. The hygiene hypothesis and the old friends hypothesis suggest that this lack of exposure resulted in allergies and other chronic inflammatory and autoimmune diseases (Lorimer 2020; Raffaetà 2022). Next to their role in the development of the immune system, the microbiota of various organs have been implicated in other noninfectious diseases, from cancer to neurological disorders. Here, the idea is that, unlike in the case of infectious disease, where it is typically a single species whose presence is seen to cause the disease, the involvement of microbiota stems from some kind of disbalance or community-level metabolic processes in the host-microbe ecosystem (dysbiosis).

The Human Microbiome Project (HMP), funded initially from 2007 to 2014 by the U.S. National Institutes of Health (NIH), focused on mapping microbiomes at various sites of the human body and estimating the diversity of microbial communities across healthy individuals, including comparisons across populations identified by conventional categories of race or ethnicity. It aimed to establish a framework for the association of microbiome composition and diseases. A second phase, the integrated HMP (iHMP), emphasized pregnancy and birth, as well as specific diseases (esp. inflammatory bowel disease and type 2 diabetes), and aimed at a functional analysis of the microbiome (The iHMP Research Network Consortium 2014). These high-profile projects were mainly focused on Western populations, with much of the work being conducted in the US context. However, humans acquire their microbiome from the environment, and it is modulated by the food they eat, physical activity, and medication. Hence, microbiome researchers became interested in comparing the microbiomes of populations who live in different environments, subsist on different food sources, and differ regarding their access to health care systems. The key distinctions in this respect are typically framed in terms of urban vs. rural environments and industrialized/developed vs. nonindustrialized/developing countries, where the latter distinction is often made with reference to the Human Development Index (HDI, e.g., Groussin et al. 2021).

Since the early 2010s, several, now well-cited studies have appeared that compare populations globally (De Filippo et al. 2010; Yatsunenko et al. 2012). These researchers mainly had backgrounds in biomedicine and genomics (partly with links to the HMP), were interested in the role of the microbiome in health and disease, and framed their work in evolutionary terms. Investigating the effects of "Westernization," researchers collected samples from children (and adults) from rural communities in Africa and South America, respectively, and compared them with populations in Europe or the United States (Yatsunenko et al. 2012, 222).[7] In a similar vein, researchers turned to populations identified as hunter-gatherers. Clemente et al. (2015) collected samples from various body sites of members of a Yanomami Amerindian village, which the researchers described as "uncontacted."[8] The main outcome of these studies was to demonstrate a marked difference in the microbial diversity between non-Western, rural populations, and those exhibiting a Westernized lifestyle. Additionally, differences in microbiome composition between populations were observed. These studies supported the notion that environmental exposure and especially diet, as well as hygiene practices,

7 For a critical, ethnographic science studies perspective on the work behind Yatsunenko et al. (2012), see Benezra (2020).

8 For a science studies perspective on this work, see Hutchison and Núñez Casal (2023).

frequency of cesarean birth, and antibiotics use, shape the human microbiome and that the loss of microbial diversity in Western populations might indeed be associated with the prevalence of certain noninfectious diseases.

At the same time, biological and evolutionary anthropologists and archaeo- and paleobiologists began to apply microbiome research in studies on hunter-gatherers (e.g., Nasidze et al. 2011). The Hadza, or Hadzabe people in Tanzania, similarly to the !Kung San, have been studied extensively by anthropologists with a great variety of questions and methods. Both groups are often treated as paradigmatic exemplars or models for hunter-gatherers. One expert on the Hadza, Frank Marlowe, can be considered a post-revisionist debate anthropologist who rejects the notion of a "living fossil" and takes the effects of a history of interactions into account, while nevertheless insisting that, with caution, contemporary hunter-gatherers can be treated as "the most relevant analogs for at least Late Pleistocene humans" to test hypotheses on human evolution (Marlowe 2005, 65). A collaborative project led by a yet younger generation of evolutionary anthropologists (Alyssa Crittenden) and paleobiologists (Amanda Henry), but also involving Marlowe, studied the gut microbiome of the Hadza. In contrast to the studies discussed above, the emphasis was on understanding human evolution, with the medical significance presented only as a secondary benefit. In one of their articles, they emphasize that the Hadza are genetically modern humans, but suggest that it is "likely the Hadza persist with a very ancient traditional lifestyle into present times" and that the Hadza gut microbiome diversity "is almost certainly the ancestral state for humans" (Schnorr et al. 2014, 9–10). In a more public-facing *Nature* piece, the lead author, Stephanie Schnorr, associates the Hadza with the "wild" and writes: "Thanks to the Hadza, we know that ancient human hunter-gatherers must have maintained a direct and persistent interface with the natural environment" (Schnorr 2015, S15). Due to their location in Eastern Africa, the Hadza are represented as inhabiting the same geographical space, and by implication the same environment, as early humans (i.e., the EEA).

Just a couple of years later, a group led by Justin Sonnenburg (who has a background in biochemistry and biomedicine), along with collaborators (some of whom were involved in the HMP and the Yanomami study), also turned to the Hadza (Fragiadakis et al. 2019; Smits et al. 2017). Though citing Marlowe, this group was less connected to hunter-gatherer studies. Their agenda focused on the gut microbiome in health and disease, but they still describe the Hadza as offering insights into the coevolution of humans and their microbiota. The strategy here, again, rests on comparing industrialized and nonindustrialized populations, with an emphasis on biomedical benefit for Western societies. The Hadza are characterized as "unperturbed" (Fragiadakis et al. 2019, 216), and while the populations in question are described as "just as genetically modern as industrialized populations," their lifestyle is referred to as "traditional" and it is assumed that "in the absence of forces associated with urbanization, it is likely that these populations harbour microbiotas that are more similar to ancestral microbiotas" (Sonnenburg and Sonnenburg 2019, 385).

This brief overview of current research serves to illustrate how the microbiome of hunter-gatherers became established as a research topic, albeit pursued by researchers with different backgrounds and goals (evolutionary anthropology vs. biomedicine). This research is often presented with a sense of urgency, based on the observations that hunter-gatherer ways of living are quickly fading out as the populations identified in these terms face various pressures to change their behaviors (Gibbons 2018). In the following, I will discuss some problematic issues in the context of this kind of research.

Hunter-Gatherer Microbiome Research and Its Discontents

Certainly, much can be learned from a comparative approach to human microbiomes, including the study of populations living in ways that are relatively unaffected by industrialization. Nonetheless, given the problematic status of the category of hunter-gatherers, the aims of some of the researchers, the assumptions they make, and the conclusions they draw, it is fair to say that the growing number of studies on the microbiomes of hunter-gatherers require further scrutiny to guarantee sound, useful, and equitable science. The critical science studies literature on microbiome research has emphasized political and ethical dimensions in terms of racism and exploitation; I will discuss these and add epistemological aspects of problematic generalizations and inference.

Referencing the microbiome studies mentioned above, science studies scholars have pointed out how microbiome research on Indigenous populations, and especially on hunter-gatherers, can "reify racist views of difference" (Rawson 2024, 5) or constitute a "reviving of colonial science" (Maroney 2017). Describing populations as "ancient" or "ancestral" (in some aspects at least), representing them as "uncontacted" or "isolated" and as living in the "wild" or closer to (an original state of) "nature," or discussing them in comparison to early hominins or even primates, places them on a linear trajectory from "primitive" (the meaning shifting between primary, basic or fundamental, and simple) to "civilized" that can be interpreted in evolutionary or valuative terms.[9] Despite the rejection of notions like "living fossils" or "missing links," this trajectory is reminiscent of colonialist social evolutionism. Furthermore, this language might be perceived as offensive or as a form of violence by populations involved in a study (Mangola et al. 2022). Even where the state of a population marked as "ancient" is valued higher than the "civilized" state in terms of health, this echoes the trope of a "noble savage" (Rawson 2024). Such framings are especially problematic when they occur in popular science, as exemplified by the activities of the publisher and entrepreneur Jeff Leach, who also participated in the studies by the Sonnenburg group (see Hobart and Maroney 2019; Lavi et al. 2024; Rowland 2020). When populations are described in biological terms, this can often amount to a naturalization of social or cultural differences. Associating populations identified as hunter-gatherers with characteristic microbiomes can lead to renewed essentialist understanding of hunter-gatherers, now as a biological category.[10]

The often implicit and sometimes explicit primordialism is thus in itself problematic. But it also has consequences. De Wolfe et al. (2021) aim to identify "ghost variables," which they define as "complex, historically loaded racial categories used in microbiome research without explicitly naming race" (ibid., 1; see also Benezra 2020). Ghost variables

9 For analysis of these notions and their effects, see the literature cited in the introduction, for a review, see Rawson (2024).

10 See Nieves Delgado and Baedke (2021) for a related discussion regarding microbiome research and race.

serve to evade the discussion of racism or racialized injustice especially regarding health disparities. The authors describe "hunter-gatherer" as ghost variable, although I would argue that the category functions in this way only where it is associated with attributes such as "pure," "pristine," "ancient," "traditional," etc., as opposed to "Western," "industrialized," etc. In the studies on the Yanomami or the Hadza, even though these populations are represented as healthy with regard to their microbiome, their framing as inhabiting an archaic environment can render invisible the effects of the interaction with neighboring groups, tourists, and state agents or NGO workers, including physicians, in postcolonial contexts and the impact of resulting potential health threats such as stress, pollution, or infectious disease. Especially, in post-settler colonial contexts, where Indigenous groups, including postforager groups, are often studied because certain health problems are especially prevalent in these populations, the structural reasons for health inequality can be masked through ghost variables. Hence De Wolfe et al. suggest making explicit the associated factors (e.g., dietary differences, pollution, variation in family structures etc.) and structural drivers (capitalism, colonialism, segregation, etc.) linked to microbiome change, instead of using ghost variables. Done right, they argue, microbiome research can also help alleviate racism and health injustice.

Rendering populations as "premodern," "pristine," or "ancient" with regard to certain biological aspects such as their microbiomes (even when emphasizing genomic modernity) presents them as somehow less human compared to Western populations and thus "obscure[s] Indigenous People's agency and rationalizes the instrumentalisation of their microbial diversity for the benefit of privileged segments of neoliberal societies" (Hutchison and Núñez Casal 2023, 21). Microbiome research can thus reproduce a colonial logic in which representational violence and exploitation are two sides of the same coin (Benezra 2020; Hutchison and Núñez Casal 2023; Maroney 2017; Rawson 2024). This then highlights another ethical and political point, concerning the risk that microbiome research on Indigenous populations will facilitate bioprospecting and exploitation. Especially studies like those of the Sonnenberg group aim at eventually improving the health of Western, mainly White populations, where diseases that are potentially related to the reduction of microbiome diversity are prevalent. This research is also connected to attempts to build biobanks that store samples from populations exhibiting lifeways that are expected to disappear.[11] This scenario is prone to reproduce a logic of (post)colonial extractivism, in which sampled populations in the Global South are treated as natural resources, whereas benefits fall to companies and affluent sections in Western countries.

Some microbiome researchers are aware of these problems. In a conversation with (and cited by) the social anthropologist Matthäus Rest that probably expresses an ethical concern as much as a sense of competition, Schnorr deplores researchers "who search for the golden shit to correct the Western diet and want to capitalize on it" (Rest 2021, S358). Crittenden describes how interaction with her Hadza informants made her change her practices (Crittenden 2020). Schnorr and Crittenden, together with a member of the Hadza and an Indigenous anthropologist, have coauthored a paper that works towards ethical principles for microbiome research with Indigenous communities (Mangola et al.

11 Benezra speaks of "salvage microbiomics" (2020, 7).

2022). They address questions of informed consent, as well as compensation and benefit sharing. And while Schnorr and Crittenden have used such terminology before, they now recommend avoiding terms such as "traditional," "ancestral," "wild/re-wild," "non-modern," "uncontacted," or "non-industrial," as their use "is based on concepts of linear evolution: the idea that there is a linear gradient of human subsistence activity" and can reproduce racist ideas or be experienced as offensive (Mangola et al. 2022, 752). However, researchers might ignore such principles, and as research is often carried out by international consortia and across varied national contexts, it is difficult to enforce agreed upon rules, even though funding institutions and journals might have some leverage to demand best practices.

I believe that a critical science studies perspective on issues of racism and exploitation in microbiome research on hunter-gatherers, or Indigenous populations in general, can benefit from considering the history of hunter-gatherer studies and its internal debates in various ways. It might help to

- understand how microbiome research is driven towards evolutionary primordialism due to a sense of crisis of late modernity, this time less related to disenchantment and the nuclear bomb, as to a perceived health crisis and anthropogenic climate change (Lavi et al. 2024; Wilmsen et al. 2009).
- trace divides between microbiome researchers and the origins of different attitudes and sensibilities, with some researchers connected to the hunter-gatherer studies tradition, while others, coming from genomics and biomedicine, are not.
- see more nuance, especially in the work of anthropologists: Schnorr et al. (2014), for instance, makes problematic ascriptions but is full of hedging and qualifications, which are certainly informed by the critical stances to treating living hunter-gatherers as isolated analogs of ancient populations.
- find resources to more explicitly articulate, for specific (post-)hunter-gatherer groups, what De Wolfe et al. (2021) call associative factors and structural drivers, as these will concern many aspects that have been studied by revisionist anthropologists. By the same token, being explicit about the way in which populations are not isolated and live in environments that are different from precolonial environments can empower Indigenous activists in their negotiation of rights related to research and other issues.

However, the critical assessment of the category of hunter-gatherer that crystalized in the revisionist controversy, while being present throughout the history of hunter-gatherer studies, was at the core about methodology, and about the kind of data that was gathered and how it could be interpreted. Hence, I suggest an additional, epistemological issue to be addressed in hunter-gatherer microbiome research in the light of the history of hunter-gatherer studies.

The main contestation of the revisionist critique was that contemporary populations and their predecessors were influenced by global and local history. To downplay these effects through a rhetoric of traditional or ancestral lifeways will lead to bad science. Even though Schnorr and Crittenden initially used this rhetoric (Schnorr et al. 2014), together with their coauthors they contend that "[e]very population that is studied in the 21st century is contemporary and influenced by the global economy as well as the nation state; to suggest otherwise is ... scientifically inaccurate (and overlooks important ecological variables that should be measured)" (Mangola et al. 2022, 752). Also here, the long-standing debates in hunter-gatherer studies can help further articulate this insight. These debates make clear that

- there can be no model population, as there is diversity geographically as well as temporally; this is due to significant differences and changes in environments—hence, there can be no such thing as the typical hunter-gatherer microbiome that could be compared with industrial microbiomes.
- there is, additionally, great variation in cultural features of groups identified as hunter-gatherers, regarding the complexity of social organization or the degree of sedentary life; furthermore, there are symbolic and religious aspects to the culture that are in turn different for each studied group. Hence microbiome research needs to take these factors into account, as they shape many aspects of hunter-gatherer lifeways, from division of labor to child rearing, food choice, and food preparation, that strongly affect the microbiome.
- hunter-gatherers have not (at least for the last few thousand years) been isolated; they have often been engaged in labor for and exchange with herders or agriculturalists, or in pursuing small-scale herding or horticulture themselves, and they have often been subject to violence, pollution, or medical interventions. Here, too, all these aspects will influence the microbiome and need to be investigated, documented, and considered in microbiome studies.

The focus on hunter-gatherers is clearly motivated by the special status of this mode of subsistence as primary in terms of human evolution. In biomedically oriented studies, the tenets of Darwinian medicine are always in the background and contemporary populations identified as hunter-gatherers are thus seen as providing information about the EEA and the respective physiological responses, including symbiotic relationships with microbiota. As debates in hunter-gatherer studies make clear, however, different Paleolithic hunter-gatherer populations have lived in diverse environments and in environments different from those inhabited by contemporary foragers. In addition to the variation in today's (post)foraging societies noted above, these issues undermine attempts to establish models of microbiome acquisition, composition, and function that represent generalized ancient conditions in which humans were well adapted to their environments, and that can inform medical interventions *on these grounds*. Moreover, the notion of maladaptedness inherent to Darwinian medicine can be contested, because it assumes that evolution is too slow to catch up with the fast pace of human cultural development. Yet in the light of the many evolutionary processes recognized today beyond changes in gene frequencies, this view cannot be upheld (Dupré 2012, Ch. 14). Most notably, for instance, microbiota evolve fast, through horizontal gene transfer (Groussin et al. 2021). Hence, the microbiomes of members of industrialized societies, even if they are less diverse, might still be adapted to the respective recent lifestyles. Low microbiome diversity is correlated with a higher prevalence of certain diseases. Nonetheless, attempts to attribute these prevalences primarily to dysbiotic microbiota are reminiscent of the earlier attempts to ascribe all disease susceptibility primarily to genetic variation that were prominent in the context of the Human Genome Project (Darrason 2017).

Studies in evolutionary anthropology make use of the analogy between contemporary and ancient hunter-gatherers to learn about the latter. While these researchers tend to be closer to hunter-gatherer studies and hence show more awareness of the long-standing debates on the legitimacy of such analogies, it is not sufficient to introduce some caveats about the uncertainties of such inferences. Such analogies have to be justified separately for each property for which inferences are made. This follows clearly from the fact that hypotheses generated by analogy need to be supported by independent evidence (Lane 2014; Wylie 1985). Tool-use in ancient or contemporary hunter-gatherer populations might exhibit different patterns of similarity and difference than, say, kinship structures, and the evidence to support each claim would be very different. Hence research on microbiomes of contemporary populations aimed at understanding the biology of ancient populations needs to be more explicit about the reasons for assuming that analogical inference would be justified in each particular case. This requires specific competencies and knowledge that need to be acquired through learning or collaboration between evolutionary and paleo-anthropologists: 1) on specificities of the evolution of symbiosis and microbial communities; 2) on cultural specificities and local histories of interactions; and 3) on how the factors in (2) might affect the microbiome.

Many in hunter-gatherer studies see the category of hunter-gatherers as having more pragmatic value, rather than carving out a natural kind. The category is useful to orient research in archaeology and anthropology on populations with little to no herding or horticulture activity. Furthermore, it provides grounds for comparing very heterogeneous populations along the lines of shared themes to find family resemblances, but also, more importantly, interesting differences in need of specific local explanations. I suggest that this is an attitude that microbiome researchers should adopt, as well. Here, too, one can expect a complex network of similarities. More worthwhile will be the study of differences in microbiome composition and function in different hunter-gatherer populations (past and present) and their correlation with differences in the respective environments, histories, and ways of life. If, instead, the only outcome, driven by a desire to generalize, is that hunter-gatherer populations in general have a more diverse microbiome than members of industrialized societies, then this branch of research does not add much to the already well-established fact that rural lifeways in less developed nations (according to the HDI) are correlated with higher diversity in the microbiome. Given that such research adds to the threats that the studied populations face from expanding agricultural, extractive land use, and tourism, such weak results might not balance the negative effects (Gibbons 2018; Mangola et al. 2022).

Conclusion

With this chapter, I complement a growing body of critical science studies literature on race in microbiome research. My focus is on the use of the category of hunter-gatherers in this area of research. Science studies scholars have pointed to political and ethical issues in this field, including racism and representational violence, as well as exploitation and bioprospecting. I have offered a brief summary of the history of hunter-gatherer studies and its internal debates on its central categories. I argue that awareness of these debates can lead to more depth and nuance in critical approaches, especially where they aim to improve the research. Additionally, it highlights another line of critique focused on epistemological questions. I have outlined this problem to set the stage for further discussion.

With regard to the questions that motivate this volume, the chapter has discussed some of the consequences of categorizing populations as hunter-gatherers (which might apply to other designations such as "pastoralists," as well). The category scheme of modes of subsistence, to which the concept of hunter-gatherer belongs, is different from race or ethnic categories. Yet the respective classification systems strongly interact. As a result, designating populations as hunter-gatherers can, under certain circumstances, contribute to racialized injustice. Additionally, using this category can lead to methodological errors in research, affecting not only the researchers but also the groups that participate in research. Hence, microbiome researchers are well advised to listen to and engage with participant communities. As everywhere, unquestioned assumptions can lead to injustice as well as to poorly justified scientific claims. Microbiome researchers certainly discuss methodological issues, and increased interaction between this group with cultural anthropologists will lead to self-correcting processes of mutual critique. Nonetheless, science studies scholars can clearly contribute to making research sounder, not least by facilitating the transfer of knowledge between hunter-gatherer studies and microbiome research.

Acknowledgments

I thank the participants and commentators of the authors' workshop, the editors, the fact-checker, and the reviewers for helpful feedback. My work was funded by the Deutsche Forschungsgemeinschaft (DFG) within the Cluster of Excellence "Precision Medicine in Chronic Inflammation" (EXC 2167), project nr. 390884018. I was also a visiting postdoctoral fellow in the MPRG "Practices of Validation in the Biomedical Sciences" at the Max Planck Institute for the History of Science.

References

Barnard, Alan. 2007. *Anthropology and the Bushman*. Oxford: Berg Publishers.

Barnard, Alan. 2014. "Defining Hunter-Gatherers: Enlightenment, Romantic, and Social Evolutionary Perspectives." In *The Oxford Handbook of the Archaeology and Anthropology of Hunter-Gatherers*, edited by Vicki Cummings, Peter Jordan, and Marek Zvelebil, 43–54. Oxford: Oxford University Press. DOI: https://doi.org/10.1093/oxfordhb/9780199551224.013.011.

Benezra, Amber. 2020. "Race in the Microbiome." *Science, Technology, & Human Values* 45 (5): 877–902. DOI: https://doi.org/10.1177/0162243920911998.

Benezra, Amber, Joseph DeStefano, and Jeffrey I. Gordon. 2012. "Anthropology of Microbes." *Proceedings of the National Academy of Sciences* 109 (17): 6378–6381. DOI: https://doi.org/10.1073/pnas.1200515109.

Berg, Gabriele, Daria Rybakova, Doreen Fischer et al. 2020. "Microbiome Definition Re-Visited: Old Concepts and New Challenges." *Microbiome* 8 (1): 103. DOI: https://doi.org/10.1186/s40168-020-00875-0.

Cannon, Aubrey. 2014. "Historical and Humanist Perspectives on Hunter-Gatherers." In *The Oxford Handbook of the Archaeology and Anthropology of Hunter-Gatherers*, edited by Vicki Cummings, Peter Jordan, and Marek Zvelebil, 92–103. Oxford: Oxford University Press. DOI: https://doi.org/10.1093/oxfordhb/9780199551224.013.016.

Chellappoo, Azita, and Jan Baedke. 2023. "Where the Social Meets the Biological: New Ontologies of Biosocial Race." *Synthese* 201 (1): 14. DOI: https://doi.org/10.1007/s11229-022-04006-0.

Clemente, Jose C., Erica C. Pehrsson, Martin J. Blaser et al. 2015. "The Microbiome of Uncontacted Amerindians." *Science Advances* 1 (3): e1500183. DOI: https://doi.org/10.1126/sciadv.1500183.

Crittenden, Alyssa. 2020. "Who Owns Poop? And Other Ethical Dilemmas Facing an Anthropologist Who Works at the Interface of Biological Research and Indigenous Rights." In *The Secret Lives of Anthropologists: Lessons from the Field*, edited by Bonnie L. Hewlett, 299–321. New York: Routledge.

Darrason, Marie. 2017. "From the Concept of Genetic Disease to the Geneticization of Diseases: Analyzing and Solving the Paradox of Contemporary Medical Genetics." In *Philosophy of Molecular Medicine: Foundational Issues in Research and Practice*, edited by Giovanni Boniolo and Marco J. Nathan, 57–78. New York: Routledge.

De Filippo, Carlotta, Duccio Cavalieri, Monica Di Paola et al. 2010. "Impact of Diet in Shaping Gut Microbiota Revealed by a Comparative Study in Children from Europe and Rural Africa." *Proceedings of the National Academy of Sciences* 107 (33): 14691–14696. DOI: https://doi.org/10.1073/pnas.1005963107.

De Lima Hutchison, Coll, and Andrea Núñez Casal. 2023. "Sustaining (Dis)Embodied Inequalities in the(ir) Eurocene: Ancient Microbes, Racial Anthropometry, and Life Choices." *Medicine Anthropology Theory* 10 (2): 1–33. DOI: https://doi.org/10.17157/mat.10.2.7105.

De Wolfe, Travis J., Mohammed Rafi Arefin, Amber Benezra, and María Rebolleda Gómez. 2021. "Chasing Ghosts: Race, Racism, and the Future of Microbiome Research." *mSystems* 6 (5). DOI: https://doi.org/10.1128/msystems.00604-21.

Dupré, John. 1993. *The Disorder of Things: Metaphysical Foundations of the Disunity of Science*. Cambridge, MA: Harvard University Press.

Dupré, John. 2012. *Processes of Life: Essays in the Philosophy of Biology*. Oxford: Oxford University Press. DOI: https://doi.org/10.1093/acprof:oso/9780199691982.001.0001.

Ellen, Roy. 1994. "Modes of Subsistence: Hunting and Gathering to Agriculture and Pastoralism." In *Companion Encyclopedia of Anthropology: Humanity, Culture and Social Life*, edited by Tim Ingold, 197–225. London: Routledge.

Ernst, Waltraud, and Bernard Harris, eds. 1999. *Race, Science and Medicine, 1700–1960*. New York: Routledge.

Fowler, Catherine S., and Nancy J. Turner. 1999. "Ecological/Cosmological Knowledge and Land Management among Hunter-Gatherers." In *The Cambridge Encyclopedia of Hunters and Gatherers*, edited by Richard B. Lee and Richard Daly, 419–425. Cambridge: Cambridge University Press.

Fragiadakis, Gabriela K., Samuel A. Smits, Erica D. Sonnenburg, et al. 2019. "Links between Environment, Diet, and the Hunter-Gatherer Microbiome." *Gut Microbes* 10 (2): 216–227. DOI: https://doi.org/10.1080/19490976.2018.1494103.

Gibbons, Ann. 2018. "Farmers, Tourists, and Cattle Threaten to Wipe out Some of the World's Last Hunter-Gatherers." *Science*, 17 May 2018, https://www.science.org/content/article/farmers-tourists-and-cattle-threaten-wipe-out-some-world-s-last-hunter-gatherers.

Groussin, Mathieu, Mathilde Poyet, Ainara Sistiaga et al. 2021. "Elevated Rates of Horizontal Gene Transfer in the Industrialized Human Microbiome." *Cell* 184 (8): 2053–2067.e18. DOI: https://doi.org/10.1016/j.cell.2021.02.052.

Guenther, Mathias. 2007. "Current Issues and Future Directions in Hunter-Gatherer Studies." *Anthropos* 102 (2): 371–388. DOI: https://doi.org/10.5771/0257-9774-2007-2-371.

Haraway, Donna J. 1989. *Primate Visions: Gender, Race, and Nature in the World of Modern Science*. New York: Routledge.

Headland, Thomas N., Lawrence A. Reid, M. G. Bicchieri, et al. 1989. "Hunter-Gatherers and Their Neighbors from Prehistory to the Present [and Comments and Replies]." *Current Anthropology* 30 (1): 43–51. DOI: https://doi.org/10.1086/203710.

Helmreich, Stefan. 2016. *Sounding the Limits of Life: Essays in the Anthropology of Biology and Beyond*. Princeton: Princeton University Press.

Hobart, Hiʻilei Julia, and Stephanie Maroney. 2019. "On Racial Constitutions and Digestive Therapeutics." *Food, Culture & Society* 22 (5): 576–594. DOI: https://doi.org/10.1080/15528014.2019.1638120.

Jordan, Peter, and Vicki Cummings. 2014. "Introduction." In *The Oxford Handbook of the Archaeology and Anthropology of Hunter-Gatherers*, edited by Vicki Cummings, Peter Jordan, and Marek Zvelebil, 1–30. Oxford: Oxford University Press. DOI: https://doi.org/10.1093/oxfordhb/9780199551224.013.063.

Kelly, Robert L. 2013. *The Lifeways of Hunter-Gatherers: The Foraging Spectrum*. Cambridge: Cambridge University Press.

Lane, Paul J. 2014. "Hunter-Gatherer-Fishers, Ethnoarchaeology, and Analogical Reasoning." In *The Oxford Handbook of the Archaeology and Anthropology of Hunter-Gatherers*, edited by Vicki Cummings, Peter Jordan, and Marek Zvelebil, 104–120. Oxford: Oxford University Press. DOI: https://doi.org/10.1093/oxfordhb/9780199551224.013.024.

Lavi, Noa, Alice Rudge, and Graeme Warren. 2024. "Rewild Your Inner Hunter-Gatherer: How an Idea about Our Ancestral Condition Is Recruited into Popular Debate in Britain and Ireland." *Current Anthropology* 65 (1): 72–99. DOI: https://doi.org/10.1086/728528.

Lee, Richard B. 1992. "Art, Science, or Politics? The Crisis in Hunter-Gatherer Studies." *American Anthropologist* 94 (1): 31–54. DOI: https://doi.org/10.1525/aa.1992.94.1.02a00030.

Lee, Richard Borshay, and Irven DeVore. 1968. *Man the Hunter.* Piscataway: Transaction Publishers.

Lorimer, Jamie. 2020. The Probiotic Planet: *Using Life to Manage Life*. Minneapolis: University of Minnesota Press.

Love, Alan C. 2009. "Typology Reconfigured: From the Metaphysics of Essentialism to the Epistemology of Representation." *Acta Biotheoretica* 57 (1): 51–75. DOI: https://doi.org/10.1007/s10441-008-9059-4.

Mangola, Shani Msafiri, Justin R. Lund, Stephanie L. Schnorr, and Alyssa N. Crittenden. 2022. "Ethical Microbiome Research with Indigenous Communities." *Nature Microbiology* 7 (6): 749–756. DOI: https://doi.org/10.1038/s41564-022-01116-w.

Marlowe, Frank W. 2005. "Hunter-Gatherers and Human Evolution." *Evolutionary Anthropology: Issues, News, and Reviews* 14 (2): 54–67. DOI: https://doi.org/10.1002/evan.20046.

Maroney, Stephanie. 2017. "Reviving Colonial Science in Ancestral Microbiome Research." *MicrobioSocial*, 10 January 2017, last accessed 30 September 2024, https://microbiosocial.wordpress.com/2017/01/10/reviving-colonial-science-in-ancestral-microbiome-research/.

Méthot, Pierre-Olivier. 2015. "Darwin, Evolution, and Medicine: Historical and Contemporary Perspectives." In *Handbook of Evolutionary Thinking in the Sciences*, edited by Thomas Heams, Philippe Huneman, Guillaume Lecointre, and Marc Silberstein, 587–617. Dordrecht: Springer. DOI: https://doi.org/10.1007/978-94-017-9014-7_27.

Nasidze, Ivan, Jing Li, Roland Schroeder, Jean L. Creasey, Mingkun Li, and Mark Stoneking. 2011. "High Diversity of the Saliva Microbiome in Batwa Pygmies." *PLOS ONE* 6 (8): e23352. DOI: https://doi.org/10.1371/journal.pone.0023352.

Nieves Delgado, Abigail, and Jan Baedke. 2021. "Does the Human Microbiome Tell Us Something about Race?" *Humanities & Social Sciences Communications* 8 (1): 1–12. DOI: https://doi.org/10.1057/s41599-021-00772-3.

Pluciennik, Mark. 2001. "Archaeology, Anthropology and Subsistence." *Journal of the Royal Anthropological Institute* 7 (4): 741–758. DOI: https://doi.org/10.1111/1467-9655.00087.

Pontzer, Herman, Brian M. Wood, and David A. Raichlen. 2018. "Hunter-Gatherers as Models in Public Health." *Obesity Reviews* 19 (S1): 24–35. DOI: https://doi.org/10.1111/obr.12785.

Raffaetà, Roberta. 2023. *Metagenomic Futures: How Microbiome Research Is Reconfiguring Health and What It Means to Be Human*. Abingdon: Routledge.

Rawson, Ariel Janaye. 2024. "Anti-Racism, Racism, and the Microbiome: A Review." *Progress in Environmental Geography* 3 (2): 137–159. DOI: https://doi.org/10.1177/27539687241245428.

Rest, Matthäus. 2021. "Preserving the Microbial Commons: Intersections of Ancient DNA, Cheese Making, and Bioprospecting." *Current Anthropology* 62 (S24): S349–S360. DOI: https://doi.org/10.1086/715810.

Robinson, Jennie. 2014. "The First Hunter-Gatherers." In *The Oxford Handbook of the Archaeology and Anthropology of Hunter-Gatherers*, edited by Vicki Cummings, Peter Jordan, and Marek Zvelebil, 177–190. Oxford: Oxford University Press. DOI: https://doi.org/10.1093/oxfordhb/9780199551224.013.029.

Rowland, Allison L. 2020. *Zoetropes and the Politics of Humanhood*. Columbus: Ohio State University Press.

Schnorr, Stephanie L. 2015. "The Diverse Microbiome of the Hunter-Gatherer." *Nature* 518 (7540): S14–S15. DOI: https://doi.org/10.1038/518S14a.

Schnorr, Stephanie L., Marco Candela, Simone Rampelli et al. 2014. "Gut Microbiome of the Hadza Hunter-Gatherers." *Nature Communications* 5 (1): 3654. DOI: https://doi.org/10.1038/ncomms4654.

Schramm, Katharina. 2016. "Casts, Bones and DNA: Interrogating the Relationship between Science and Postcolonial Indigeneity in Contemporary South Africa." *Anthropology Southern Africa* 39 (2): 131–144. DOI: https://doi.org/10.1080/23323256.2016.1168267.

Smits, Samuel A., Jeff Leach, Erica D. Sonnenburg, et al. 2017. "Seasonal Cycling in the Gut Microbiome of the Hadza Hunter-Gatherers of Tanzania." *Science* 357 (6353): 802–806. DOI: https://doi.org/10.1126/science.aan4834.

Sonnenburg, Erica D., and Justin L. Sonnenburg. 2019. "The Ancestral and Industrialized Gut Microbiota and Implications for Human Health." *Nature Reviews Microbiology* 17 (6): 383–390. DOI: https://doi.org/10.1038/s41579-019-0191-8.

Sterling, Kathleen. 2014. "Man the Hunter, Woman the Gatherer? The Impact of Gender Studies on Hunter-Gatherer Research (A Retrospective)." In *The Oxford Handbook of the Archaeology and Anthropology of Hunter-Gatherers*, edited by Vicki Cummings, Peter Jordan, and Marek Zvelebil, 151–174. Oxford: Oxford University Press. DOI: https://doi.org/10.1093/oxfordhb/9780199551224.013.032.

The Integrative HMP (iHMP) Research Network Consortium. 2014. "The Integrative Human Microbiome Project: Dynamic Analysis of Microbiome-Host Omics Profiles during Periods of Human Health and Disease." *Cell Host & Microbe* 16 (3): 276–289. DOI: https://doi.org/10.1016/j.chom.2014.08.014.

Wilmsen, Edwin N. 1989. *Land Filled with Flies: A Political Economy of the Kalahari*. Chicago: The University of Chicago Press.

Wilmsen, Edwin N., Alan Barnard, Megan Biesele, Akira Takada, and Owen B. Sichone. 2009. "To See Ourselves as We Need to See Us: Ethnography's Primitive Turn in the Early Cold War Years." *Critical African Studies* 1 (1): 20–95. DOI: https://doi.org/10.1080/20407211.2009.10530743.

Winterhalder, Bruce. 2001. "The Behavioral Ecology of Hunter-Gatherers." In *Hunter-Gatherers: An Interdisciplinary Perspective*, edited by Catherine Panter-Brick, Robert H. Layton, and Peter Rowley-Conwy, 12–38. Cambridge: Cambridge University Press.

Wylie, Alison. 1985. "The Reaction against Analogy." *Advances in Archaeological Method and Theory* 8 (1): 63–111.

Yatsunenko, Tanya, Federico E. Rey, Mark J. Manary et al. 2012. "Human Gut Microbiome Viewed across Age and Geography." *Nature* 486 (7402): 222–227. DOI: https://doi.org/10.1038/nature11053.

New Categories but Old Meanings

Ethnicity Disguised as Migration Background?

Technologies of Difference in German Epidemiology

Andrea zur Nieden, Laura Schnieder, Isabelle Bartram, Nils Ellebrecht, Tino Plümecke[1]

Abstract *A new body of medical and epidemiological research on human diversity is emerging in Germany. Unlike in the United States, the terms "race" or "Rasse" are largely considered taboo as classifications in German research. Instead, alternative concepts such as "migration background" play a more prominent role. In this literature study, we present findings from a systematic review of 546 papers examining concepts of classification in German life sciences research, followed by a qualitative, grounded theory–based analysis of a subset of 77 publications from the discipline of epidemiology. Our primary focus is on how the concept of "migration background," as an administrative technology, is enacted in epidemiological publication practices. Epidemiologists aim to facilitate research on health disparities by introducing this new label. At the same time, however, the pursuit of greater heterogeneity gives rise to new (or, on closer inspection, not-so-new) forms of homogenization, particularly of the vulnerable, hard-to-reach Other. We find that "migration background" is frequently conflated with "ethnicity" in an essentializing way and identify different modes in which this occurs. We argue that examining these processes of othering in epidemiological research is a crucial prerequisite for developing meaningful and inclusive public health measures.*

Migration, Ethnicity, and German Epidemiological Research

In the United States, political efforts to address health inequalities have led to the inclusion of supposedly relevant ethnic, racial, gender, and age groups in medical research.

1 The authors are part of the SoSciBio research group (Human Diversity in the New Life Sciences: Social and Scientific Effects of Biological Differentiations), funded by the German Ministry of Education and Research (BMBF, grant number 01GP1790), at the Institute of Sociology at the University of Freiburg. Andrea zur Nieden and Laura Schnieder conducted the qualitative study and prepared the draft manuscript, with Andrea zur Nieden responsible for its revisions. We would like to thank Charlotte Schulze-Marmeling, Felix Fink, Isabel Schön and Jan Weber for their support throughout our work. We are also grateful to the participants of the authors' workshops for this edited volume, the anonymous reviewers for their valuable comments on earlier versions of this article and Michael Thomas Taylor for his excellent linguistic revision. The quantitative study was carried out by all group members (see Bartram et al. 2023).

This practice was formalized by the National Institutes of Health Revitalization Act of 1993, which mandated such representation in all NIH-sponsored clinical trials; the requirement was later extended to all drugs seeking approval from the Food and Drug Administration (FDA). Both the NIH and FDA recommend using the U.S. Census system of racial and ethnic classification, which defines ethnicity as either Hispanic/Latino or not Hispanic/Latino, and race according to the following categories: American Indian or Alaska Native, Asian, Black or African American, Native Hawaiian or Other Pacific Islander, and White (Food and Drug Administration 2016, 9–10).[2] Steven Epstein has traced the history of what he terms the "inclusion-and-difference paradigm" in the United States:

> Characteristic of this way of thinking is the assumption that social identities correspond to relatively distinct kinds of bodies—female bodies, Asian bodies, elderly Hispanic male bodies, and so on—and that these various embodied states are medically incommensurable. (Epstein 2007, 2)

Multiple researchers from sociology and science studies have critically examined these classifications (Shim et al. 2014; Fujimura and Rajagopalan 2020; Fujimura and Rajagopalan 2011; Pollock 2012). In Germany, however, no comparable formalization of difference exists to date. Instead, a growing interest in human diversity is emerging in medical, epidemiological, and other life sciences research. This development appears to be a response to international debates on rectifying health disparities.

Of course, some fields of medicine are highly internationalized, and a great deal of research is conducted through international collaborations. However, national differences become particularly apparent in studies involving data collected and analyzed in Germany—especially in epidemiology, which is closely tied to country-specific policies and national biopolitics. Unlike in the United States—and in part due to Germany's Nazi history—the term "Rasse" (the German word for "race") is considered taboo (Chin 2016; zur Nieden 2014; Plümecke 2010; Lipphardt et al. 2018). Our systematic literature review of 546 life science articles confirmed that the concept of race is used very rarely in German-based research (Bartram et al. 2023). Still, what might be called an "absent presence" persists, a phenomenon observed in other European countries as well (M'charek, Schramm, and Skinner 2014). Countries that perceive themselves as postracial and avoid explicit references to race often maintain spaces where race is subtly embedded rather than overtly articulated—where it is inscribed and reproduced, for instance, within technical apparatuses. Moreover, alternative categories that appear politically less problematic are in use, such as "Migrationshintergrund" ("migration background" or "immigrant background") instead of race. Yet these categories may still carry racializing implications in the sense described by Wacquant:

---

2 The census system has since been updated to using a combined race/ethnicity question instead of the two-question format and adding a new "Middle Eastern or North African" (MENA) category (United States Census Bureau 2024), but the FDA has not yet revised its recommendations in this regard.

> To racialize means to naturalize, to turn history into biology, cultural differences into dissimilarities of essence; to eternalize, to stipulate that those differences are enduring if not unchanging across time, past, present and future; and to homogenize, to perceive and picture all members of the racialized category as fundamentally alike, as sharing a permanent essential quality that warrants differential treatment of its members in symbolic, social and physical space. (Wacquant 2022, 78)

In this article, we examine how the pursuit of inclusion and diversity for historically underrepresented groups in (bio)medical studies is integrated into epidemiological research based in Germany. We are especially interested in how the concept of migration background, as an administrative technology, is enacted in epidemiological (publication) practices and how it is frequently conflated with the concept of ethnicity. We draw on approaches by Weber, Wacquant, M'charek et al., and Mecheril as heuristics to analyze potential processes of othering in epidemiological research. We argue that such an examination is a crucial prerequisite for developing meaningful and inclusive public health measures. Our analysis is based on fieldwork conducted in Germany as part of the research group SoSciBio at the University of Freiburg.[3]

Materials and Methods

Our empirical material is derived from an original dataset generated by our research group (see Bartram et al. 2023). The full dataset comprises 546 articles published between 2018 and 2020 that categorize their research subjects using terms related to ethnicity, race, ancestry, or migration. These articles were identified through a systematic literature search of PubMed and Web of Science (WoS). Based on our preliminary research, we determined that "race," "ethnicity," "migration background," and "ancestry" (or related terms) are the most frequently applied concepts in the life sciences to represent human diversity. Consequently, we selected these terms—combined with (German*)—as our search criteria.[4] This dataset includes a subset of 135 publications from epidemiology, of which 77 articles contain at least one variant of the term "migra*" as a category describing study subjects. In addition to the research group's quantitative content analysis, we applied a grounded theory approach to generate initial hypotheses from these texts (Charmaz 2012; Berg and Milmeister 2011). Beginning with the 77 articles, we conducted open coding to develop a coding system and carried out detailed qualitative analyses (Kruse 2015), recording codes and memos using MAXQDA Standard 2020 20.1.0 (Verbi GmbH). In a second step, these codes were refined into more focused categories, which were then, in a third step, aligned with larger portions of data from the full dataset. Here, we focus primarily on one key finding: the gradual conflation of the category "migration background" with "ethnicity" and its analytical consequences.

3 In another part of the case study on epidemiology, we conducted ethnographic fieldwork on a large German health study.

4 Details on the selection of studies and the content analysis can be found in Bartram et al. 2023.

Results

In the epidemiology subsample examined below, the category "migration background/migrant" is the most frequently applied, appearing in 58.5 percent of papers in this discipline. As noted above, national differences are particularly evident in this discipline, since researchers collaborate internationally less often than in other fields. Specifically, in 51.9 percent of the epidemiology publications, all authors were affiliated with a German institution—slightly more than the 43.6 percent observed across the entire dataset of 546 publications. Additionally, 66 percent of the studies collected their data within Germany, a proportion notably higher than the 50.6 percent in the overall sample. Our quantitative content analysis revealed a significant impact of *authors' national affiliations* on the *categories* chosen: when all authors of a paper were affiliated with a German institution, terms such as "migration background," "migrant," and their derivatives were the most frequently applied to describe study subjects (78.8 percent). By contrast, other categories became more prevalent in studies involving international collaboration. "Ethnicity," for example, was the most commonly used term (56 percent) when either the first *or* last author was affiliated with a German institution. Moreover, the more authors were based at German institutions, the less frequently the term "race" appeared. It was not used in any publication where all authors were affiliated with German institutions but appeared in nearly 30 percent of publications where either the first *or* last author had a German institutional affiliation.

Migration Background as an Administrative Technology of Difference

Before presenting more of our results, we find it necessary to provide some background on the term "(people with a) migration background"—in German: "(Menschen mit) Migrationshintergrund." Over the past few decades, this term has become widely used in public and political discourse to refer to immigrants to Germany and their descendants. German epidemiological studies have also begun to adopt this category, often citing a model developed by Liane Schenk (Schenk et al. 2006; Schenk 2007). Schenk et al. (2007) advocate for the use of *migration background* instead of *race* or *ethnicity*: they argue, first, that the latter terms are both controversial and difficult to operationalize; and second, that Germany is "an immigration country without postcolonial migration and without numerically relevant autochthonous ethnic minorities, i.e., a country with a relatively young immigration history" (Schenk et al., 88, our translation).[5] As this quote suggests, unlike in the United States, Germany lacks a deeply rooted self-perception as a country of immigration and has no established ethnic or racial census categories.

The largest migrant group in the postwar period came from Turkey. The standard terms for these immigrants in German used to be "foreigners" and "guest workers" ("Ausländer" and "Gastarbeiter"), reflecting the perception that they were not part of German society but rather temporary guests. However, many chose to remain after years of labor

5 During the time of our research, this model was revised through new recommendations, which we cite in our discussion (Kajikhina et al. 2023).

in German industries. Since the late 1990s, Germany has reformed its previously restrictive citizenship laws, making it possible for immigrants and their descendants to acquire German citizenship. To statistically account for individuals who were previously categorized as *foreigners* (*Ausländer*) but have since obtained German citizenship, the term "people with migration background" was introduced. Since 2005, this category has been used in the German Microcensus, a representative annual household survey that collects sociodemographic data from about 1 percent of the population. After several revisions, the Federal Statistical Office (Statistisches Bundesamt) defined the term in 2016 as follows for the microcensus: "A person has a migration background if they or at least one of their parents did not possess German citizenship at birth" (Statistisches Bundesamt 2017, 4, our translation). It further explained that the purpose of this new definition was

> to keep those groups of persons identifiable that have always been associated with migration in public debate and official statistics, such as: foreigners, naturalized, displaced, resettlers, late resettlers or asylum seekers. … Justifiable questions should not have to be left unanswered because the affected population groups had been "defined out"; on the other hand, only those people should be included who at least in principle have a need for integration. (Ibid. 4)

The reference here to people "who at least in principle have a need for integration" highlights how the term "migration background" is deeply intertwined with ongoing political debates in Germany about the "integration" of immigrants and their descendants. Religion (and especially *Muslim* religion), national loyalty, and "culture" have often served as markers for drawing boundaries between groups and individuals considered well-integrated and those seen as lacking integration.

The second-largest group of immigrants or people with a migration background are the so-called "resettlers." Resettlers are understood as descendants of German settlers who had lived in Russian regions since the twelfth century and later "resettled"—that is, immigrated to Germany—after the collapse of the Soviet Union. They were granted German citizenship with relative ease because German law classifies them as "ethnic Germans." This practice offers important insights into Germany's concept of nationhood: historically, and before the shift in citizenship law that took effect in the late 1990s, the German national identity was primarily based on ancestry (*jus sanguinis*) rather than place of birth (*jus soli*). However, despite being legally recognized as German, resettlers were assumed to face typical challenges of integration. Ensuring that these individuals remained visible in statistical and policy frameworks became another key rationale for the creation and implementation of the category of migration background (Schenk et al. 2006, 854).

Like *race* and *ethnicity* in US medical discourse, *migration background* in Germany functions as an administrative biopolitical technology of difference. However, it arises from a distinct national political context. In the United States, *race* and *ethnicity* are census categories shaped by a long history of colonialism, slavery, and liberation movements of disadvantaged and minoritized groups, along with the accompanying political struggles (Epstein 2007; Thompson 2016). By contrast, in Germany, *migration background* is a more recent, top-down administrative technology designed to distinguish and quantify certain population groups that would otherwise become statistically invisible for purposes of government regulation. Moreover, Germany's classification system is more rigidly formalized than that of the United States. While US racial and ethnic categories—used in both the census and FDA guidelines—are based on self-identification and may change over a person's lifetime, Germany determines *migration background* through citizenship data, specifically whether an individual or their parents held German citizenship at birth.

Despite the relatively formalized nature of this approach, the definition of migration background has been criticized for its ambiguity. On the one hand, it appears to focus on the social experience of migration and its potential discriminatory consequences. On the other hand, it also carries associations with ethnicity (Will 2019, 2018). For instance, so-called ethnic German refugees and their descendants who fled to the territory of present-day Germany during or after the Second World War but before 1950 are not classified as having a migration background—unlike "ethnic German" resettlers. The same applies to children of German citizens born abroad.

Another criticism is that, beyond its official definition, the category is often used in a stigmatizing way to describe individuals who are perceived as problematic (see, for instance, Neue Medienmacher). For example, while citizens of other Western European Union countries and their descendants living in Germany are officially considered to have a migration background, the term is rarely applied to them in public discourse. Instead, it is more commonly used for people perceived as "visibly" different or as culturally or religiously "other."[6]

Despite its association with ethnicity, however, the formal category of migration background in Germany differs significantly from how ethnicity is conceptualized in other countries. In the United States, for example, a fourth-generation Hispanic individual could still be classified as Hispanic—or self-identify as such. In Germany, by contrast, the classification of migration background ends after the third generation: only individuals who were not born with German citizenship or whose parents were not born as German citizens are included. As a result, a fourth-generation person of Turkish origin would no longer be officially categorized as having a Turkish migration background, even if they continue to experience discrimination or racism in practice.

Ethnicity (Lurking) in the Background

One of the key empirical findings from our analysis of the publications was that the category of migration background frequently conveys elements of ethnicity, often blurring conceptual distinctions between the two. In roughly one-quarter (18) of the 77 epidemiology publications in our sample that use the terms "migration background" or "migrant" to describe their research subjects, the term ethnicity is also employed to characterize

6 Since 2022, the Federal Statistical Office has responded to these criticisms by introducing the new category "immigrants and their descendants," which more specifically includes immigrants and their children but no longer the grandchildren's generation. This category emphasizes migration experience: only those who have immigrated to Germany themselves or whose parents have immigrated are classified within this group. To date, both categories are being used in the microcensus on a trial basis (Statistisches Bundesamt 2024).

the analyzed groups. An additional 18 articles incorporate terms with the word stems "migra*" and "ethn*" in the main text—often to frame a particular issue or to reference comparable studies. In the following sections, we examine two examples in greater detail, highlighting recurring themes or "central motives" (Kruse 2015) that emerged from our qualitative analysis of the entire publication sample.

The Blurring of Classifying Concepts

In their study on the prevalence of dementia among people with a migration background in Germany, Monsees, Hoffmann, and Thyrian (2019) introduce their focus in a way that is typical of the publications we analyzed:

> In 2013, 16.5 million people with a migration background were living in Germany, 1.5 million of them older than the age of 65. For this population, there are no reliable figures on dementia prevalence, which constitutes a challenge for the healthcare system that is difficult to assess in its full scope. (Monsees, Hoffmann, and Thyrian, 654, our translation)

By quantifying the supposed "challenge for the healthcare system," the authors emphasize both the scale of the issue and the urgency of further research. The urgency is reinforced by another common argumentative move we observed in our material: *framing these populations as difficult to reach within the healthcare system*. The authors note:

> A significant proportion of people with a migrant background is difficult to reach through the healthcare system, and where this is successful, communication problems often arise. (Ibid., 654, our translation)

Many texts in our sample expressed similar concerns, citing a lack of knowledge about these populations, "underdiagnosis," or "diminished use of existing services" (ibid.). Such framings contribute to an *othering* discourse that constructs these groups as *closed, homogeneous communities* that need to be reached.

The authors estimate the dementia prevalence among people with migration background by referring to "prevalences in the countries of origin of the different ethnic groups" (Monsees, Hoffmann, and Thyrian 2019, 654, our translation). Here, ethnicity is defined and presented in a chart according to country of origin or nationality (e.g., "Poland"). This illustrates how migration background is not only conflated with ethnicity but also with nationality—despite the fact that the data stem from statistics on people with migration background, many of whom may hold German citizenship. Following this homogenization through processes of ethnicization and nationalization, the authors proceed to calculate dementia risks for these groups based on dementia prevalence data from their countries of origin. For instance, if Turkey has an overall dementia prevalence of 4.25 percent, this figure is used to estimate the risk for people with a Turkish migration background living in Germany.

Crucially and unfortunately, the authors leave open the question of how these supposed country-specific or "ethnic" differences could be further explained. They acknowledge that significant variations—such as 4.25 percent for Turkey versus 8.24 percent for Austria—could result from differences in data collection methods. However, they also suggest the possibility of "ethnic differences," citing studies on higher dementia risks among Hispanics and African Americans in the United States. This evidence appears sufficient for them to assume that the overall dementia prevalence in Poland or Turkey can predict the dementia risk of an individual who immigrated from Poland or Turkey in early childhood. They even concede in the limitations that this practice might be inappropriate, and yet they still consider their results to be valid approximations. In our view, this approach reflects an essentializing notion—implying that people from the same country share a substantial, unchanging characteristic that influences their dementia risk. A more reasonable alternative would be to assume that people with migration background in Germany have the same dementia prevalence as those without migration background, unless specific factors are identified to justify claims of "ethnic differences."

The authors conclude their paper by introducing another recurring theme in our dataset, and additional contested categories: They emphasize the need for a "culture-sensitive approach" in dementia care and highlight the role of societal "integration" in shaping healthcare needs and access for people with a "migration background." On the other hand, this move reintroduces a degree of heterogeneity—such as generational differences or varying levels of integration—within the groups being studied.

Monsees, Hoffmann, and Thyrian (2019) serve as an example of what happens when distinct classifications of difference—migration background, ethnicity, nationality, country of origin (along with culture and integration)—become so entangled that they lose conceptual clarity. The internal inconsistencies in their text suggest an underlying reliance on both biological and essentialist cultural notions of ethnicity.

The conflation of concepts becomes even more confusing when, as in other publications, the second-largest group of immigrants—the so-called resettlers—is addressed. Hagenfeld et al. (2019), for instance, present "a comparison of Germans and two migrant groups" in their study on "periodontal health and use of oral health services." Once again, migration background is conflated with ethnicity: "migrant groups" and nonmigrant groups are redefined as "ethnic groups." This is particularly evident in Table 1, where "Turks," "resettlers," and "Germans" are listed as three distinct ethnic groups. Moreover, the term "German" is used in at least two different ways throughout the text. The group without migration background is simply called "Germans," implying that these individuals are somehow more German than people with a migration background who hold a German passport. So if "German" does not refer here to citizenship, does it instead indicate an ethnic identity? The answer is not straightforward, as resettlers are also described as "ethnic German immigrants from the former Soviet Union," while still being distinguished from the "German group." As a result, the three so-called ethnic groups being compared are "Germans," "ethnic German resettlers," and "Turks." The latter, who would more accurately be described as people with a Turkish migration background, are thus completely excluded from being considered "German."

Other examples in our dataset take a seemingly more sensitive approach to the topic while still representing another mode of conceptually mixing analytically distinct terms. Zhou et al. (2018), for example, focus on migration background and the prevalence of overweight and obesity. One of their findings is that "Turkish migrant children showed a higher overweight prevalence compared to their peers" (Zhou et al., 761). The authors advocate for self-reporting in human classification, which raises an important point for discussion, but their inconsistent use of distinct concepts makes it difficult to assess their proposal. They note, for instance: "Migration background is more a matter of self-perception than an objective fact. However, epidemiological research on ethnic disparities in health is inconsistent in the use of terms" (ibid., 759). Here, the authors directly link "migration background" to "ethnic disparities." They go on to argue that using parents' birthplace as a classification criterion "might result in inaccuracies when considering cultural implications." They then propose: "In our study, ... the classification referred to a self-identified migration background targeting personal aspects and cultural lifestyle ... using a self-defined ethnicity may improve the ability to capture more meaningful groups that share language, religion, and traditions" (ibid.).

As in these examples, throughout the paper it is impossible for the reader to disentangle migration background and ethnicity, as the terms appear interchangeable. The authors categorize children "into six groups based on the main ethnic groups in the German population (non-migrant, Turkish migrant, Polish migrant, other Eastern European migrant, Russian migrant, and others)" (ibid., 755). This classification leads to references to children of immigrants—who may never have migrated themselves—as, for example, "Turkish migrants," reinforcing a distinction between them and "ethnic non-migrants."

Inclusion or Integration?

In the discussion section, Zhou et al. (2018) note that "[a]mong Turkish mothers in Germany, a shorter length of breastfeeding and higher smoking prevalence during pregnancy were reported, both of them known to be risk factors for childhood obesity," and suggest that "culturally tailored interventions starting in pregnancy and early childhood may be more effective, and further research should focus on its development and implementation" (759).

This exemplifies another recurring pattern we identified in our material: health differences are attributed to *cultural* differences and to immigrants' lack of behavioral integration in Germany, including adherence to a "healthy" lifestyle. This reinforces our impression that aspects of German public health discourse align with broader political debates on the proper "integration" of people with a migration background. As a result, they may be shaped more by an *integration paradigm* (zur Nieden 2014; Bartram et al. 2023) than by an *inclusion paradigm* or *biomulticulturalism*, as Epstein has described for the US context (Epstein 2007). The latter, closely linked to ethnic identity politics in the United States, seeks to incorporate presumed ethnic groups in their (biological) diversity to create a more inclusive health system. By contrast, the integration paradigm views health disparities as a consequence of a lack of awareness or nonconformance with preventive strategies. The implicit goal seems to be assimilation—for example, that Turkish mothers adopt German behavioral norms, such as avoiding smoking during pregnancy and breastfeeding.

While Monsees, Hoffmann, and Thyrian (2019) appeared to treat ethnicities as *biologically* distinct, in this case, the ethnicization of migration background reflects a *culturalist* form of essentialization. It attributes ethnic differences to "culture" rather than considering structural disadvantages that may result from migration status. The study does control for one factor—parental education level—but does not consider income or experiences of discrimination. Both forms of essentialization ultimately fail to sufficiently consider socioeconomic explanations for health inequalities.

Discussion

We have presented some of the results of our empirical analysis in more detail. What has become quite clear is that the use of migration background as a classifying principle adds an additional layer of complexity to epidemiological research. This complexity is addressed in very different ways across the publications we analyzed.

Since the introduction of the category into German epidemiology, the concept has been increasingly criticized for various reasons—not only by migrant self-organizations but also by epidemiologists, including those from the German government institution Robert Koch Institute (Kajikhina et al. 2023). Their critique culminated in the recommendation that the concept should no longer be used, accompanied by revised "Recommendations for Collecting and Analyzing Migration-Related Determinants in Public Health Research." One of the main reasons cited was the inconsistent operationalization of migration background in studies, where country of birth and current citizenship are often conflated, despite the Federal Statistical Office defining it based on one's own and/or parental citizenship at birth. Our findings also highlight the problems arising from this conflation.

A comparable qualitative analysis in the German-speaking context, focusing on research in education and psychology (Moffitt and Juang 2019), supports our findings by emphasizing the inconsistent and heterogeneous use of terminology in these fields. Among the key issues they identify is the conflation of "im/migrant" and/or "migration background" with "ethnic minority." Similarly, in our empirical material, migration background is frequently used as an indicator of—or even as a synonym for—ethnicity, and vice versa. This shows that migration background is not only associated with the social experience of migration and its potential consequences, such as legal and socioeconomic disadvantages, discrimination, and language barriers, but is also being reinterpreted—that is to say, retranslated—in terms of older concepts of ethnic difference. While migration background as a formal concept includes ethnic elements (Will 2018, 2019), it also differs significantly from ethnicity as it is used, for example, in the United States.

Despite originally being a formalized administrative category that lacks the subjective elements and ambiguity of race or ethnicity, the way migration background is enacted in practice leads to similar problems, with wide-ranging practical and analytical consequences:

1) In epidemiological research, the alignment of migration background with ethnicity in many studies may be an implicit or explicit attempt to make research involving mi-

gration background compatible with international research—and research databases.[7] Here, ethnicity appears to function as the preferred mediation category or boundary object (Ellebrecht et al. 2023), one that is (at least to some extent) understood across borders. However, classification systems such as race and ethnicity are anything but easily translatable: because of their social nature, they carry specific national meanings and differ in their operationalization. The blurring of terms further reduces reproducibility and translatability, which may have serious consequences—potentially rendering results ineffective for the implementation of meaningful public health measures.

2) Migration background, as a seemingly "more innocent" biopolitical technology of difference and classification, may in fact perpetuate the *absent presence of race* we cited in the introduction through its unquestioned (re)attachment to ethnicity. We agree with Max Weber's (1990) definition that ethnicities are groups that share a subjective belief in their common origin (237). Weber, like most sociologists after him[8], thus emphasizes the socially constructed nature of ethnicity while also pointing out that it attributes unchangeable qualities to a group, such as a shared origin. For Weber, race is simply a form of ethnicity that further emphasizes common *biological traits*. In our material, as exemplified by Monsees, Hoffmann, and Thyrian (2019), ethnic groups are sometimes attributed more essential characteristics—such as a shared risk for dementia—that resemble the innate essential group qualities found in racial constructions. Such racializing notions (Wacquant 2022, 78) of distinct kinds of bodies also inform the "inclusion-and-difference paradigm" described by Epstein (2007).

The administrative category of migration background, as defined in the microcensus, also refers to an unchangeable quality—being born with or without German citizenship. However, this classification differs in two key ways. First, it is inherited by only one generation and disappears in the next, meaning it does not create an enduring group. Second, it lacks the subjective element of belief that fosters a sense of community.

3) As we have discussed in the case of Zhou et al. (2018), the category of migration background can still have a boundary-drawing effect by keeping the children or grandchildren of immigrants—people who have never migrated themselves but may experience the social consequences of their ancestors' immigration—within a distinct "not-quite-German" group. These individuals are often simply labeled as "migrants" or "Turkish mothers," even though they may be German citizens. In this publication, the subjective element of belief is already embedded in the definition of migration background as a "matter of self-perception," making it more comparable to how ethnicity is used in the United States. However, unlike in the US and its inclusion-and-difference paradigm (Epstein), where ethnic groups are considered part of a larger national identity, the German case illustrated here—as well as in Monsees, Hoffmann, and Thyrian (2019)—that it constructs "German" as one ethnicity that is contrasted with the "migrant" Other. Migration background thus functions as a "natio-ethno-cultural" framework for defining belonging and non-belonging, or non-Germanness (Mecheril 2002, 2003).

---

7 Authors of papers that we interviewed in another part of our study confirmed that they used the category to make it comparable with the international literature.

8 For an overview of the state of research see the article by Bartram and Plümecke in this volume.

Part of this construction is the reference to cultural differences while simultaneously demanding adaptation to the dominant health behaviors of the majority—a pattern that may stem from an integration paradigm. For example, Turkish mothers are to be approached with "culturally tailored interventions," but the goal remains for them to conform to the idealized behaviors of German mothers, such as breastfeeding and not smoking.

Our research supports the observations of Kajikhina et al.:

> [E]thnicity, origin, culture, or tradition are often spoken of as individual or group-related characteristics, without taking into account the legal and structural disadvantages, belonging, and discrimination. In this way, relevant explanatory mechanisms of health inequality are obscured and the 'otherness' of the supposed holders of these characteristics is emphasised and essentialised, i.e., substantiated as inherent. (Kajikhina et al. 2023, 64)

In German epidemiology, as in other scientific fields and practices, it becomes evident that the pursuit of greater heterogeneity always carries the risk of generating new (or, on closer inspection, not so new) forms of homogenization—particularly of the vulnerable, hard-to-reach Other. Further research is needed to examine the processes of othering at work here in more detail (Akbulut and Razum 2021) and, in the long run, to ensure that public health measures based on epidemiological data truly count people in rather than merely adding another layer of counting them out. The new recommendations by Kajikhina et al. for "a mutual and differentiated consideration of migration-related and social determinants of health," which include variables that account for discrimination, may contribute to this goal. However, we join the authors in their assessment that diversity research always "entails the risk of external attributions, discrimination, and misinterpretations" (Kajikhina et al. 2023, 64).

References

Akbulut, Nurcan, and Oliver Razum. 2021. "Considerations on the Relationship Between Othering and Public Health." *Social Epidemiology Discussion Papers (SEDiP)*, no. 1. DOI: https://doi.org/10.4119/unibi/2950920.

Bartram, Isabelle, Laura Schnieder, Nils Ellebrecht, Florian Ruland, Tino Plümecke, and Andrea zur Nieden. 2023. "Categorizing People in the German Life Sciences: A Systematic Literature Review of Classifications of Human Diversity." *Discover Social Science and Health* 3 (1). DOI: https://doi.org/10.1007/s44155-023-00033-5.

Berg, Charles, and Marianne Milmeister. 2011. "Im Dialog mit den Daten das eigene Erzählen der Geschichte finden: Über die Kodierverfahren der Grounded-Theory-Methodologie." In *Grounded Theory Reader*, edited by Günter Mey and Katja Mruck, 303–332. Wiesbaden: VS Verlag für Sozialwissenschaften.

Charmaz, Kathy C., ed. 2012. *Constructing Grounded Theory*. 2nd ed., repr. Los Angeles: Calif. Sage.

Chin, Rita. 2016. "Thinking Difference in Postwar Germany: Some Epistemological Obstacles Around "Race."" In *Migration, Memory, and Diversity: Germany from 1945 to the Present*, edited by Cornelia Wilhelm, 206–230. New York: Berghahn Books.

Ellebrecht, Nils, Hannah Schönberger, Isabelle Bartram, Tino Plümecke, and Andrea zur Nieden. 2023. "Tabu, Problem oder Standard? Die Ethnisierung von Race-Klassifikationen in der Deutschen Medizin- und Gesundheitsforschung." In *Polarisierte Welten: Verhandlungen Des 41. Kongresses der Deutschen Gesellschaft für Soziologie in Bielefeld 2022*, edited by Paula-Irene Villa Braslavsky. Accessed February 4, 2025. https://publikationen.soziologie.de/index.php/kongressband_2022.

Epstein, Steven. 2007. *Inclusion: The Politics of Difference in Medical Research*. Chicago: University of Chicago Press.

Food and Drug Administration. 2016. "Collection of Race and Ethnicity Data in Clinical Trials: Guidance for Industry and Food and Drug Administration Staff." Accessed February 4, 2025. https://www.fda.gov/media/75453/download.

Fujimura, Joan H., and Ramya Rajagopalan. 2011. "Different Differences: The Use of 'Genetic Ancestry' Versus Race in Biomedical Human Genetics Research." *Social Studies of Science* 41 (1): 5–30.

Fujimura, Joan H., and Ramya M. Rajagopalan. 2020. "Race, Ethnicity, Ancestry, and Genomics in Hawai'i." *Historical Studies in the Natural Sciences* 50 (5): 596–623. DOI: https://doi.org/10.1525/hsns.2020.50.5.596.

Hagenfeld, Daniel, Heiko Zimmermann, Katja Korb et al. 2019. "Periodontal Health and Use of Oral Health Services: A Comparison of Germans and Two Migrant Groups." *International Journal of Environmental Research and Public Health* 16 (16). DOI: https://doi.org/10.3390/ijerph16163000.

Kajikhina, Katja, Carmen Koschollek, Navina Sarma et al. 2023. "Recommendations for Collecting and Analysing Migration-Related Determinants in Public Health Research." *Journal of Health Monitoring* 8 (1): 52–72. DOI: https://doi.org/10.25646/11144.

Kruse, Jan. 2015. *Qualitative Interviewforschung: Ein Integrativer Ansatz.* 2., überarb. und erg. Aufl. Weinheim, Basel: Beltz Juventa.

Lipphardt, Veronika, Anna Lipphardt, Amade M'charek et al. 2018. "Lost in Translation: Why the US American Concept of "Race" Is Not to Be Translated Using the German Term "Rasse."" Accessed December 19, 2024. https://cpb-us-e1.wpmucdn.com/sites.ucsc.edu/dist/4/415/files/2018/04/Lost_in_translation_english-10.21.18-Final-21wstie.pdf.

M'charek, Amade, Katharina Schramm, and David Skinner. 2014. "Technologies of Belonging." *Science, Technology, & Human Values* 39 (4): 459–467. DOI: https://doi.org/10.1177/0162243914531149.

Mecheril, Paul. 2002. "Natio-kulturelle Mitgliedschaft – Ein Begriff und die Methode seiner Generierung." *Tertium Comparationis* 8 (2): 104–15.

Mecheril, Paul. 2003. *Prekäre Verhältnisse. Über natio-ethno-kulturelle (Mehrfach-)Zugehörigkeit.* Münster: Waxmann.

Moffitt, Ursula, and Linda P. Juang. 2019. "Who Is 'German' and Who Is a 'Migrant?' Constructing Otherness in Education and Psychology Research." *European Educational Research Journal* 18 (6): 656–674. DOI: https://doi.org/10.1177/1474904119827459.

Monsees, Jessica, Wolfgang Hoffmann, and Jochen René Thyrian. 2019. "Prävalenz von Demenz bei Menschen mit Migrationshintergrund in Deutschland.". *Zeitschrift für Gerontologie und Geriatrie* 52 (7): 654–660. DOI: https://doi.org/10.1007/s00391-018-01469-0.

Neue Medienmacher. "NdM Glossar: Menschen mit Migrationshintergrund." Accessed December 19, 2024. https://glossar.neuemedienmacher.de/glossar/menschen-mit-migrationshintergrund-mh/#easy-footnote-bottom-1-97.

Plümecke, Tino. 2010. "Die neuen Differenzen der Lebenswissenschaften: ‚Rasse', Genetik und die ungenutzten Potentiale der Soziologie." In *Ethnowissen: Soziologische Beiträge zu ethnischer Differenzierung und Migration*, edited by Marion Müller and Dariuš Zifonun, 423–449. Wiesbaden: VS.

Pollock, Anne. 2012. *Medicating Race: Heart Disease and Durable Preoccupations with Difference*. Durham, NC: Duke University Press.

Schenk, Liane, Anne-Madeleine Bau, Theda Borde et al. 2006. "Mindestindikatorensatz zur Erfassung des Migrationsstatus. Empfehlungen für die epidemiologische Praxis." *Bundesgesundheitsblatt, Gesundheitsforschung, Gesundheitsschutz* 49 (9): 853–860. DOI: https://doi.org/10.1007/s00103-006-0018-4.

Schenk, Liane. 2007. "Migration und Gesundheit – Entwicklung eines Erklärungs- und Analysemodells für epidemiologische Studien." *International Journal of Public Health* 52 (2): 87–96. doi.org/10.1007/s00038-007-6002-4.

Shim, Janet K., Sara L. Ackerman, Katherine Weatherford Darling, Robert A. Hiatt, and Sandra Soo-Jin Lee. 2014. "Race and Ancestry in the Age of Inclusion: Technique and Meaning in Post-Genomic Science." *Journal of Health and Social Behavior* 55 (4): 504–518. DOI: https://doi.org/10.1177/0022146514555224.

Statistisches Bundesamt. 2017. "Fachserie 1 Reihe 2.2 Ergebnisse des Mikrozensus 2016." Accessed February 6, 2025. https://www.statistischebibliothek.de/mir/receive/DEHeft_mods_00070829.

Statistisches Bundesamt. 2024. "Hintergrundpapier für die Veröffentlichung von Ergebnissen zur Bevölkerung nach Einwanderungsgeschichte." Accessed December 19, 2024. https://www.destatis.de/DE/Themen/Gesellschaft-Umwelt/Bevoelkerung/Migration-Integration/Methoden/Erlauterungen/einwanderungsgeschichte-hintergrundpapier.html.

Thompson, Debra E. 2016. *The Schematic State: Race, Transnationalism, and the Politics of the Census.* Cambridge: Cambridge University Press.

United States Census Bureau. 2024. "What Updates to OMB's Race/Ethnicity Standards Mean for the Census Bureau." Accessed February 5, 2025. https://www.census.gov/newsroom/blogs/random-samplings/2024/04/updates-race-ethnicity-standards.html.

Wacquant, Loïc. 2022. "Resolving the Trouble with 'Race'." *New Left Review* 133/134: 67–88.

Weber, Max. (1921) 1990. *Wirtschaft und Gesellschaft: Grundriss der Verstehenden Soziologie*. Studienausg., 5., rev. Aufl., Nachdruck 1990. Tübingen: J. C. B. Mohr (Paul Siebeck).

Will, Anne-Kathrin. 2018. "Wie Werden Zuwanderer und ihre Nachkommen in der Statistik erfasst?" *Mediendienst Integration*, 1–14. Accessed December 19, 2024. https://mediendienst-integration.de/fileadmin/Info-Papier_Mediendienst_Integration_Migrationshintergrund_2020.pdf.

Will, Anne-Kathrin. 2019. "The German Statistical Category 'Migration Background': Historical Roots, Revisions and Shortcomings." *Ethnicities* 19 (3): 535–557. DOI: https://doi.org/10.1177/1468796819833437.

Zhou, Yusheng, Thomas von Lengerke, Ulla Walter, and Maren Dreier. 2018. "Migration Background and Childhood Overweight in the Hannover Region in 2010–2014: A Population-Based Secondary Data Analysis of School Entry Examinations." *European Journal of Pediatrics* 177 (5): 753–763. DOI: https://doi.org/10.1007/s00431-018-3118-x.

zur Nieden, Andrea. 2014. "The Presence of the Past: 'Ashkenazi BRCA Mutations' and Transnational Differences in Categories of 'Race' and 'Ethnicity': The German Case." In *Breast Cancer Gene Research and Medical Practices: Transnational Perspectives in the Time of BRCA*, edited by Galen Joseph, Jessica Mozersky, Andrea zur Nieden, and Sonja Palfner, 17–34. Oxon (UK): Taylor & Francis.

Biologies of Ethnicity

Artifacts of Politics in German Life Sciences

Tino Plümecke and Isabelle Bartram

Abstract *Ethnic classifications are among the most utilized concepts in the life sciences to categorize people. This paper critically examines the diverse aims and approaches characterizing their use in life science research in Germany. For this purpose, we conducted a systematic literature review and a quantitative and qualitative content analysis of 249 studies by authors affiliated with German institutions, published between 2018 and 2020. Our analysis meticulously explores how these studies assign ethnicity to research subjects, as well as the types of differences they examine and the explanations they provide for these differentiations. Our findings reveal that most articles do not comply with the guidelines of leading science journals regarding definitions of ethnicity and classification of research subjects. Most articles we reviewed employ ethnicity to investigate biological and health-related disparities caused by social inequalities. However, they also use the concept to identify biological or genetic differences, with many studies offering genetic explanations alongside social and cultural ones for the differences they found. This highlights several problematic aspects inherent to the concept of ethnicity, which is often used in ways that are fuzzy, with a spectrum of social, biological, and genetic meanings. Ethnicity is not a clearly defined entity in the studies examined; rather, it emerges as a social-biological construct influenced by various political demands, hopes, expectations, and specific cultural-temporal research frameworks. Consequently, human diversity and social inequity are transformed into a methodologically and theoretically underdeveloped artifact.*

Introduction

The emergence of advanced genetic and computational technologies within the last twenty-five years has enabled and fostered a vast amount of research in the fields of human genetics, precision medicine, and public health, as well as anthropology and history, by facilitating the statistically mapping of groups of people, which has also increased the interest in group-based differentiations to investigate health inequities between minorities and the majority population and to trace historical events such as histories of migration, pandemics, and population growth. To capture the diversity of research subjects on a level beyond individuals, classifications referring to the origin, descent, and social-cultural group differentiation are now, again, being increasingly applied in life science research—so much so that one can speak of a current renaissance of biological differentiation according to ethnic and racial concepts (Duster 2015). This recent expansion in the identification of race and ethnicity can also be attributed to a growing awareness of, and research interest in, social inequalities among various parts of the population (Krieger 2020; Epstein 2007). Both research goals—investigating biological differences and social inequality—are used to justify certain categories for sorting humans into groups. Moreover, even if researchers rely on different terms and concepts, these words often convey similar meanings (Malinowska and Żuradzki 2023; Fujimura and Rajagopalan 2011).

Internationally, "race," "ethnicity," "population," and "ancestry" are the concepts used most often to distinguish people into groups (Byeon et al. 2021). Nonetheless, other terms are used in specific national and disciplinary contexts.[1] It is important to note that each of these concepts is imprinted in ways that are specific to context, culture, time, and discipline in its meanings and implications. Particularly noteworthy are the different meanings that have been constructed around the concept of race as a result of the extensive critiques that have taken shape over many decades (Duster 1990; Reardon 2005; Plümecke 2010). Social scientists and scholars in the humanities have criticized the concept for its a priori assumptions regarding its supposed precision as a tool of classification, its inherent essentialism, and the way it carries forward ideas of biological determinism (Koenig, Lee, and Richardson 2008; Roberts 2011). From within the life sciences, many have argued that concepts of race are not empirically substantiated, and that they hinder the investigation of biological and anthropological questions (Lewontin 1972; Marks 1995, 2008; Fischer et al. 2019). Not only have concepts of race faced criticism; the practice of labeling human groups more broadly has also been the subject of extensive political and ethical deliberations (Bowker and Star 2000; Fujimura and Rajagopalan 2011; National Academies 2023).

In Germany, the inclusion of models for human classification in bioscientific research has become the focus of a renewed debate. Alongside references to international research—where distinctions are predominantly based on categories such as race and ethnicity—there is a growing emphasis on arguments advocating for the differentiation of research subjects into population groups. For us, the question is how human classifications are dealt with in the German context, against the background of a Nazi past and the various contemporary criticisms and problematizations of classification models. A recent comprehensive analysis our research group conducted of publications by German research institutions revealed that terms related to ethnicity—such as "ethnic group," "ethnic minority," "ethnic background," or "ethnic ancestry"—are the most frequently used in the German life sciences (Bartram et al. 2023). In contrast to the relatively

1 In Germany, for example, the term "migration background" is frequently applied by researchers in the life sciences, especially in the field of epidemiology (Schenk 2007; zur Nieden and Bartram 2020; Bartram et al. 2023). In the Netherlands, the terms "autochtoon" and "allochtoon" are widely used (Yanow and van der Haar 2013). Furthermore, in some countries, immigration status is the primary factor in making differentiations, which is why terms such as "migrant" or "refugee" are preferred (see e.g., Mangrio, Paul-Satyaseela, and Strange 2020).

infrequent use of the term "race" (13.9%), "ethnicity" appeared in nearly half (45.5%) of the 546 articles from the period 2018 to 2020 we analyzed. This finding from Germany is consistent with trends observed in international studies. For instance, an analysis by Byeon et al. (2021) of classification terms in the *American Journal of Human Genetics* from 1949 to 2018 found that the use of ancestry and ethnicity increased significantly over the time period, while the use of "race" declined. Similarly, a study by Afshari and Bhopal (2010) examining terms used in international and US-based articles from 1965 to 2005 found that the concept of ethnicity surpassed the use of race between 1976 and 1980 in the international context and between 1991 and 1995 in the United States.

The increasing use of ethnicity may seem surprising given the apparent dominance of racial classification in the debate. However, several authors have advocated for this approach since the early twentieth century, and leading institutions have long supported ethnicity as a more innocuous alternative to race (Deniker 1900; Weber 1922; Huxley and Haddon 1935). Most prominently, the UNESCO statement on race recommended that one "drop the term 'race' altogether and speak of *ethnic groups*" (UNESCO 1950, 6, emphasis in the original).[2] From these early drafts and statements, it took time for the concept to become part of a broader international political debate and conceptual shifts within the scientific community. For instance, as recently as 1999, a committee of the US Institute of Medicine (now the National Academies of Medicine) advised "an emphasis on ethnic groups rather than on race in the NIH's cancer surveillance and other population research" (Institute of Medicine 1999, 4).[3] In Germany, it was only in 2006 that the German Society for Anthropology, representing human biologists, formally urged its members to avoid the term "Rasse" (race) and instead use "the most fitting term for the observed group of people, namely the geographical variant, the ethnicity or the respective population" (Niemitz, Kreutz, and Walter 2006, 463). It is important to note that ethnicity has long served as a substitute for race in German-speaking countries, particularly in the life sciences. As one of the nations that sees itself as a "post-racial" society, Germany has increasingly removed the concept of race from everyday language, legal texts, media, and academic discourse (Liebscher 2021). However, it is crucial to recognize that racism and racialized attributions do not simply vanish as a result, but may reconfigure and reassign with the concept of ethnicity.

Lack of Research on the Concept of Ethnicity

Given the rise in its use, it is surprising that the utilization of "ethnicity" in the life sciences does not receive the same level of critical historical, ethical, or social scrutiny typically directed toward "race." This lack of attention is even more surprising, especially considering the associations tied to the term "ethnos," such as its links to nationalism and the ethnicization of social conflicts—connections extensively documented in research

---

2 The recommendation to replace the concept of "race" with that of "ethnicity" was not repeated in the second UNESCO Statement on the Nature of Race and Racial Differences from 1950.

3 However, this is countered by the fact that the U.S. Food and Drug Administration, as the regulatory agency for the NIH, has been recording race (as well as ethnicity, sex, and age) since 1993, for purposes such as clinical trials for the approval of drugs.

on racism and discrimination. Race theorists such as Joseph Deniker also used the term "ethnic groups" (Deniker 1900), and Adolf Hitler—who primarily used the concept of race—occasionally employed "ethno" as a term, as in a speech he gave to the Reichstag on October 6th 1939, when he spoke of a "new ordering of ethnographic relations" that would follow the destruction of the Polish state (Domarus 1973, 1863). Finally, given that ethnic groups are mostly conceptualized as sociocultural entities, the use of this term to describe biological, genetic, or natural traits should give rise to serious questions of clarification. All of this raises the important question of how ethnicity is conceptualized and applied and whether ethnicity can effectively serve as a scientific classification tool for identifying social and group-based differences, as well as whether it offers a meaningful alternative to race for such a purpose.

To address these questions, this article focuses on three key areas. First, we offer a *brief overview of the historical use* of the concept of ethnicity and its interpretations. This seems particularly important to us, as there is little literature to date that systematically documents the historical and disciplinary negotiations that form the basis of our investigation. Second, we analyze *how ethnicity is produced as a biologically meaningful concept* in life science research in Germany. With the term of "biologies of ethnicity," our aim is to explore how ethnicity is used in life science research in Germany, i.e. *how it is defined, applied, and how its variations of meaning are established*. This includes examining *how research subjects are classified* into ethnic categories, what *kind of differences* researchers claim to identify between ethnic groups, and *what reasons* they propose for these differences. Third, we aim to foster a more critical examination of the topic.

Our empirical findings will ground a "critical interrogation on the present" (Foucault 1984, 49–50, 2010, 21). Foucault's concept refers to the critical practice of interrogating and analyzing the historical conditions and power structures that underpin contemporary realities. Central to this endeavor is an examination of how current norms, truths, and practices have been historically constructed, revealing the contingent mechanisms through which human beings are classified and governed. Rather than seeking a return to the past or positing an idealized future, this critique aims to unsettle established certainties, thereby creating space for alternative modes of existence and thought. The aim of our interrogation is to uncovering the politics embedded within the use of the concept "ethnicity." Treating ethnicity as an artifact of politics, we examine how politics and scientific knowledge is constructed and sustained together through specific practices, norms, standards, and discourses. Our findings reveal that there is nothing neutral or merely descriptive about the concept of ethnicity.

Methods

To investigate ethnicity in German life sciences, we conducted a systematic review of articles from German life science research institutions. Our review focused on studies published between 2018 and 2020 that classified people into ethnic, racial, national, or migration-related groups (Bartram et al. 2023).[4] Specifically, we searched the databases PubMed and Web of Science (WoS) to identify all German life sciences studies that applied potentially racializing classifications to human research subjects. We then used document analysis and descriptive statistics to explore how concepts of human differentiation were applied and implemented throughout the research process (Mayring 2014).

Our investigation uses this descriptive analysis to ground our inquiry of the ways in which ethnicity is given its meanings. With reference to Foucault's critical interrogation on the present, we seek to draw out the relationships between power and knowledge—and particularly the explicit and implicit political effects operating at different levels of politics—that concepts of ethnicity express. Along with Foucault, we consider power not as an abstract force but as something that shapes and drives scientific questions, research designs, chosen concepts, analyses, and forms of presentation. Human classification, in our view, provides a particularly clear example of how scientific knowledge production is intertwined with sociocultural questions, policies, and frameworks. Against this backdrop, we do not merely ask whether "artifacts have politics" (Winner 1980); instead, we approach ethnicity as an artifact shaped by a series of political interpellations.

For this study, we followed the PRISMA-P 2020 guidelines for systematic reviews and meta-analyses (Page et al. 2021). Using variations of the terms "race," "ethnic," "migration background," and "ancestry," and filtering for articles authored by individuals affiliated with German institutions (either as first or last authors), we identified 546 articles relevant to our review. We analyzed the content of these articles both quantitatively and qualitatively using MAXQDA Standard 2020 (version 20.1.0, VERBI GmbH, Berlin). Our coding system included the following categories: authorship (subcodes: all authors, first and last authors, or first or last author affiliated with German institutions), discipline (based on the research institutions of first and last authors), terminology (terms used to classify or define groups of subjects), and study location (the country where samples or data were collected). In our content analysis, we examined not only the classifications described in the methods sections to define or stratify research subjects but also the terms used in the discussion sections. For this article, we focused on the 249 studies that used the term "ethnicity" or its derivatives (e.g., ethnic group, ethnic minorities) to describe research subjects.

We further analyzed how research subjects were assigned to an ethnicity (e.g., based on country of birth, genetics, self-identification, etc.) and the types of differences identified between ethnic groups (e.g., anthropometric variables, health status, genetics). Finally, we investigated the explanations provided for these observed differences, examining whether and how study authors attributed them to cultural or lifestyle factors, health service access, biological factors, or other causes.

Focusing on the methods, results, and discussion sections, we carefully read and coded all articles. When uncertainties arose, our team discussed the publications and coding decisions until reaching a consensus.

---

4 Details on our selection of studies and analysis of their content can be found in this previous publication.

Brief Overview of the Concept of Ethnicity

The word "ethnicity" originates from the Greek term "ethnos" and its corresponding adjective "ethnikos," which denotes a form of we/they distinction within a nonspecific group or collective. In French, the noun "etnique" has been in use since the thirteenth century, meaning "pagan." The adjective "ethnicque" with the same meaning came into usage in the sixteenth century (Académie française 2024). In English, the adjective "ethnic" has been recorded since the fifteenth century, and in German "ethnisch" has been used since the sixteenth century. European scholars and colonial administrators applied these terms to describe the diverse groups of people they encountered in colonized territories who were not considered Christians or Jews (Oxford English Dictionary 2024; DWDS 2024). It was only in the nineteenth century, with advances in historical and ethnographic scholarship, that these terms gained broader usage to denote belonging to a (foreign) people, tribe, or nation characterized by diverse customs, religions, and appearances (DWDS 2024; Bös 2005, 2015). The development of ethnic terms has been closely tied to their relationships with nationality, culture, and, above all, race (Hattam 2004).

Ethnicity in the Social Sciences

While the concept of race was relatively widespread in the early development of social science—encompassing a spectrum of biological, Social Darwinist, and sociocultural meanings—ethnic terms appeared only sporadically and lacked analytical precision until the 1920s. They were often used synonymously with race. One example is found in the work of Polish-Austrian sociologist Ludwig Gumplowicz, who, at the end of the nineteenth century, wrote about social groups and communities and warned that states and their people should not be misunderstood as "permanent ethnic elements," as these themselves consist of "heterogeneous ethnic components of populations" (Gumplowicz 1883, 183, our translation). At the beginning of the twentieth century, American sociologist William Graham Sumner introduced the term "ethnocentrism" (Sumner 1906, 14; see also Bös 2020). In the second decade of the century, some American Zionists in New York began describing themselves as ethnic groups (Hattam 2004, 45). Jewish immigrants had to grapple with issues of group membership and national origin in different ways than other immigrants, as the common national origin labels of "hyphenated Americans" did not provide sufficient identification for them (Banton 2015, 96). In the following decades, debates around ethnicity in the United States were significantly fueled for various reasons: the expanding immigration from countries that were no longer seen as easily integrated into the "melting pot," an increasing criticism of the category of "race" in the context of debates about anti-Semitism in France and Germany, and the social and political controversies surrounding the so-called "race question" in the United States (Plümecke 2013, 81–84).

Importantly, ethnicity was not originally conceptualized as a counterpoint to race. Instead, the two were intricately interconnected in their relationships with nation and culture. The term ethnicity became more important as European immigrant groups (and only these) in the United States were increasingly be seen not as different races, but as different ethnic groups. People previously recorded as biologically distinct races in public discourse and administrative registries were now redefined based on factors such as nationality, language, religion, and behavior.

One of the first—and still widely used—definitions of "ethnicity" was developed by Max Weber. In *Economy and Society* (1922, 389), he defined ethnic groups as "human groups that entertain a subjective belief in their common descent because of similarities of physical type or of customs or both, or because of memories of colonization and migration." Weber conceptualized ethnicity as a form of social relationship, that united group members through a shared belief in common origins, whether based on physical or cultural characteristics. "It does not matter," he emphasized, "whether or not an objective blood relationship exists" (ibid.). Importantly, Weber viewed race as a subset of ethnicity, while other social scientists emphasized their distinctions. Some of these defined race as based on biological or phenotypical characteristics, contrasting it with ethnicity, which they saw as rooted in cultural attributes (van den Berghe 1967; Omi and Winant 1986; Feagin and Feagin 2011). Others highlighted differences in identification processes. They linked ethnicity to positive tendencies of self-identification and inclusion, and argued race primarily linked to negative attributions of exclusion by others (Banton 1977, 136).

Contemporary empirical research on ethnicity mirror back this diversity of theoretical and methodological approaches. Thus, some scholars argue that there is no universally agreed meaning of the term within sociology (Banton 2015, 119; Gabbert 2015, 185). However, certain elements are common to all definitions. First, ethnicity involves a distinction between "we" and "the other." Second, this distinction is based on perceptible or imagined characteristics, such as physical features or cultural practices such as language, religion, or traditions. Third, these distinctions go beyond the (putative) characteristics themselves to create socially relevant differences—differences that carry social consequences. Fourth, ethnicity is often linked to power dynamics, structural advantages or disadvantages, and relationships of social marginalization or dominance. Many researchers see ethnicity as deeply intertwined with issues of income inequality, wealth disparities, and poverty, underscoring its intersection with class (Gordon 1964; Stein 2004).

Administrative definitions of ethnicity vary significantly across countries, with varied and often conflicting meanings and usages across political, legal, medical, and activist contexts. Many European nations do not officially record ethnicity, while in the United States, the census recognizes only Hispanics or Latinos as an ethnic group. The most recent census in the United Kingdom, by contrast, identified nineteen distinct ethnic groups. In Canada, ethnicity is an official category primarily for First Nations people, with the 2021 census offering over 580 examples to account for diverse identities (Statistics Canada 2024). Moreover, as the social anthropologist Frederik Barth demonstrated, ethnic boundaries are not necessarily based on cultural content but on strategically drawn demarcations. Ethnicity is not the sum of "objective" differences, he argues, but of those differences deemed significant by the actors involved: some characteristics are emphasized as markers of distinction, while others are downplayed or denied (Barth 1969, 14). Such demarcations are therefore not fixed; they are continually shaped through interactions, with individuals crossing into different ethnic groups and changing their identities over time. Surveys in the United States, for instance, have shown that respondents often change their reported race or ethnicity. In a comparison of data from the First National Health and Nutrition Examination Survey and the Epidemiologic Follow-up Survey (published in 1975 and 1984, respectively), only 58 percent of respondents identified with the same ethnicity in both surveys (Hahn, Truman, and Barker 1996). These findings highlight the fluidity of ethnicity (and race) as self-constructed categories rather than fixed, objective markers of group membership (Jugert et al. 2022, 4).

All contemporary approaches agree that ethnicity is a social construction. This means it is a socioculturally generated classification often associated with cultural essentialism or biodeterminism. In the social sciences, ethnicity is thus understood not as a property of a group or a natural entity, as something in groups, but as a relationship *between* people (Barth 1969; Eriksen 2010). Ethnicity is thus seen as "highly situational," dependent on the "relations to other groups and persons in a given situation," and subject to frequent change (Bös 2015, 137). It is a process of boundary making and boundary maintenance (Barth 1969, 10), and more a way of "perceiving, interpreting, and representing the social world" (Brubaker 2004, 17). Work in cultural studies often emphasizes the diversity of "cultural landscapes," while disciplines such as anthropology, political science, and sociology focus on the persistence and reformation of cultural boundaries. Many studies in the social sciences aim to identify commonalities in ethnic phenomena, while others investigate construction practices, processes of othering, comparisons of national contexts, and the effects of discrimination against ethnic groups. Research on the impact of social inequality on ethnic groups, in particular, increasingly connects social science studies with bioscientific investigations.

Ethnicity in the Life Sciences

In their seminal work *We Europeans* (1935), evolutionary biologist Julian Huxley and anthropologist Alfred Haddon rigorously challenged the scientific validity of the concept of "race," advocating instead for the adoption of "ethnic group" as a more precise and constructive term (Huxley and Haddon 1935, 136). Their proposal had a profound and lasting impact, notably influencing the UNESCO statement "The Race Question" (UNESCO 1950), which similarly recommended to drop the term "race" altogether and speak of "ethnic groups." The statement and this recommendation encountered significant opposition; however, over the following decades, its perspective gained traction among a number of bioscientists, who criticized the scientific basis of the usual concepts used to classify humans (e.g., Livingstone 1962; Lewontin 1972; Marks 2008; UNESCO 1995).

With the emergence of new molecular genetic methods from the 1980s onwards, criticism of racial concepts also intensified. The molecularization of race—where the concept is increasingly defined through genetics and "reinscribed" or "refashioned" in terms of DNA (Duster 1990; Fullwiley 2007; El-Haj 2007)—has been extensively problematized. How the concept of ethnicity has evolved in the course of this development, on the other hand, is largely unclear. For example, no studies exist to date that have examined the molecularization or genetization of ethnicity apart from its connection to race. Studies show, for example, that the use of the combination term "race/ethnicity" is often used as a "cipher for innate genetic differences" (Outram and Ellison 2006, 83). Similarly, Catherine Lee's content analysis of 204 biomedical research publications (2009) found that in many cases where "race and ethnicity" were invoked to explain observed differences, the "authors relied on biological or genetic explanations" (Lee 2009, 1188). However, these findings cannot be generalized, as the elaborations under the term ethnicity predominantly refer to race. The fact that there is also no scientific consensus on a definition was already pointed out by the director of the National Human Genome Research Institute in 2004, for example, who stated that ethnicity as well as race "are poorly defined terms that serve as flawed surrogates for multiple environmental and genetic factors in disease causation" (Collins 2004, S13).

The lack of clarity in the use of terms has already been documented in a couple of reviews. For instance, in an analysis of reporting on "race/ethnicity" in the leading academic journals for general medicine, surgery, and oncology from 2007 to 2018, the physician Therezia Bokor-Billman and colleagues found that of 995 studies, only 4.5 percent provided a formal definition of the concepts used (Bokor-Billmann, Langan, and Billmann 2020, 2). Similar reviews of surgical research (Maduka et al. 2021) and ophthalmology literature (Moore 2020) revealed widespread failure to define race/ethnicity among biomedical researchers. This lack of definition for key concepts, while otherwise unusual in scientific research, should not be dismissed as mere oversight or imprecision but should itself be the subject of a critical investigation.

Instead of focusing on definitions, we can examine the varying meanings attributed to the concept. An overview shows that these range from perspectives emphasizing its social construction, fluidity, and complexity to those presenting clear biological definitions. One example of a clear biological conceptualization of ethnicity can be seen in the work of the geneticist George Rédei, who defines ethnicity as a "human population related by biological characteristics and identifiable by anthropometric, cultural, linguistic, biochemical, serological, molecular characteristics, SNPs and gene frequencies" (Rédei 2008, 636). In contrast, medical scientist Jessica Cerdeña and colleagues emphasize ethnicity's reference to "cultural, socioeconomic, religious, linguistic, and political qualities of groups that establish cohesion and order through membership, rather than their population genetics" (Cerdeña, Grubbs, and Non 2022, 2147). Psychologist Kevin Cokley further defines ethnicity as "a characterization of a group of people who see themselves and are seen by others as having common ancestry, common history, common traditions and common cultural traits such as language, beliefs, values, music, dress and food" (Cokley 2007, 225). The phrase "see themselves and are seen by others" notably does not assert that ethnicity is objectively linked to a common origin but echoes Weber's concept of "subjective belief," introduced in the previous chapter. Nonetheless, ancestry is still presented here in a fuzzy way: it can be interpreted as something only "seen" (as subjective belief) or, also possibly, as a recognition of an at least probable "common ancestry" (as a biological origin).

Geneticist Charmaine Royal offers a perspective in which biological and cultural attributes are used to determine ethnicity. She initially describes ethnicity as a "legitimate descriptor for human groups, referring to a recognized degree of shared *linguistic, religious* or other *socio-cultural* attributes" but then adds that it "will likely also exhibit (based on an average of its members) relatively appreciable frequencies of some *physical* and/or *biological* characteristics, resulting from a combination of the effects of common geographic ancestry, physical environment, diet, socio-political environment, etc." (Royal

2006, 326, emphasis added, compare also Bhopal 2004, 443). Such descriptions intertwine biological characteristics and ancestry with cultural elements such as language, religion, and traditions.

In recent years, a plethora of suggestions and recommendations for the application of classification terms have been published. A scoping review, for example, identified 121 articles published between 2000 and 2021 that provided normative guidelines on the use of population categories, particularly in genetics research (Mauro et al. 2022). The authors of the review note a broad consensus across the articles on most points. However, they highlight substantial disagreement of opinion regarding the "appropriate definitions of population categories and contexts for use" (2116). For instance, the recommendations contain very different views on how strictly population categories should be defined, with proposals ranging from a rigorous differentiation between the definitions of race and ethnicity to proposals to define them jointly (ibid.: 2019). In summary, they state that the "considerable disagreement among these recommendations" indicates that there is a lack of "clear, centralized definitions of population categories" and that there is also "confusion and disagreement" as to when these should be used (ibid.: 2120).

In our view, the varied interpretations and lack of consensus on the definition of ethnicity highlights how important a careful approach to classifications of ethnic group is in research and publications. In any case, existing standards from institutions such as the International Committee of Medical Journal Editors (ICMJE) and the European Journal of Medical and Health Research (EJMHR) consistently emphasize that "race and ethnicity are social and not biological constructs" and that "authors should interpret results associated with race and ethnicity in this context" (ICMJE 2022; EJMHR 2024). Most importantly, they require that "authors should define how they have determined race or ethnicity and justify their relevance" (ibid.). *The Lancet*, one of the most influential medical journals, similarly mandates that authors "explain the definitions, categories, or conceptual framework used and how they were assigned (e.g., self-report, census or registry data)" in the methods section (The Lancet 2024). Given the critical importance of human classifications in the life sciences, providing empirical insights on this issue is essential. Our analysis, based on a systematic literature review, seeks to address these concerns, as we will explore in detail below.

Ethnicity in the German Life Sciences

To our knowledge, no study has yet examined how ethnicity is conceptualized by researchers at German life science institutions.[5] Nor are we aware of any investigations into how the concept is applied within specific disciplines at German research institutions, such as epidemiology, genetics, or medicine, or in human research more broadly. While treatises have been published on terms such as "race" and "migration background" (Schenk 2007; Kajikhina et al. 2023)—the latter being a term frequently used in the

5 The conceptualization of migration background and race by life science researchers in Germany has been the focus of other studies within our research group SoSciBio; see Ellebrecht and zur Nieden et al. in this volume.

German-speaking world, often accompanied by guidelines for its use in scientific studies—no such work exists for the concept of ethnicity. Even strong critiques of the concept of race by representatives of the life sciences, with statements such as "[n]ot using the term race should be part of scientific decency" (Fischer et al. 2019) or "[t]here is no scientific reason to continue using the term 'race'" (UNESCO 1995: 2), fail to specify how research should be done in its absence.

It is therefore necessary to systematically examine the publications of German research institutions in order to determine how ethnicity is defined and applied. Among the studies identified, the majority were authored by multidisciplinary teams, reflecting extensive collaboration across various fields. In terms of disciplinary affiliation, the distribution is as follows: The most studies originate from the field of medicine (194), followed by epidemiology (79), psychology (11), health services & economics (8), the biomedical and pharmaceutical industry (8), archaeogenetics (4), and anthropology (2).

In the sections that follow, we provide a brief overview of the terminology commonly used to describe ethnic distinctions before analyzing how ethnicity was operationalized by study authors, how individuals were assigned to specific ethnic groups, the differences observed between these groups, and the explanatory approaches authors offered for these differences.

The Most Commonly Used Terms

Of all the classification terms recorded and coded in our search, ethnicity-related terms are the most frequently used. The one notable exception is the frequent use of the term "migration background" in German-speaking countries, which is also reflected in research publications—even when most are written in English. Among the 238 articles authored entirely by researchers affiliated with German institutions, the term "migration" (including derivations such as migration background, migrants, etc.) appears in more than half (57.1%) to describe the research subjects. Notably, 23 percent of these texts also use terms like ethnicity or ethnic group interchangeably with migration (for an examination of the category migration background, see zur Nieden et al. in this volume).

It is also significant that the term race is used far more often by international author teams (28.6%) than by exclusively German-based teams (4.6%). This suggests that for many German researchers, terms such as ethnicity, population, or ancestry are often employed as translations or substitutes for race.

The specifics of the German context become even clearer when we analyze the use of terms based on sampling locations. For instance, the term migration is used in 58.5 percent of studies involving participants sampled in Germany but only in 6.6 percent of studies conducted in the United States. In contrast, the terms race or ethnicity appear in 63.9 percent of studies on participants sampled in the United States. Nevertheless, among studies on participants in Germany, the term ethnicity is still used in 31.8 percent of the articles (Bartram et al. 2023).

Figure 1: Frequency of classification terms used in relation to affiliation of author teams to German institutions. More than half of the studies (54.8%) applied several terms to the subjects under investigation. For more detail, see Bartram et al. 2023.

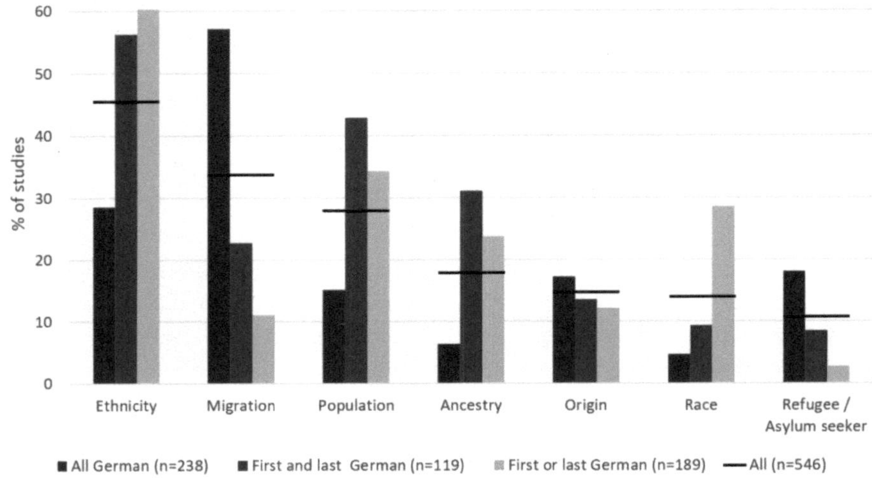

There are also disciplinary differences in the use of ethnicity. However, since the majority of articles analyzed come from the fields of medicine, epidemiology, and psychology—with similar frequencies of ethnicity use (48.3%, 35.6%, and 41.8%, respectively)—and the number of articles from other disciplines (e.g., archaeogenetics, human genetics, pharma, and biotech) is relatively small, these differences are not the focus of our analysis here (see Bartram et al. 2023). Instead, we concentrate on the wide range of meanings attributed to the term "ethnicity," as reflected in the diversity of its operationalization.

Hardly Any Explanation of the Meaning of Ethnicity

In accordance with the principles of sound scientific practice, it is essential that the definition and operationalization of any key classificatory term be clearly and explicitly articulated in scientific literature. This requirement is not only a cornerstone of scholarly integrity but is also emphasized in the editorial standards of numerous leading scientific journals (ICMJE 2022; The Lancet 2024). In the structured format of life science publications, the methods section plays a pivotal role in detailing the concepts, procedures, and analytical steps undertaken to produce the study's findings. Such clarity is indispensable for ensuring transparency, facilitating reproducibility, and enabling meaningful comparisons across studies. This focus on terminological precision is particularly important when dealing with complex and socially loaded categories such as ethnicity or race, where common usage often diverges significantly from academic definitions, leading to potential misunderstandings or misapplications in research contexts.

In sharp contrast to these established requirements, our analysis reveals that most of the reviewed literature fails to adequately define the classification methods used for the people studied. Specifically, 137 articles—almost 55 percent of the studies examined—either completely omitted any information about how subjects were classified (30.5%) or merely referenced data sources or previous studies without providing details on the classification process (25.3%).[6] While some sources, such as the U.S. Cancer Registry Surveillance, Epidemiology, and End Results (SEER), offer accessible online documentation, other cases present significant challenges. For instance, classification information may not be available in English (or German), may be hidden behind a paywall, or may simply not exist in the referenced article.

Articles that do not specify their classification methods typically only list the number of subjects in different groups. For example, Wuestemann et al. (2021, 483), in their study on bone morphology, state that to investigate "ethnic differences," they included "817 Caucasian femora," "245 Japanese," "240 Non-Japanese Asian," "30 African," and "13 Middle Eastern" in their dataset. However, the authors provide no information on the classification criteria used and, moreover, assign the femora studied to disparate and inconsistent classification schemes, including categories such as African, the extremely vague category of Middle Eastern (which overlaps with Asia and Africa), and nationality-based classifications such as Japanese. The article offers no explanation for how individuals were assigned to these groups or why nationality, race, and geography are collectively referred to as "ethnic."

An example of referencing the original database used (but still without explaining how subjects were classified) is the article by Amare et al. (2021). The authors study clinical data and genetic information from 2,586 "multi-ethnic" patients of the "International Consortium on Lithium Genetics" to investigate the genetics of lithium response in patients with bipolar disorder, dividing them into a "European sample" and an "Asian sample" (ibid., 2459) Neither the referenced website ("www.ConLiGen.org") nor the articles cited in the methods section provide clear information on how the study subjects were classified by "ethnicity" as "European" or "Asian." In the supplementary file of one referenced article, we found a statement that "EIGENSOFT was used to identify population outliers [...], which were removed" (Hou et al. 2016). This suggests ethnicity was somehow genetically validated to exclude certain individuals, but how the subsamples were originally labeled as European or Asian remains unexplained.

Both examples, from our perspective, contradict the guidelines of most major journals, such as those formulated by the International Committee of Medical Journal Editors (ICMJE). In its guidelines, the ICMJE requires authors to "define how they determined race or ethnicity and justify their relevance" (ICMJE 2022). Similarly, *Nature*, one of the most prestigious science journals, requires authors to specify "who provided the classification terms (the participants, the researchers or third parties)" and "how racial/ethnic identity was determined (by the participants, the researchers or third parties)" (Nature 2024).

The failure to specify how research subjects were classified runs counter to good scientific practice and the guidelines of numerous journals. However, this issue is not uniformly present across all life science disciplines. Interestingly, it is much less common in

---

6 There is an overlap consisting of two studies because these articles used several cohorts and mentioned a data source for one but not the other.

epidemiology, where only 6.1 percent of studies failed to specify how ethnicity was determined (Figure 2). Nevertheless, 32.7 percent of epidemiological studies referenced a data source without providing further details, making them quite similar to other disciplines in the life sciences in this regard.

Figure 2: How ethnicity is specified according to different disciplines. Some of the studies were carried out by interdisciplinary teams and were accordingly assigned to multiple disciplines by us. Only the three most common disciplines (n>30 items) are shown.

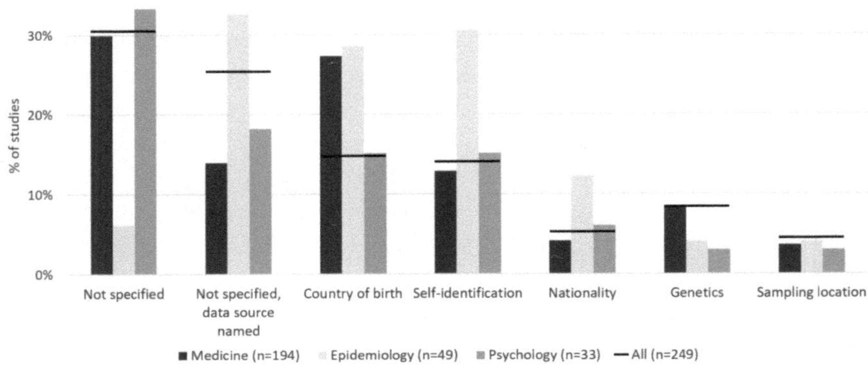

In the remaining 112 studies that provided information on the classification process, we identified a wide variety of approaches to assigning ethnicity. The most common methods were based on the country of birth of the research subject (or their parents or grandparents) and self-identification, used in 14.9 percent and 14.1 percent of the articles, respectively. Notably, twenty-one articles (8.4%) used genetic analyses to attribute ethnicity to the study subjects. For example, Schote et al., in a study on hyperhidrosis (excessive sweating), conducted a genetic analysis to confirm that the samples were "of European origin" (Schote et al. 2020, 12) Based on this analysis, the authors excluded samples from further study if they "were not ethnically matched" (ibid., 13). Similarly, Toepfer et al. (2019, 159) referred to a "genotype-based ethnicity" in their analysis of DNA regulation changes in the oxytocin gene locus for mothers after childbirth. Other studies used information about nationality (5.2%) or sampling location (4.4%) to assign ethnicity to study subjects.

Differences in the disclosure of classification methods also correlate with the degree of affiliation authors have with German research institutions. As shown in Figure 3, while the differences are not dramatic, author teams entirely affiliated with German institutions are more likely to use the country of birth (30.4%) to assign ethnicity compared to more international teams (7.6%/9.6%). This trend can be explained by the interchangeable use of ethnicity and migration background in many articles authored by "all German" author teams. Migration background is frequently operationalized using the country of birth of the subject or their parents.

It is also noteworthy that several studies provide difficult-to-understand or even confusing information regarding the attribution of ethnicity. For example, a transfusion medicine study by Flesch et al. (2020) aimed to determine blood group allele frequencies in blood donors from "Arabian countries and Iran" to identify "donors with rare blood groups," referring to blood with genetic markers uncommon in Central Europe (Flesch et al. 2020, 397). The authors employed a combination of methods to assign ethnicity. First, donors were asked for their "mother country" (ibid.). However, this term is ambiguous and could refer to the country of birth, the parents' or mother's origin, or even that of more distant ancestors. The authors then compared the collected blood samples with a sample from the German Red Cross, describing the latter as predominantly of "Middle European origin," while assuming a small number of donors could have a different "ethnic background" (ibid.: 398). In this context, the terms "origin" and "ethnic background" appear synonymous—at least where "another ethnic background" means "not of Middle European origin" (ibid., 400). The discussion section further complicates matters. The authors state that they know "only the *mother country* but not the *ethnicities*" and speculate that many donors with a specific blood type might be "mostly of African black *ancestry*" (Flesch et al. 2020, 405, emphasis added). This phrasing suggests that the researchers differentiate between ethnicity and origin, while also introducing ancestry as a third classification. Additionally, the authors assign some individuals to "African or Arabian ancestry" based solely on surnames, estimating that around half of those with a specific rare blood group fall into this category (ibid., 405).

Overall, the studies largely fail to meet journal guidelines and standards of scientific integrity. Many provide no information—or vague, ambiguous, or even contradictory statements—about their methods for assigning ethnicity.

No Differences Investigated or a Spectrum of Differences Found

In numerous settings, ethnicity serves as a seemingly straightforward concept of differentiation. This impression is reinforced by the widespread availability of ethnic classifications in many databases and data collections. Here, it is noteworthy that in public discourse ethnic assignments are typically employed to emphasize distinctions *between* groups—often highlighting differences in an externalized population (the "minority group" or the "foreigners") in contrast to the majority group. Conversely, a significant portion of the studies we examined—27.3 percent of the publications—focus exclusively on a single population identified as an ethnic group without conducting comparative analyses with other groups (see Figure 3).

Our analysis suggests various reasons for the high prevalence of "no comparisons" in studies that classify by ethnicity. A large proportion of the texts use the term merely as a descriptor for the single sample of people studied. For instance, in a study on "predictors of heart-focused anxiety," Wedegärtner et al. (2020) examined 107 patients (87.9% male) at the Saarland University Medical Center in Homburg, Germany. The authors note in their limitations that their sample "comprised mainly Caucasian men"[7] and that no "further ethnicity" was included (Wedegärtner et al. 2020, 385). From a race-critical perspective, it may appear peculiar that a term originally coined as racial and predominantly

7 The term "Caucasian" appears frequently in studies in our corpus.

used as such in international literature is framed here as "ethnicity." Such a terminological—but not conceptual—substitution is a common attempt to circumvent the problematic connotations of the concept of race (see also Ellebrecht in this volume). The issues in the authors' approach extend beyond conflating racial ("Caucasian") and ethnic categories. A significant problem arises when psychological and social characteristics, such as anxiety, are linked to continental origins, implying a genetic basis for observed differences. Furthermore, it is unlikely that this classification reflects self-identification, as the term "Caucasian" is uncommon in German and would more likely be interpreted as referring to individuals from the Caucasus region. Lastly, it seems implausible that none of the participants would identify with a different ethnic group, given that the population of the city of Homburg is far from as homogeneous as the study suggests. This approach, likely intended to align the study with international discourse, employs racializing terminology and uses ethnicity as a broad, continental proxy that oversimplifies and misrepresents the complex social, biological, and biosocial realities of human diversity.

Figure 3: The significance of differences investigated between ethnic groups. In some studies, differences were described in several categories, resulting in multiple entries.

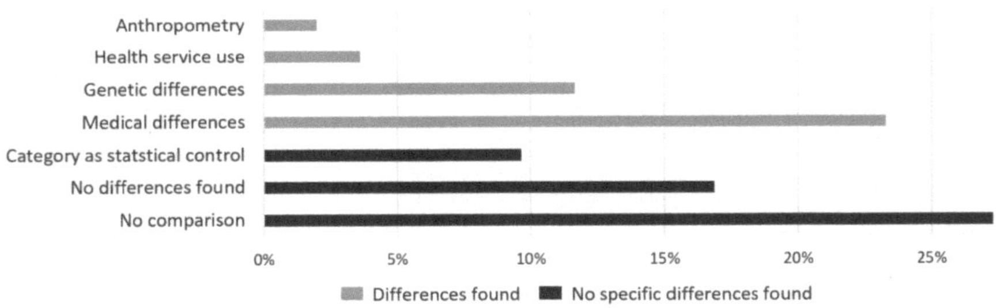

Beyond the local classifications observed in studies like Wedegärtner et al. (2020), other studies use indicators such as nationality, migration status, refugee experience, language, or religion to attribute ethnicity. Often, these attributions are made to align with the prevailing practices of the international (primarily US-dominated) scientific community, particularly in medical and epidemiological research, or are even mandated by scientific journals (Rubin 2021). Some studies also appear to assume a meaningful difference between ethnic groups without conducting comparative analyses that could confirm or refute this hypothesis. Furthermore, in pharmacological research and the biotechnological sector, studies sometimes seem to follow the objective of marketing drugs to the broadest possible patient population. As a result, there is often a vested interest in ensuring that no significant differences between population groups are identified.

Within our database, of the studies that conducted a comparative analysis across various ethnic groups, 23.3 percent identified significant medical disparities, evident in aspects such as disease incidence, disease burden, mental health status, and mortality rates. Among these studies, 16.9 percent found no meaningful differences in their comparisons: 11.6 percent reported discovering genetic differences between individuals categorized into different ethnic groups. For instance, Degenhardt et al. (2019) conducted a genetic study at the Institute of Clinical Molecular Biology at Kiel University. The authors stated that in their "multi-ethnic reference panel" on immune system genes, the various "subpopulations" clustered well with the "respective ethnicities of the 1000 Genomes population" (Degenhardt et al. 2019, 2078, 2080). However, it is important to note that the 1000 Genomes Project does not address ethnic groups or races but instead refers to twenty-six populations (internationalgenome.org/1000-genomes-summary).

It is also worth noting that most studies identifying genetic differences between ethnic groups focus on very rare genetic variants. For instance, Jacobi et al. (2019) reported the discovery of a rare genetic variant in a patient described as being of "Asian ancestry." Participants in this study were children with diabetes, whose blood samples were collected at outpatient clinics at Charité University Hospital in Berlin, with the public health "Study of Health in Pomerania" serving as a control.[8] As is common in many of the articles we reviewed, the authors employed multiple classification frameworks for human diversity interchangeably, noting that a specific genetic variant associated with obesity and insulin resistance "has not been detected in individuals of European ancestry but is a known variant in other *ethnicities*" (Jacobi et al. 2019, 1171, emphasis added). In another study, genetic differences were used to trace "patterns of diversity," suggesting that "human genetic diversity can reflect physical and cultural geography" (Peter, Petkova, and Novembre 2020, 943, 946). This research aimed to map out geographical zones that exhibit not only "genetic but also linguistic and ethnic differentiation" (Peter, Petkova, and Novembre 2020, 944).

Overall, our findings indicate that the concept of ethnicity is used both to identify existing differences and to highlight the absence of such differences. Furthermore, it often serves as a seemingly self-evident category for the (frequently vague) description of the sampling population. To better understand the role of this concept in the search for and detection of differences, it is necessary to take a closer look at the explanations given—or presumed—by the studies for the differences identified.

Reasons Given for the Differences

Among the 181 articles that reported differences between groups categorized by ethnicity, 16.0 percent (29 articles) presented their findings without offering any explanation or hypothesis to account for the differences they observed (see Figure 4). Nevertheless, the majority of studies elucidate their findings, proposing at least potential explanations, encompassing factors including social, biological, or measurement artifacts.

8 For a detailed analysis of the "Study of Health in Pomerania," see Ellebrecht in this volume.

Figure 4: Stated or presumed reasons for differences found. Results from the 181 studies where differences between research subjects grouped by ethnicity. Many articles propose multiple hypotheses to explain the identified differences, leading to overlap. Only explanations appearing in more than 5 percent of the studies are listed.

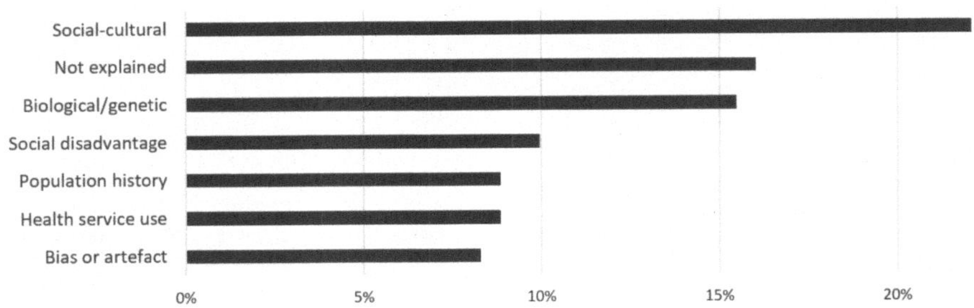

In a total of 40.9 percent (74 articles) of the studies, the measured differences were explained socially, for example by referring to socio-cultural reasons such as lifestyle, cultural or religious diversity (22.1%), social disadvantage (9.9%), and differences in the use of healthcare services (8.8%). For example, a study on the prediction of attention deficit hyperactivity disorder (ADHD) in children found that, among other factors, "mixed or Asian ethnicity was a negative predictor" of ADHD, which means that Asian children were less likely to be diagnosed with ADHD than white children (Brandt, Patalay, and Kernerauch Koerner 2021, 880). Citing previous research, the authors suggested that these variations might arise from disparities in the "perception of hyperactivity by the parents," potentially shaped by "cultural factors" (ibid., 882).

Biological or genetic differences between ethnic groups were referenced in 24.3 percent of the studies (44 articles). In 15.5 percent (28 articles), such differences were suspected as potential cause for medicial differences under investigation. Another 8.8 percent (16 articles) of the studies explicitly searched for genetic differences between different groups based on population genetics questions. When biological or genetic explanations for measured differences were suggested, they were often mentioned within a list of several different hypothesis, typically without further elaboration. This was also evident in studies that emphasized social factors. For instance, a study on eating disorders among girls with type 1 diabetes mellitus reported "lower odds of eating disorders in girls with *migration background*," suggesting that "*cultural*, *social*, or *genetic* factors" might influence prevalence (Reinehr et al. 2019, 210, emphasis added). The authors further speculated that "ethnicity and cultural background are more important factors than migration itself" (ibid.).

It is also noteworthy that among studies authored entirely by researchers affiliated with German institutions, only two proposed genetic reasons for differences. Of the studies that identified genetic differences, more than half (15) linked their findings to population history. These were mostly studies focused on population genetics rather than medical aims. For example, Chaichoompu et al. (2020, 46) examined the genetic "population structure of Western African populations" by distinguishing individuals "from 25 ethnic groups" on the basis of more than 300,000 genetic variants.

Social disadvantage, which includes explanations related to socioeconomic inequalities, was mentioned in only 9,9 percent (18) of the studies. Differences in "health service use," such as those caused by language barriers, were cited in 8.8 percent (16) of studies. In the studies that indicated sociocultural reasons, 11 (6%) mentioned discrimination as a factor, while only two studies mentioned experiences of racism or structural racism as potential causes for differing health outcomes.

When comparing disciplines, few significant differences emerged, except that no psychological studies in our corpus linked ethnicity to genetic explanations for group differences. Thus, while many authors attributed differences between ethnic groups to sociocultural factors, biological or genetic explanations were also frequently hypothesized.

Conclusions

Ethnicity is a very complex and multifaceted term. As a scientific concept, ethnicity aims to capture relationships between groups distinguished by perceived or assumed physical and/or cultural differences. The publications we examined reveal that, in Germany, the concept of race has nowadays largely been replaced by the concepts of ethnicity and migration background in analyses of biological differences between population groups, minorities, and the majority society. However, a thorough reflection on the limitations and dangers of the concept of ethnicity—including its historical continuities and inadequacies as a classification tool for assessing biological differences—remains lacking.

As demonstrated in numerous other studies, merely substituting terms without enacting conceptual change fails to address the fundamental problems inherent in broad group classifications. The interchangeable use of terms such as ethnicity, race, population, continental designations, or migration background additionally exacerbates the "terminological mess" (Malinowska and Żuradzki 2023) and leads to an imprecise and therefore unscientific use of the concept of ethnicity. But it is not just a matter of scientifically inadequate use. Rather, broad and unspecific classifications tend to run the risk of overemphasizing genetic differences while neglecting sociocultural factors (Morning 2011; Kaufman, Merckx, and Cooper 2021). In this way, the publications reviewed here often exhibit misleading simplifications and bias toward biological or genetic determinisms that can lead them to misinterpret health outcomes. Ethnic classifications can be a relevant category in the study of health-related effects of social inequalities. But when applied out of the context of other social factors, they are insufficient and misguiding in addressing the diverse health needs of various groups.

As used in the studies examined, ethnicity emerges as an artifact shaped through social, institutional, discursive, and political negotiations. This means that ethnicity, far from being a life science entity awaiting discovery, is the product of a complex web of interpellations directed at researchers, authors, and journal editors. These interpellations arise from sociocultural classifications, directives from funding bodies, journal guidelines, and the prestructuring of the data collected through forms and sociodemographic surveys.

Our analysis of life science studies conducted by authors affiliated with German research institutions revealed that while ethnicity is frequently invoked, it is rarely specified. Its use reflects a broad range of meanings, with diverse connotations. Moreover, we identified overlapping and synonymous use of classification terms such as race, migration, ancestry, nationality, and origin. This imprecision contradicts journal guidelines, undermines established standards of sound scientific practice, and highlights a clear discrepancy between guidelines and the practices of editorial supervision and peer review. Standards for the application and reporting of human diversity classifications are inconsistently enforced, and violations often go unnoticed. When we sought clarification from the ICMJE regarding whether simply naming the data source without detailing how ethnicity was determined aligns with their guidelines, they responded by assigning responsibility to journal editors: "The level of detail published in the article is at the discretion of the editor, but our guidance encourages transparency" (personal email correspondence). Future research could explore whether this lack of adherence to guidelines reflects broader issues within peer review and the scientific publication process.

It is important not to dismiss the conceptual and definitional ambiguity in authors' work as trivial or careless. This vagueness often serves a strategic purpose, allowing the category of ethnicity to remain open to essentialist and biologistic interpretations. While there is widespread acknowledgment in the life sciences that ethnic categories are socially constructed, the concepts often remain internally contradictory, with a persistent tendency to imply or explicitly assume a biological-genetic foundation. As an artifact of politics, ethnicity is applied across diverse contexts, often within methodologically and theoretically underdeveloped frameworks.

Politics influence the processes of categorizing human diversity across multiple levels, ranging from individual preferences within the research process to institutional, national, and international guidelines and standards. Normative frameworks shaped by biopolitical and identity-political discourses play a crucial role in regulating life, health, and the body, especially within marginalized communities, the inherently political nature of categorizing human diversity is often obscured or denied in scientific practices, which claim to objectively measure human differences while attributing any politicization to external social forces rather than acknowledging their internal role in shaping these categories.

Thus, while ethnicity is frequently employed in scientific research, it must be recognized as an artifact shaped by norms, standards, and expectations to replace the concept of race. However, ethnicity does not appear to offer a viable solution to the challenges posed by race as a category, leaving the debate about its use in the life sciences unresolved. A crucial question is whether ethnicity, as an inevitably constructed category, can function effectively as a scientific tool for capturing human biological diversity. Furthermore, it remains uncertain whether the current conceptualization of ethnicity can mitigate issues of essentialization and stigmatization faced by marginalized communities. In fact, the vague and inconsistent use of ethnicity may perpetuate and reinforce the biologization of group categories. For example, research linking ethnic labels with genetics has stigmatized the European Roma as "inbred" and "genetic high-risk," with tangible effects on policy decisions (Lipphardt et al. 2021).

While ethnicity, as a proxy, can point to underlying phenomena that are difficult to measure, it also risks obscuring or even concealing relevant inequities. A more specific categorization of people based on their social and socioeconomic status, tailored to the particular context and research questions, may ultimately render the concept of ethnicity redundant and replace it with more precise and contextually relevant classifications.

Acknowledgments

The research was funded by the German Federal Ministry of Education and Research (BMBF, grant number 01GP1790) at the Institute of Sociology at the University of Freiburg, Germany. We would like to thank Felix Fink, Paula Steinbrenner, and Levin Nikolai for their support in coding the research publications. We are indebted to the participants of the authors' workshops for this edited volume, the anonymous reviewer, as well as Andrea zur Nieden and Nils Ellebrecht for their valuable comments on earlier versions of this article. Michael Thomas Taylor contributed substantially to improving the linguistic quality of the text.

This article is a revised and expanded version of "Ethnizität in den deutschen Lebenswissenschaften: Spektrum an Bedeutungen und Implikationen" (Plümecke and Bartram, forthcoming).

References

Académie française. 2024. "Ethnique." Accessed December 16, 2024. https://www.dictionnaire-academie.fr/article/A9E2884.

Afshari, Reza, and Raj S. Bhopal. 2010. "Ethnicity Has Overtaken Race in Medical Science: MEDLINE-Based Comparison of Trends in the USA and the Rest of the World, 1965–2005." *International Journal of Epidemiology* 39 (6): 1682–1683. DOI: https://doi.org/10.1093/ije/dyp382.

Amare, Azmeraw T., Klaus Oliver Schubert, Liping Hou, et al. 2021. "Association of Polygenic Score for Major Depression with Response to Lithium in Patients with Bipolar Disorder." *Molecular Psychiatry* 26 (6): 2457–2470. DOI: https://doi.org/10.1038/s41380-020-0689-5.

Banton, Michael. 1977. *The Idea of Race*. London: Tavistock.

Banton, Michael. 2015. *What We Now Know About Race and Ethnicity*. New York: Berghahn Books.

Barth, Fredrik. 1969. *Ethnic Groups and Boundaries: The Social Organization of Culture Difference*. Long Grove: Waveland Press.

Bartram, Isabelle, Laura Schnieder, Nils Ellebrecht, Florian Ruland, Tino Plümecke, and Andrea zur Nieden. 2023. "Categorizing People in the German Life Sciences: A Systematic Literature Review of Classifications of Human Diversity." *Discover Social Science and Health* 3 (1). DOI: https://doi.org/10.1007/s44155-023-00033-5.

Bhopal, Raj. 2004. "Glossary of Terms Relating to Ethnicity and Race: For Reflection and Debate." *Journal of Epidemiology and Community Health* 58 (6): 441–445. DOI: https://doi.org/10.1136/jech.2003.013466.

Bokor-Billmann, Therezia, Ewan A. Langan, and Franck Billmann. 2020. "The Reporting of Race and/or Ethnicity in the Medical Literature: A Retrospective Bibliometric Analysis Confirmed Room for Improvement." *Journal of Clinical Epidemiology* 119: 1–6. DOI: https://doi.org/10.1016/j.jclinepi.2019.11.005.

Bös, Mathias. 2005. *Rasse und Ethnizität: Zur Problemgeschichte zweier Begriffe in der amerikanischen Soziologie*. Wiesbaden: VS.

Bös, Mathias. 2015. "Ethnicity and Ethnic Groups: Historical Aspects." In *International Encyclopedia of the Social & Behavioral Sciences*, 2nd ed., edited by James D. Wright, 136–141. Amsterdam: Elsevier.

Bös, Mathias. 2020. "Migration, Ethnizität und 'Rasse'." In *Handbuch Migrationssoziologie*, edited by Antje Röder and Darius Zifonun, 193–225. Wiesbaden: Springer VS.

Bowker, Geoffrey C., and Susan Leigh Star. 2000. *Sorting Things Out: Classification and Its Consequences*. Cambridge, MA: MIT Press.

Brandt, Valerie, Praveetha Patalay, Julia Kerner auch Koerner. 2021. "Predicting ADHD symptoms and diagnosis at age 14 from objective activity levels at age 7 in a large UK cohort." *European Child & Adolescent Psychiatry* 30 (6): 877–884. DOI: https://doi.org/10.1007/s00787-020-01566-9.

Brubaker, Rogers. 2004. *Ethnicity without Groups*. Cambridge, MA: Harvard University Press.

Byeon, Yen Ji Julia, Rezarta Islamaj, Lana Yeganova, et al. 2021. "Evolving Use of Ancestry, Ethnicity, and Race in Genetics Research: A Survey Spanning Seven Decades." *American Journal of Human Genetics* 108 (12): 2215–2223. DOI: https://doi.org/10.1016/j.ajhg.2021.10.008.

Cerdeña, Jessica P., Vanessa Grubbs, and Amy L. Non. 2022. "Racialising Genetic Risk: Assumptions, Realities, and Recommendations." *The Lancet* 400 (10368): 2147–2154. DOI: https://doi.org/10.1016/S0140-6736(22)02040-2.

Chaichoompu, Kridsadakorn, Fentaw Abegaz, Bruno Cavadas, et al. 2020. "A Different View on Fine-Scale Population Structure in Western African Populations." *Human Genetics* 139 (1): 45–59. DOI: https://doi.org/10.1007/s00439-019-02069-7.

Cokley, Kevin. 2007. "Critical Issues in the Measurement of Ethnic and Racial Identity: A Referendum on the State of the Field." *Journal of Counseling Psychology* 54 (3): 224–234. DOI: https://doi.org/10.1037/0022-0167.54.3.224.

Collins, Francis S. 2004. "What We Do and Don't Know About 'Race,' 'Ethnicity,' Genetics and Health at the Dawn of the Genome Era." *Nature Genetics* 36 (11): S13–S15. DOI: https://doi.org/10.1038/ng1436.

Degenhardt, Frauke, Mareike Wendorff, Michael Wittig, et al. 2019. "Construction and Benchmarking of a Multi-Ethnic Reference Panel for the Imputation of HLA Class I and II Alleles." *Human Molecular Genetics* 28 (12): 2078–2029. DOI: https://doi.org/10.1093/hmg/ddy443.

Deniker, Joseph. 1900. *The Races of Man: An Outline of Anthropology and Ethnography*. London: Walter Scott.

Domarus, Max, ed. 1973. *Hitler: Reden und Proklamationen 1932–1945. Kommentiert von einem deutschen Zeitgenossen, 1939–1940*. Vienna: R. Löwit.

Duster, Troy. 1990. *Backdoor to Eugenics*. New York: Routledge.

Duster, Troy. 2015. "A Post-Genomic Surprise. The Molecular Reinscription of Race in Science, Law and Medicine." *The British Journal of Sociology* 66 (1): 1–27. DOI: https://doi.org/10.1111/1468-4446.12118.

DWDS Digitales Wörterbuch der Deutschen Sprache. 2024. "Ethnisch." Accessed December 16, 2024. https://www.dwds.de/wb/ethnisch.

EJMHR European Journal of Medical and Health Research. 2024. "Ethical Considerations." Accessed December 14, 2024. https://ejmhr.com/index.php/journal/ethical-considerations.

El-Haj, Nadia Abu. 2007. "The Genetic Reinscription of Race." *Annual Review of Anthropology* 36 (1): 283–300. DOI: https://doi.org/10.1146/annurev.anthro.34.081804.120522.

Epstein, Steven. 2007. *Inclusion: The Politics of Difference in Medical Research*. Chicago: University of Chicago Press.

Eriksen, Thomas Hylland. 2010. *Ethnicity and Nationalism: Anthropological Perspectives*. 3rd ed. London: Pluto Press.

Feagin, Joe R., and Clairece Booher Feagin. 2011. *Racial and Ethnic Relations*. 9th ed. Boston: Pearson.

Fischer, Martin S., Uwe Hoßfeld, Johannes Krause, and Stefan Richter. 2019. "Jena Declaration: The Concept of Race Is the Result of Racism, Not Its Prerequisite." Accessed December 19, 2024. https://www.uni-jena.de/en/190910-jenaererklaerung-en.

Flesch, Brigitte Katharina, Vanessa Scherer, Burkhard Just, et al. 2020. "Molecular Blood Group Screening in Donors from Arabian Countries and Iran Using High-Throughput MALDI-TOF Mass Spectrometry and PCR-SSP." *Transfusion Medicine and Hemotherapy* 47 (5): 396–408. DOI: https://doi.org/10.1159/000505495.

Foucault, Michel. 1984. "What Is Enlightenment?" In *The Foucault Reader*, edited by Paul Rabinow, 32–50. New York: Pantheon Books.

Foucault, Michel. 2010. *The Government of Self and Others: Lectures at the Collège de France 1982–1983*. Translated by Graham Burchell. New York: Palgrave Macmillan.

Fujimura, Joan H., and Ramya Rajagopalan. 2011. "Different Differences: The Use of 'Genetic Ancestry' Versus Race in Biomedical Human Genetic Research." *Social Studies of Science* 41 (1): 5–30. DOI: https://doi.org/10.1177/0306312710379170.

Fullwiley, Duana. 2007. "The Molecularization of Race: Institutionalizing Human Difference in Pharmacogenetics Practice." *Science as Culture* 16 (1): 1–30. DOI: https://doi.org/10.1080/09505430601180847.

Gabbert, Wolfgang. 2015. "Ethnicity in History." In *Ethnicity as a Political Resource: Conceptualizations Across Disciplines, Regions, and Periods*, edited by University of Cologne Forum "Ethnicity as a Political Resource," 183–200. Bielefeld: transcript.

Gordon, Milton M. 1964. *Assimilation in American Life: The Role of Race, Religion, and National Origins*. New York: Oxford University Press.

Gumplowicz, Ludwig. 1883. *Der Rassenkampf: Sociologische Untersuchungen*. Innsbruck: Wagner.

Hahn, Robert A., Benedict I. Truman, and Nancy D. Barker. 1996. "Identifying Ancestry: The Reliability of Ancestral Identification in the United States by Self, Proxy, Interviewer, and Funeral Director." *Epidemiology* 7 (1): 75–80. DOI: https://doi.org/10.1097/00001648-199601000-00013.

Hattam, Victoria. 2004. "Ethnicity: An American Genealogy." In *Not Just Black and White: Historical and Contemporary Perspectives on Immigration, Race, and Ethnicity in the United States*, edited by Nancy Foner and George M. Fredrickson, 42–60. New York: Russell Sage Foundation.

Hou, Liping, Urs Heilbronner, Franziska Degenhardt, et al. 2016. "Genetic Variants Associated with Response to Lithium Treatment in Bipolar Disorder: A Genome-Wide Association Study." *The Lancet* 387 (10023): 1085–1093. DOI: https://doi.org/10.1016/S0140-6736(16)00143-4.

Huxley, Julian S., and Alfred C. Haddon. 1935. *We Europeans: A Survey of "Racial" Problems*. London: Jonathan Cape.

ICMJE International Committee of Medical Journal Editors. 2022. "Recommendations for the Conduct, Reporting, Editing, and Publication of Scholarly Work in Medical Journals." Accessed December 03, 2024. https://www.icmje.org/news-and-editorials/icmje-recommendations_annotated_may22.pdf.

Institute of Medicine, ed. 1999. *The Unequal Burden of Cancer: An Assessment of NIH Research and Programs for Ethnic Minorities and the Medically Underserved*. Washington: National Academies Press.

Jacobi, Simon F., Noushafarin Khajavi, Gunnar Kleinau, et al. 2019. "Evaluation of a Rare Glucose-Dependent Insulinotropic Polypeptide Receptor Variant in a Patient with Diabetes." *Diabetes, Obesity and Metabolism* 21 (5): 1168–1176. DOI: https://doi.org/10.1111/dom.13634.

Jugert, Philipp, Marie J. Kaiser, Francesca Ialuna, and Sauro Civitillo. 2022. "Researching Race-Ethnicity in Race-Mute Europe." *Infant and Child Development* 31 (1). DOI: https://doi.org/10.1002/icd.2260.

Kajikhina, Katja, Carmen Koschollek, Navina Sarma, et al. 2023. "Empfehlungen zur Erhebung und Analyse migrationsbezogener Determinanten in der Public-Health-Forschung." *Journal of Health Monitoring* 8 (1): 55–77. DOI: https://doi.org/10.25646/11093.

Kaufman, Jay S., Joanna Merckx, and Richard S. Cooper. 2021. "Use of Racial and Ethnic Categories in Medical Testing and Diagnosis: Primum Non Nocere." *Clinical chemistry* 67 (11): 1456–1465. DOI: https://doi.org/10.1093/clinchem/hvab164.

Koenig, Barbara A., Sandra Soo-Lin Lee, and Sarah S. Richardson, eds. 2008. *Revisiting Race in a Genomic Age*. New Brunswick: Rutgers University Press.

Krieger, Nancy. 2020. "Measures of Racism, Sexism, Heterosexism, and Gender Binarism for Health Equity Research: From Structural Injustice to Embodied Harm: An Ecosocial Analysis." *Annual Review of Public Health* 41: 37–62. DOI: https://doi.org/10.1146/annurev-publhealth-040119-094017.

Lee, Catherine. 2009. "'Race' and 'Ethnicity' in Biomedical Research: How Do Scientists Construct and Explain Differences in Health?" *Social Science & Medicine* 68 (6): 1183–1190. DOI: https://doi.org/10.1016/j.socscimed.2008.12.036.

Lewontin, Richard C. 1972. "The Apportionment of Human Diversity." *Evolutionary Biology* 6: 381–398.

Liebscher, Doris. 2021. *Rasse im Recht – Recht gegen Rassismus: Genealogie einer ambivalenten rechtlichen Kategorie*. Berlin: Suhrkamp.

Lipphardt, Veronika, Mihai Surdu, Nils Ellebrecht, Peter Pfaffelhuber, Matthias Wienroth, and Gudrun A. Rappold. 2021. "Europe's Roma People are Vulnerable to Poor Practice in Genetics." *Nature* 599: 368–371. DOI: https://doi.org/10.1038/d41586-021-03416-3.

Livingstone, Frank B. 1962. "On the Non-Existence of Human Races." *Current Anthropology* 3 (3): 279–281.

Maduka, Richard C., Michael Broderick, Erin M. White, et al. 2021. "The Reporting of Race and Ethnicity in Surgery Literature." *JAMA Surgery* 156 (11): 1036–1041. DOI: https://doi.org/10.1001/jamasurg.2021.3752.

Malinowska, Joanna K., and Tomasz Żuradzki. 2023. "Reductionist Methodology and the Ambiguity of the Categories of Race and Ethnicity in Biomedical Research: An Exploratory Study of Recent Evidence." *Medicine, Health Care and Philosophy* 26 (1): 55–68. DOI: https://doi.org/10.1007/s11019-022-10122-y.

Mangrio, Elisabeth, Maneesh Paul-Satyaseela, and Michael Strange. 2020. "Refugees in Sweden During the Covid-19 Pandemic: The Need for a New Perspective on Health and Integration." *Frontiers in Public Health* 8: 574334. DOI: https://doi.org/10.3389/fpubh.2020.574334.

Marks, Jonathan M. 1995. *Human Biodiversity. Genes, Race, and History*. New York: Aldine de Gruyter.

Marks, Jonathan. 2008. "Race: Past, Present, and Future." In *Revisiting Race in a Genomic Age*, edited by Barbara A. Koenig, Sandra Soon-Jin Lee, and Sarah Richardson, 21–38. New Brunswick: Rutgers University Press.

Mauro, Madelyn, Danielle S. Allen, Bege Dauda, Santiago J. Molina, Benjamin M. Neale, and Anna C. F. Lewis. 2022. "A Scoping Review of Guidelines for the Use of Race, Ethnicity, and Ancestry Reveals Widespread Consensus but Also Points of Ongoing Disagreement." *American Journal of Human Genetics* 109 (12): 2110–2125. DOI: https://doi.org/10.1016/j.ajhg.2022.11.001.

Mayring, Philipp. 2014. *Qualitative Content Analysis*. Klagenfurt: Beltz.

Moore, Daniel B. 2020. "Reporting of Race and Ethnicity in the Ophthalmology Literature in 2019." *JAMA Ophthalmology* 138 (8): 903–906. DOI: https://doi.org/10.1001/jamaophthalmol.2020.2107.

Morning, Ann. 2011. *The Nature of Race. How Scientists Think and Teach About Human Difference.* Berkeley: Univ. of California Press.

National Academies of Sciences, Engineering, and Medicine. 2023. *Using Population Descriptors in Genetics and Genomics Research: A New Framework for an Evolving Field*. Washington, DC: National Academies Press.

Nature. 2024. "Research Ethics: Race, Ethnicity and Racism." Accessed December 13, 2024. https://www.nature.com/nature-portfolio/editorial-policies/ethics-and-biosecurity#race-ethnicity-racism.

Niemitz, Carsten, Kerstin Kreutz, and Hubert Walter. 2006. "Wider den Rassenbegriff in Anwendung auf den Menschen." *Anthropologischer Anzeiger* 64 (4): 463–464. DOI: https://doi.org/10.1127/anthranz/64/2006/463.

Omi, Michael, and Howard Winant. 1986. *Racial Formation in the United States: From the 1960s to the 1980s*. New York: Routledge.

Outram, Simon M., and George T. H. Ellison. 2006. "Anthropological Insights into the Use of Race/Ethnicity to Explore Genetic Contributions to Disparities in Health." *Journal of Biosocial Science* 38 (1): 83–102. DOI: https://doi.org/10.1017/S002193200500921.

Oxford English Dictionary. 2024. Ethnic. DOI: https://doi.org/10.1093/OED/1011146976.

Page, Matthew J., Joanne E. McKenzie, Patrick M. Bossuyt, et al. 2021. "The PRISMA 2020 Statement: An Updated Guideline for Reporting Systematic Reviews." *PLoS Medicine* 18 (3): e1003583. DOI: https://doi.org/10.1371/journal.pmed.1003583.

Peter, Benjamin M., Desislava Petkova, and John Novembre. 2020. "Genetic Landscapes Reveal How Human Genetic Diversity Aligns with Geography." *Molecular Biology and Evolution* 37 (4): 943–951. DOI: https://doi.org/10.1093/molbev/msz280.

Plümecke, Tino. 2010. "Die neuen Differenzen der Lebenswissenschaften. 'Rasse', Genetik und die ungenutzten Potentiale der Soziologie." In *Ethnowissen. Soziologische Beiträge zu ethnischer Differenzierung und Migration*, edited by Marion Müller and Darius Zifonun, 423–450. Wiesbaden: VS.

Plümecke, Tino. 2013. *Rasse in der Ära der Genetik: Die Ordnung des Menschen in den Lebenswissenschaften*. Bielefeld: transcript.

Plümecke, Tino, and Isabelle Bartram. forthcoming. "Ethnizität in den deutschen Lebenswissenschaften: Spektrum an Bedeutungen und Implikationen."

Reardon, Jenny. 2005. *Race to the Finish. Identity and Governance in an Age of Genomics*. Princeton, NJ: Princeton University Press.

Rédei, George P. 2008. *Encyclopedia of Genetics, Genomics, Proteomics, and Informatics*. 3rd ed. Dordrecht: Springer.

Reinehr, Thomas, Barbara Dieris, Angela Galler, et al. 2019. "Worse Metabolic Control and Dynamics of Weight Status in Adolescent Girls Point to Eating Disorders in the First Years After Manifestation of Type 1 Diabetes Mellitus: Findings from the Diabetes Patienten Verlaufsdokumentation Registry." *The Journal of Pediatrics* 207: 205–212.e5. DOI: https://doi.org/10.1016/j.jpeds.2018.11.037.

Roberts, Dorothy. 2011. *Fatal Invention: How Science, Politics, and Big Business Re-Create Race in the Twenty-First Century*. New York: The New Press.

Royal, Charmaine D. M. 2006. "'Race' and Ethnicity in Science, Medicine and Society." *BioSocieties* 1 (3): 325–328. DOI: https://doi.org/10.1017/S1745855206003048.

Rubin, Eric. 2021. "Striving for Diversity in Research Studies." *The New England Journal of Medicine* 385 (15): 1429–1430. DOI: https://doi.org/10.1056/NEJMe2114651.

Schenk, Liane. 2007. "Migration und Gesundheit: Entwicklung eines Erklärungs- und Analysemodells für epidemiologische Studien." *International Journal of Public Health* 52 (2): 87–96. DOI: https://doi.org/10.1007/s00038-007-6002-4.

Schote, Andrea B., Florian Schiel, Benedikt Schmitt, et al. 2020. "Genome-Wide Linkage Analysis of Families with Primary Hyperhidrosis." *PLoS One* 15 (12): e0244565. DOI: https://doi.org/10.1371/journal.pone.0244565.

Statistics Canada. 2024. "List of Ethnic or Cultural Origins 2021." Accessed December 02, 2024. https://www.statcan.gc.ca/en/subjects/standard/otherclass-alpha/origins2021spchar.

Stein, Stuart. 2004. "Ethnicity." In *Encyclopedia of Race and Ethnic Studies*, edited by Ellis Cashmore, 142–146. New York: Routledge.

Sumner, William Graham. 1906. *Folkways: A Study of the Sociological Importance of Usages, Manners, Customs, Mores, and Morals*. Boston: Ginn and Company.

The Lancet. 2024. "Information for Authors." Accessed December 03, 2024. https://www.thelancet.com/pb-assets/Lancet/authors/tl-info-for-authors-1717152770277.pdf.

Toepfer, Philipp, Kieran J. O'Donnell, Sonja Entringer, et al. 2019. "Dynamic DNA methylation changes in the maternal oxytocin gene locus (OXT) during pregnancy predict postpartum maternal intrusiveness." *Psychoneuroendocrinology* 103: 156–162. DOI: https://doi.org/10.1016/j.psyneuen.2019.01.013.

UNESCO. 1950. "The Race Question." Unesco Publication 791. Accessed November 27, 2024. https://unesdoc.unesco.org/ark:/48223/pf0000128291.

UNESCO. 1995. "Declaration of Schlaining Against Racism, Violence, and Discrimination." Accessed December 15, 2024. https://uol.de/f/5/inst/biologie/ag/didaktik/Ulrich/Resengl.pdf?v=1517221751.

van den Berghe, Pierre L. 1967. *Race and Racism: A Comparative Perspective*. New York: John Wiley & Sons.

Weber, Max. 1922/1978. *Economy and Society: An Outline of Interpretive Sociology*. Berkeley: University of California Press.

Wedegärtner, Sonja Maria, Igor Schwantke, Ingrid Kindermann, and Julia Karbach. 2020. "Predictors of Heart-Focused Anxiety in Patients with Stable Heart Failure." *Journal of Affective Disorders* 276: 380–387. DOI: https://doi.org/10.1016/j.jad.2020.06.065.

Winner, Langdon. 1980. "Do Artifacts Have Politics?" *Daedalus* 109 (1): 121–136.

Wuestemann, Thies, Samuel G. Hoare, Andreas Petersik, et al. 2021. "Bone Morphology of the Proximal Femoral Canal: Ethnicity Related Differences and the Influence on Cementless Tapered Wedge Stem Designs." *HIP International* 31 (4): 482–491. DOI: https://doi.org/10.1177/1120700019895458.

Yanow, Dvora, and Marleen van der Haar. 2013. "People out of Place: Allochthony and Autochthony in the Netherlands' Identity Discourse. Metaphors and Categories in Action." *Journal of International Relations and Development* 16 (2): 227–261. DOI: https://doi.org/10.1057/jird.2012.13

zur Nieden, Andrea, and Isabelle Bartram. 2020. "Categories of Diversity in Precision Medicine and Epidemiology: Contrasting Germany and the US." Annual Meeting of the Society for Social Studies of Science (4S) 2020, Virtual Prague, unpublished manuscript.

Translation Matters

Racial Classification in South Korean Genetic and Genomic Research

Jaehwan Hyun

Abstract *This chapter explores the use of human group classifications in scientific articles related to genetic research in South Korea. Building on a systematic literature review of human group categories in South Korean genetic research, it reveals that while South Korean geneticists and genomicists rarely use the English term "race," they frequently rely on its Korean equivalent, "injong," as well as the more ambiguous category "minjok," which is often used in a similar sense. The widespread use of injong and minjok in South Korean genetic and genomic research not only reinforces biological distinctions between Koreans and non-Koreans but also risks creating a public misconception that genetic and genomic research supports the concept of biological race. With this case study, I argue that historians of science and technology, as well as science and technology (STS) scholars, should critically examine the translation of group categories across different national contexts.*

Introduction

In the fall of 2018, when I interviewed Jong Hwa Park, a leading scholar in South Korean genomic research, he insisted that the idea of "race" ("injong" in Korean, 人種 인종) was totally refuted by genomic research on human diversity and that only "ethnic groups" ("jongjok," 種族 종족 or "injok," 人族 인족 in Korean) were a valid grouping category (Park 2018). In another interview that I carried out with Han-jun Jin, one of the pioneers in the genetic history of Koreans, this geneticist also agreed that concepts of race are totally outdated and not used in genetic history research anymore (Jin 2017). According to those geneticists, the idea of race had already receded from South Korean genetic and genomic research by the 1990s. South Korean geneticists' belief in the absence of racial categories more or less resonates with the wider public's belief that racism is lacking in South Korea, since the majority of the country is made up of ethnic Koreans, with only a few foreign residents mostly from other Asian countries (Lie 2014, 8–15). This belief holds that Koreans do not use racial categories in everyday life, resulting in a weak awareness of race as a category.[1]

To scrutinize the views of these geneticists, this chapter examines the use of human group classifications in scientific articles related to genetic research in South Korea. Based on a systematic literature review on the use of human group categories in South Korean genetic research on Koreans, it aims to uncover the complex use of these categories in both English and Korean. South Korean geneticists and genomicists mostly do not use the English term "race" in their English abstracts or in English papers. However, in Korean papers, they still frequently use its Korean translation "injong" and an ambiguous group category "minjok" (民族, 민족) that is similar in some senses to that of race.[2]

This widespread use of the Korean categories of "injong" and "minjok" in South Korean genetic and genomic research reifies biological differences between Koreans and non-Koreans. Although further study is required, this categorizing practice in science might also risk misinforming the public that genetic and genomic research supports the idea of biological race. Throughout this case study, I argue that historians of science and technology and STS scholars should pay attention to the translation of group categories in different national contexts.

This chapter consists of four sections. The first section offers a historical background concerning the group classification in the sciences of human heredity in South Korea. It particularly focuses on the historical origins of Korean categories—"injong" and "minjok"—and their inclusion in human heredity research from a *longue-durée* perspective. The second section explains the method and database used for the systematic literature review. In the two final sections, I analyze the review result and discuss its social impact.

Historical Backgrounds

Korean historians and historical sociologists have come to a consensus that "injong" translates to the English term "race," while "minjok" translates to "nation." Furthermore, "jongjok" is considered the most accurate translation of "ethnic group" (Kim 2005; Kang 2022).[3]

Among those terms, "minjok" is often considered the most important in understanding modern Korean history, because it was the term that nationalist Korean intellectuals and political leaders chose to foster political aspirations of a Korean people in var-

1 South Korean historians and sociologists have recently criticized the widespread belief in a "race-free Korean society" by arguing that "racism without race" is prevalent in this country. See Na (2022) and Park (2019).

2 In South Korea, "injong" is a Korean translation of the English concept of race, while "minjok" corresponds to the concept of nation. However, "minjok" is used to indicate an ethnonational entity that shares history, language, culture, and sometimes blood, so it easily conflates other English concepts of race, ethnic group, and nation (Shin 2006).

3 South Korean historians and sociologists studying nationalism differentiate between ethnic nationalism, which is based on biological ethnicity, and civic nationalism, which relies on cultural integration (Kang 2019). In some cases, however, scholars have chosen to retranslate minjok as "ethnic nation" in English in order to emphasize its biological connotations (Shin 2006).

ious contexts. During the late Chosŏn dynasty (1392–1910), Koreans were pressured by the Western imperial countries and by the Japanese empire to open up the country to foreign trade. When it seemed evident that Korea would be colonized by the Japanese empire in the early twentieth century, Korean intellectuals used "minjok" to define the Korean people as a political unity regardless of having lost their state as a political entity. The trope of *Tanil minjok*—encompassing the idea that Koreans were nationally "pure" as they originated from a single ancestor and shared a singular history, language, and culture—played a large role in maintaining nationalistic independence movements during the colonial period and in advancing nation-building projects in the two Koreas during the postliberation period (Shin 2006). In this political history-centered narrative, "injong" quickly lost its importance among the nationalist intellectuals at the turn of the century with the rise of the term "minjok." Although "injong" was used by the colonial government and pro-Japanese intellectuals in encouraging cultural assimilation and claiming a "yellow-race" alliance against "white predation" during the colonial period, the concept lost its politico-ideological space in the postliberation period because the political leaders of the two Koreas each aimed to establish a nation-state based on the idea of Korean minjok (Tikhonov 2015).[4]

In contrast to this political history-centered narrative, the history of the concepts of *injong* and *minjok* and the ways in which they were adopted in science shows that there were no rigid boundaries between their translations, and this interchangeability was maintained in specific professional areas, especially in the field of human heredity, even during the post–WWII period. In East Asia, Japanese scholar Kazan Watanabe first introduced Johann Friedrich Blumenbach's five concepts of race in 1838, while enlightenment thinker Yukichi Fukuzawa devised the term "jinshu" (人種), written in Chinese characters, to explain Blumenbach's racial classification in Japanese in 1869. The same Chinese-character term, pronounced as *injong* in Korean, first appeared in Korea's first modern newspaper the *Hanseong Sunbo* (漢城旬報) in 1883 to explain Blumenbach's theory of five races (Na 2022: 81–82). In the 1890s, the term "injong" was also widely used to define the Korean people as an independent group that was politically and biologically distinct from neighboring groups. "Chosŏn [Korean] injong," for instance, was regarded as different from "Ilbon [Japanese] injong" and was "the top injong among the Oriental yellow races" (東洋黃種中第一等人種) (Na 2022: 90–93).

Meanwhile, the introduction of the term "nation" complicated the group classification in Korea. In 1876–1879, Japanese legal scholar and politician Hiroyuki Katō coined the term "minshu" (民種) to explain the difference between the German concepts of "Nation" and "Volk" in his Japanese edition of Swiss jurist and politician Johann Caspar Bluntschli's *Allgemeines Staatsrecht* (1868) (Park 2011). According to the classification presented by Katō, "Volk" indicated a political unit of a state, named "kokumin" (国民) in Japanese, while "Nation," translated as "minshu," meant a group of people who shared cultural commonalities and biological lineages. Katō's translations were widespread in East Asia, and Chinese politician and social activist Qichao Liang, who studied in Meiji

4 Racial categories translated from the Korean, such as white, black, and yellow, as well as caucasian and mongoloid., will be spelled lowercase in this article to reflect their status as translations, setting aside questions pertaining to their capitalization in English.

Japan, became a propagator of Katō's classifications. In the late 1890s. Liang modified Katō's term "minshu" to "minzu" (民族) by replacing the Chinese character "種" (shu) with "族" (zu) and used this term in his political writings. Liang arbitrarily used the term to indicate different grouping levels; for instance, Germans, Teutons, and white people were all referred to as "minzu" (Kang 2006).

In the same way, the same term written in Chinese characters, "minzu," pronounced as "minjok" in Korean, was also used interchangeably with "injong" in the early 1900s when Korean nationalists were strongly influenced by Liang's texts. For instance, in a news article published in 1900, "Tongbang [Oriental] minjok" was mentioned as being different from "Paegin [white] minjok." However, from 1905 onwards, the term "minjok" increasingly replaced "injong" in the political arena to indicate the Korean people as a political collective, and it became a core concept of Korean nationalist intellectuals opposed to the Japanese colonization of Korea (Kwon 2007).

In the field of anthropology, a similar change occurred in the 1920s and 1930s. While in Japan by the 1910s physical anthropologists used "jinshu" (人種) when classifying the Japanese and other neighboring people, e.g., "Nippon [Japanese] jinshu" and "Chōsen [Korean] jinshu." In the 1920s, the term "Nippon jinshu" was increasingly replaced with "Nippon minzoku" in the anthropological field (Sakano 2005). Similar changes can be observed in colonial Korea with Japanese anthropologists in colonial research institutions. In the late 1920s and the 1930s, Japanese anthropologists at Keijō Imperial University Medical School in Keijō (now Seoul) carried out extensive fieldwork in Korea and Manchuria, to study the biological characteristics of "Chōsen minzoku." According to Tsunekichi Ueda, who had been trained in anthropometry at Ludwig Maximilian University in Munich and was the leader of this colonial anthropological enterprise, Korean *minzoku* was "not the same Volk, but the same Rasse" as Japanese *minzoku* (Hyun 2023). His statement implied the biological proximities and cultural differences between the Japanese and Korean people but did not deny the presence of biological differences between the two groups. This small biological difference became the target of Japanese research in human heredity during the 1930s and the 1940s (Hyun 2023: 269–270). Korean scientists, who were trained under the Japanese colonial education system, internalized these group categorizations. For instance, Korean physical anthropologist Sejin Na, who was a member of Keijō's extensive survey team of Korean *minzoku*, came to consider the Korean *minjok* (*Hanminjok*) as a principal unit for his postwar research (Hyun 2019a).

After the liberation from Japanese imperialism as a result of the end of the Pacific War (1941–1945) and after the end of the Korean War (1950–1953), South Korean anthropologists and geneticists, including Na, sought to define the Korean people as a biological population distinct from its neighboring people, especially from their former colonizers (Hyun 2019a: 503–505). This was one way for them to contribute to building South Korea as a new nation-state. For instance, Yung-Sun Kang, a founder of South Korean genetics, in 1954 initiated a decade-long project focused on "genetic research on the Korean population," stating that its aim was to identify the biological nature of "Korean minjok" (Hyun 2019a: 505). In 1965, serological geneticist Samyŏl Yi more clearly claimed that his research on the blood group distribution in South Korea—indicating the "pure blood of Korean minjok" distinct from Japanese and other neighboring *minjok*—contributed to the "scientific independence" of his new nation-state from Japanese colonialism (Hyun 2019b).

So far, the usage of *injong* in human heredity research in South Korea remains unexamined, while that of *minjok* has only been partially examined by a few case studies (Hyun 2018). In this respect, the systematic literature review of genetic and genomic research articles written in Korean after the Korean War can shed some light on the usage of the two Korean classificatory terms in South Korean genetic and genomic research. In the following sections, this chapter focuses on the scientific literature review in terms of classification use and its implications.

Method

I conducted a systematic literature review on the use of human group categories and their Korean translation in South Korean genetic research on Koreans. To do this, I utilized two national research databases: RISS (Research Information Sharing Service) and Korea Institute of Science and Technology Information's (KISTI) ScienceOn. RISS is the most comprehensive national database, containing 6,304,375 research articles (as of 2022) produced in all South Korean research institutions, including universities and colleges. ScienceOn has a science and technology focus and is helpful as it offers 328,615 national R&D reports (as of 2022) excluded from the RISS database. In both databases, I used the keyword "한국인 집단" ("han'gugin chiptan," Korean population) for the literature review search. In RISS, I searched for the term with the field limitation of "자연과학" ("chayŏn'gwahak," natural sciences) and "의약학" ("ŭiyak'ak," medical and pharmaceutical sciences) and with the time limitation from 1945 to 2022. In ScienceOn, I searched for the term with the search limitation of "보고서" ("pogosŏ," national R&D reports) with the time limitation from 1945 to 2022.[5]

As a result, I found 379 articles from 1963 to 2022 in RISS and 177 R&D reports from 1985 to 2022 in ScienceOn. Then, I excluded articles and reports that (1) were written only in English without a Korean title or abstract; (2) were nongenetic research (e.g., demography, public health, social psychology, or nutrition studies); or (3) were duplicates. After excluding those articles, I was left with 172 articles (1973–2022) from the RISS database and 97 R&D reports (1985–2022) from the ScienceOn database. Then, I classified all articles and R&D reports by discipline: (1) population genetics, (2) medical genetics, (3) forensic genetics, and (4) biological anthropology. Finally, I analyzed the English group categories and their Korean translations in those articles and reports.

5 National R&D reports are papers that are not peer reviewed and are used to report research results to government funders.

Analysis

Scientific Articles

Out of 172 scientific articles found in the RISS database, 105 articles were written in Korean, so I could identify the Korean translation of the English grouping categories: fifty-six articles, which were published from 1973 to 2022, used "minjok" as the Korean translation of "population," "race," "ethnic group," "geographic origin," and "ancestry." Nineteen papers, which were published from 1982 to 2021, used "injong" as the Korean translation for "population," "race," and "ethnic group." Twenty-nine papers (1981–2021) used "chiptan" as a Korean translation of "population" and "ethnic group," and among them, nine articles adopted racial classification, including "white" and "black" (see Table 1). Finally, only three articles (1994–1997) used "jongjok," and among them, two articles also used different group categories such as "minjok" and "injong." Even the one article that only used the term "jongjok" also adopted the racial classification of "white" and "black."

Figure 1: The Korean translation of the English grouping categories in scientific articles from 1972 to 2022 (n=105).

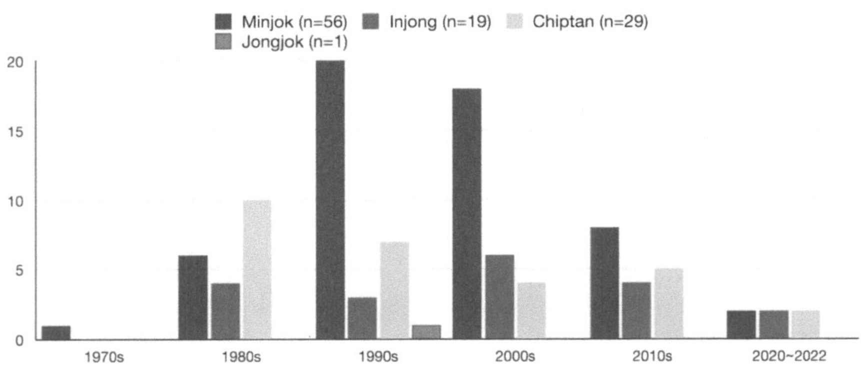

Table 1: Racial classifications used in scientific articles using the Korean translation of "chiptan."

Author, Year, Title	**Group Classification**
Lee, KY, 1980, Genetic study of phenylthiocarbamide (PTC) taste in a Korean population (1)	Mongoloid, Caucasoid
Lim, NY, 1981, Abstracts of lectures and papers presented at the 4th Congress of the Korean Genetics Society	Caucasians
Lee, HK, and HW Shin, 1996, Genetic characterization of a DXYS17 locus hypomorphism in a Korean population	white population (*paegin chiptan*)

Author, Year, Title	Group Classification
Lee, HK, and HW Shin, 1996, Genetic characterization of hTPO locus hypomorphism in a Korean Population	white population (*paegin chiptan*)
Han, JS, et al., 1997, Cloning of the variable region of D8S210 locus and its application to the forensic individual identification for Korean population	whites (*paegin*)
Kim, KS, et al., 1999, Genetic analysis of three short tandem repeat loci and one variable number of tandem repeat locus in the Korean population	Caucasian population
Hong, KJ, et al., 2000, Allele frequency distribution of Two PCR-amplified loci in the Korean population	Caucasian population
Kim, DS, et al., 2014, Morphometric Analysis of the Skull by Moiré contourography	whites (*paegin*), blacks (*hŭgin*), Hispanic (*Hisŭp'aenik*)
Ahn, GS, et al., 2014, Polymorphism of ADRB2 genes, beta-2 adrenoceptor related to atopic dermatitis in Koreans	white lineages (*paegin kyeyŏl*)

The remaining articles (n=67) were written in English, and in most cases, it was not possible to identify Korean translations of grouping categories because of the short length of the Korean abstracts. Despite such limitations, the data offers a glimpse of which English group categories were used. Among them, twelve articles defined "chiptan" as equivalent to ethnic groups, while nine of them adopted racial classifications such as "Mongoloid," "Caucasoid," and "Negroid," or "Caucasian" and "Negroid." Seven papers used "race" to translate the Korean term "chiptan" into English, and among them, one paper used "race" and "ethnic group" interchangeably. Forty-eight articles used the term "population," and among them, thirty papers either adopted the aforementioned racial classification or defined population as "racial population."

Throughout all these papers, written either in English or Korean, "injong" was consistently used from 1982 to 2021. In contrast, its English equivalent "race" was used only from 1985 to 2004. The term "ethnic group" and its Korean equivalent "jongjok" were first used in 1987, but they were not widely used: the English term was not widely used after 2009, and "jongjok" had a much shorter life span, as it was only used three times in 1994 to 1997. Meanwhile, the English term "population" continued to be used from 1985 to 2014 while "chiptan" was used from 1981 to 2021. Among the Korean terms, "minjok" was used most consistently throughout all periods to translate all the English categories—race, ethnic group, and population—in fifty-six articles from 1973 to 2022.

Figure 2: A comparison of the word frequency between the English term "race" and the Korean term "injong" in scientific articles from 1972 to 2022.

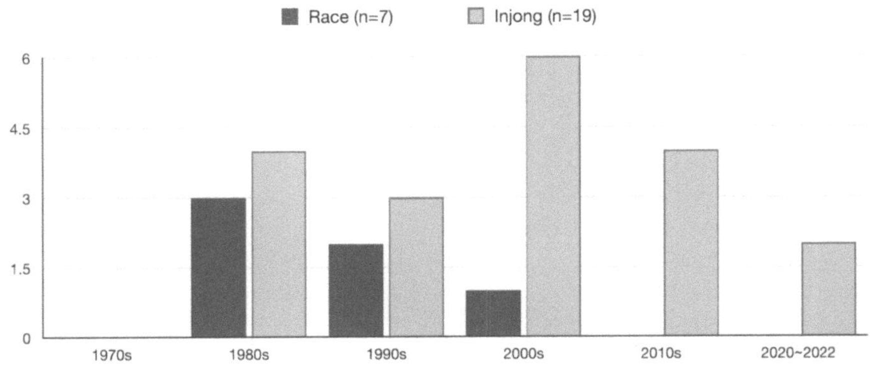

There are disciplinary differences in the use and translation of group categories. Population genetics articles used "race" and "injong" from 1982 to 1995 but changed them to "ethnic group" in 1997 and used that term until 2002. After that, "population" and "chiptan" became the dominant categories, although "minjok" was also used consistently. Forensic genetics was the field that used "minjok" most extensively throughout the first and second decades of the twenty-first century. Medical genetics papers used "race" and "injong" more than other genetic fields did. From 1996 to 2021, except for two articles, all scientific articles employing the categories "race" and "injong" came from the field of medical genetics. Biological anthropology used all grouping categories, but from 2013 onward, they started using "population" and "(ingu) chiptan" more often.

National R&D Reports

Although all the national R&D reports (n=92) found in the ScienceOn database were written in Korean, they also used English group categories in their abstracts and main parts. All papers used the English term "population" when translating "chiptan" into English, while fifteen reports also mentioned different English grouping categories: one report published in 1988 used the term "racial population," "race," and "racial characteristics" while mentioning the racial classification of "Caucasoid," "Mongoloid," and "Negroid" (Kim 1988). Between 1998 and 2020, four reports used the English term "ethnic group" interchangeably with "population," while ten reports categorized people as "Caucasian" or "Mongoloid" and "white" or "black" without explicitly mentioning the term "race."

Figure 3: The Korean translation of the English grouping categories in R&D reports from 1985 to 2022 (n=97). Nineteen articles used both "minjok" and "injong," while five articles used either/both "minjok" or "injong" with "jongjok." One article used "chiptan" and "injong" to criticize their use for human classification.

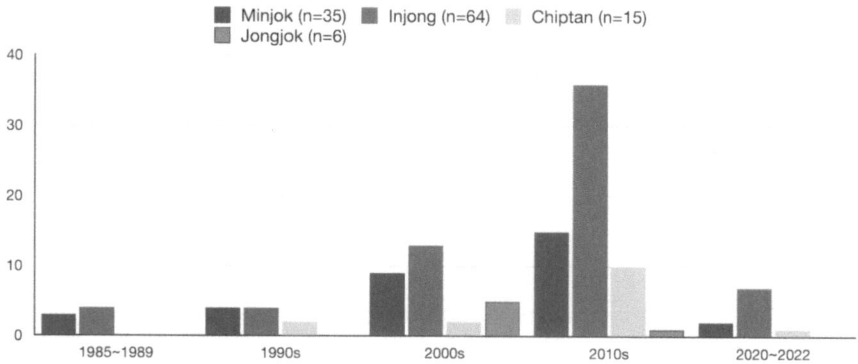

Only one report, published in 2018, problematized the mixture of diverse grouping ideas and conceptions, including "chiptan, injong, sŏnjo [ancestry], and kukka [nation-state]" (Kim 2018). Sixty-four papers, published between 1985 and 2022, used the term "injong" to translate "population," "race," "ethnic group," and "ancestry"; thirty-five reports, published from 1985 to 2020, used the term "minjok" to translate "population," "race," "ethnic group," and "ancestry." Among both groups of papers, nineteen reports used both "minjok" and "injong."

Meanwhile, the term "jongjok" was used only in six papers, and among them, five papers used that term along with other Korean translations, "minjok" and "injong." Fifteen papers used "chiptan" to translate the English word "population" while using the dichotomic category of "Westerners" (서구인, "sŏguin," or 서양인, "sŏyangin") and "Orientals" (동양인, "tongyangin").

Table 2: Racial classifications used in national R&D reports using the Korean translation of "jongjok."

Author, Year, Title	**Group Classification**
Kim, MS, 2002, National center for genomic research, part of the national institutes of health	*jongjok*, *injong*, and *minjok*
Shin, JK, 2007, Korean drug metabolizing enzyme and drug transporter protein genomics for personalized drug therapy	Asian *jongjok*, *minjok*, Westerners (서구인, *sŏguin*), and Africans (아프리카인, *ap'ŭrik'ain*)
Shin, MH, 2008, International cooperation cohort project I: overseas Korean American cohort project (3rd year)	*injong*, *minjok*, and *jongjok*

Author, Year, Title	Group Classification
Kim, YJ, 2008, Korean haplotype information development project; development of Korean HapMap Database	Korean, Chinese, and Japanese *injong*, Asian *injong*, and *jongjok*
Choi, BY, 2009, Rural community-based multiagency cohort projects	*jongjok*
Park, SK, 2020, Tracking Korean genomic epidemiology survey project and developing performance strategy	*jongjok*, *injong*, and *minjok*

Out of eighty reports which used "injong" and/or "minjok," twelve came from the population genetics field. Ten population genetics reports, published from 1985 to 1998, adopted both terms, but two reports, published in 2010, only used "minjok." Fifty-two reports published between 1994 and 2022 were categorized into medical genetics. Among them, sixteen reports adopted the term "minjok" and twelve of those sixteen reports also used "injong." The preference for "injong" was strong in the field of medical genetics: forty-eight medical genetics–related reports used that term. Of fifteen reports classified into the field of forensic genetics, "minjok" and "injong" were almost equally preferred: nine reports used "minjok" and ten used "injong," while four used both terms. Lastly, there was only one paper, published in 2018, that could be classified into the field of biological anthropology, and it explicitly objected to using "minjok" or "injong" as categories (Kim 2018: 3). The paper also cautioned about the mixture of the diverse grouping categories "chiptan, injong, sŏnjo, and kukka."

Discussion

The systematic literature review has revealed four key points. First, despite the decline in the use of the English term "race" since the mid-2000s, its Korean translation "injong" has continued to be frequently used in scientific articles (n=19, 1982–2021) and has even significantly increased in national R&D reports (n=64, 1985–2022). Second, the Korean term "minjok" is the most widely used translation for various English grouping categories in scientific articles (n=56, 1973–2022) and the second most common in national R&D reports (n=35, 1985–2020), following "injong." Third, the two Korean translations of "injong" and "minjok" have been used interchangeably for different English grouping categories such as "race," "ethnic group," "ancestry," and "population." Finally, there are disciplinary differences in the adoption of English grouping categories and their translations.

Table 3: Examples of the use of "injong" for translating the English words "ethnic group" and "population" after the 2000s.[6]

Author, Year, Title	Classification, Translation	Report Type	Field
Lee, YJ, 2004, Characterization of microsatellite markers covering chromosome 1 in the Korean and Japanese populations	ethnic group = *injong*	S	M
Chang, EH, 2012, Y haplogroup distribution in Korean and other populations	Western three *minjok* (African Americans, Caucasians, Mexican Americans) and East Asian *minjok* (Koreans, Japanese, and Chinese), population and ancestry = *minjok*	S	F
Kim, YJ, 2020, Study compares genetic effects of risk factors associated with diabetes and hypertension	Europeans and Orientals, ethnic group = *injong*	N	M
Chang, IJ, 2011, Study on the feasibility of sharing drug approval data between China, Japan, and Korea	Korean, Chinese, and Japanese injong, population = *injong*	N	M

The first two points indicate that, in contrast to humanities scholars and social scientists' usage of translations, "jongjok" was not used in scientific articles and R&D reports for translating the English term "ethnic group"; instead, in both cases, "injong" and "minjok" was mostly used to translate all English group categories, including "ethnic group." The result is in stark contrast to the explanations commonly found in South Korean social science and humanities literature. This literature describes "injong" as a social construct demystified by natural scientists, and "minjok" as a cultural grouping category unrelated to the biological characteristics of a specific human group (Kim 2020; Lee 2017; Kang 2022). However, as I showed earlier, the terms are still used in scientific articles to express biological differences between various human groups.

In terms of the third point, it is worth noting that both "injong" and "minjok" are used in various ways: South Korean geneticists and genomicists occasionally employ the term "injong" to differentiate between individuals of "Japanese," "Korean," and "Chinese" descent, although these groups are typically considered subgroups of the broader "Asian" race from an English-language perspective. This categorization is consistent with the early twentieth-century categorizations employed by Korean nationalists. The term "injong" is also used to distinguish between "Asians" and "whites" (Kim 2005). The term "minjok" is commonly used to describe populations from different Asian countries including

6 Report type: S=scientific article, N=national R&D report. Field: M=medical genetics, F: forensic genetics.

"South Korea, China, Indonesia, Japan, Philippines, and Thailand" (E.S. Kim 2020). However, that term is also used to classify "European whites" and "Asians," who are grouped based on racial categories from the English-language perspective (Zhang et al. 2012). This mixed use of *injong* and *minjok* resulted from translating diverse English terms into those two Korean terms. For instance, scientific articles and R&D reports related to the International HapMap Project and the 1000 Genomes Project adopted "injong" to translate the English terms "ethnicity" and "ancestry" (Park 2016; Choi 2020).

The disciplinary differences found in scientific articles and R&D reports indicate how the preferred terms in human grouping vary by discipline. In the field of population genetics, one can observe the decreased use of the English term "race" and its Korean translation "injong" simultaneously with the consistent preference for the Korean term "minjok." Biological anthropology papers stuck to the use of the English term "population" and the Korean translation "chiptan" while occasionally problematizing other Korean grouping categories. In the two applied fields, medical genetics and forensic genetics, the terms "injong" and "minjok" were consistently used in the 2010s and early 2020s, even though the English term "race" was not used at that time. Both Korean terms were often used in the same way as "race," alongside other racial classifications such as "Caucasians" and "whites and blacks." Further research is needed to fully understand the extent to which each field takes into account the political implications of using human group descriptors. However, it seems that biological anthropology is the most careful in the use of these terms—as some papers in this field explicitly problematize using racial categories—followed by population genetics. This heightened sensitivity might be due to physical anthropologists' recent self-critical reflection on their past involvement in racial science (Pak 2004; Woo 2021). On the other hand, researchers in medical genetics and forensic genetics appear to be less reflective about their not-scientifically-sound usage of their group classification.

Although the English term "race" is not used, the widespread use of "injong" and "minjok" in South Korea reinforces the biological reification of race. In response to the growing concern over "multicultural crime" (다문화범죄, "tamunhwa pŏmjoe" in Korean), which is a xenophobic expression of fears of foreigners committing crimes due to the influx of Asian migrant laborers and the rise of international marriages, forensic geneticists have conducted research on "injong/minjok identification (식별, 'shikpyŏl' written in Korean)." They have attempted to identify "Korean minjok" and distinguish it from other "Asian injong" by searching for more precise biological ancestry markers and algorithm systems, presuming biological disparities between Koreans and immigrants from other parts of Asia (Hyun 2024).

Forensic efforts made to identify the "injong" and "minjok" of criminal suspects heavily involve devising forensic DNA phenotyping. While it is recognized that identifying "injong/minjok" within Asian populations through conventional pigmentation markers is difficult, researchers are focusing on finding alternative markers. They are identifying single nucleotide polymorphisms (SNPs) related to "hair morphology" and "behavioral characteristics, such as smoking and alcohol consumption behaviors" (Seo et al. 2017). Their work has defined certain groups as biological "injong" or "minjok" possessing certain physical and behavioral characteristics. In other words, this research reifies biological race in Korean terms.

When scientific articles and R&D reports introduce "injong" and "minjok" as human group descriptors in media outlets, they contribute to the biological reification of race with those terms. The Genome Asia 100K initiative, for example, began in 2016 with the aim of creating a comprehensive pan-Asian genome database by sequencing 100,000 individuals from various Asian countries through a consortium of genome companies and academics. In South Korea, biotech company Macrogen Inc. and Seoul National University Bundang Hospital participated in the pan-Asian initiative, and the first sequencing result was reported in *Nature* in December 2019. In the English news release provided by Marcogen Inc., the implication of its publication was explained (Macrogen 2019):

> The research team performed whole-genome sequencing for a total of 1,739 individuals ... The study found that the genetic diversity of the approximately 142 ethnic groups living in Asia was far greater than what was found in previous studies. On the basis of this study, it was also recognized that the Asian ethnic groups differed in terms of their response to key drugs ... The research showed that Warfarin sensitivity is more likely in individuals from Korea, China, Japan, and Mongolia, as well as other persons of *North Asian descent*. (Italics added by the author.)

The news release by Macrogen Inc. was distributed to Korean newspapers, as well. In the Korean version, the terms "injong" and "jongjok" were used to translate "ethnic groups" and "descent" respectively (Lim 2019):

> The 1,739 individuals [who participated in the study were] Asians from 142 *jongjok* in 24 countries. This is the largest number of injong studied in any Asian-focused genomic research project to date. The Singaporean government-led initiative announced the completion of its [Asian] genome database including the data of 5,000 individuals but [they were] only from three injong ... We discovered that Warfarin, an anticoagulant prescribed to patients with cardiovascular disease, causes serious side effects, including allergies, to *Northeast Asian injong* such as Koreans, Chinese, and Japanese." (Italics added by the author.)

Other news articles based on the same news release of Macrogen Inc. explained the implication of the research as "the inclusion of the largest number of Asian regions and *injong* in the world's publicly available Asian genomic data" and "the identification of different pharmacokinetic responses of Asian *minjok* to key medications" [italics added by the author] (Ham 2019). Again, in those articles and news reports, they do not use the English term "race," but they use "injong" and "minjok" in explaining human diversity.

Those news reports lead the South Korean public to believe that biological differences among "injong" are real. It is beyond the coverage of this chapter, but some far-right online communities use those newspapers related to genetic and genomic research on the Korean population as a biological basis to make racist and sexist arguments.[7]

7 See, for instance, "Haplogroup Gallery," DC Inside, https://gall.dcinside.com/mgallery/board/lists/?id=haplogroup. For a detailed analysis of the use of genetic knowledge for online far-right activism, see Claude Oliver Doron's chapter in this volume. See also Panofsky and Donovan (2019).

Conclusion

This chapter examined the use of human grouping categories in South Korean genetic and genomic research by carrying out a systematic literature review of South Korean scientific articles and R&D reports. It shows that, although the English term "race" has not been used since the mid-2000s, the Korean terms "injong" and "minjok" are still consistently used in South Korean papers. The "injong" and "minjok" categories are used in the same sense as "race" in those papers. This has led to the biological reification of race in forensic research and science news. Future research might need to investigate the co-productive relationship between this biological reification of race in genetic and genomic research to shape and be shaped by broader public conceptions of race and national identity in South Korea.

The South Korean experience indicates a need for decentralizing the English-centered approach in the global regulation of human grouping descriptors in genetic and genomic research. The current regulation efforts of group descriptors in genetic and genomic research are mostly American-centered, and as a result, its focus is mainly on English terms, especially "race." The most recent and influential publication in this field is a report titled "Using Population Descriptors in Genetics and Genomics Research: A New Framework for an Evolving Field" (National Academies of Sciences et al., 2023). This report was prepared by the Committee on the Use of Race, Ethnicity, and Ancestry as Population Descriptors in Genomics Research of the US National Academies of Sciences, Engineering, and Medicine (NAS) and was published by NAS in 2023. Although the report contains various useful recommendations, it also starts with the statement that "researchers should not use race as a proxy for human genetic variation" (NAS et al., 2023: 8). The South Korean case shows that racial research can be conducted in non-English settings without the use of the English term "race." If we do not consider the matter of translation in different national contexts beyond the Northern American concerns about the English term race, we will continue to overlook different forms of racial science across the globe.

References

Choi, Moo-rim. 2020. *Tabuch'ŏ yujŏnch'e chŏngbo hwaryongŭl t'onghan han'gugin taejogun yujŏnch'e pyŏni chido chaksŏnggwa konggae p'ŭllaetp'om kaebal* [The National Research Foundation Final Report on the Establishment of Healthy Korean Variant Profile and Sharing through Utilizing Data from Post-genome Projects]. Seoul: Seoul National University.

Ham, Young Hoon. 2019. "Han'guk chudo ashia kenom... Neich'ŏji k'ŏbŏsŭt'ori [The 'Asian Genome' Research Is Led by Koreans... Nature Cover Story]." *Herald Kyungje*, 5 December 2019.

Hyun, Jaehwan. 2018. "Yujŏnjŏk minjok mandŭlgi: Han'gugŭi illyu yujŏn yŏn'gu, minjok chŏngch'esŏng, kŭrigo ch'ogukchŏk kwahak kyoryu [The Making of a Genetic Nation: Human Heredity, National Identity, and Transnational Scientific Exchange in South Korea, 1926–2009]." PhD dissertation, Seoul National University, Seoul.

Hyun, Jaehwan. 2019a. "Racializing Chōsenjin: Science and Biological Speculations in Colonial Korea." *East Asian Science, Technology and Society* 13 (4): 489–510. DOI: https://doi.org/10.1215/18752160-8005053.

Hyun, Jaehwan. 2019b. "Blood Purity and Scientific Independence: Blood Science and Postcolonial Struggles in Korea, 1926–1975." *Science in Context* 32 (3): 239–260. DOI: https://doi.org/10.1017/S0269889719000231.

Hyun, Jaehwan. 2023. "Race and Ethnicity." In *The Interwar World*, edited by Andrew Denning and Heidi J. S. Tworek, 263–278. New York: Routledge.

Hyun, Jaehwan. 2024. "'Multicultural Genes in Our Blood'? Genetic Governance and Biocultural Purity in South Korea." In *Ordering the Human: The Global Spread of Racial Science*, edited by Dorothy Roberts, Eram Alam, and Natalie Shibley, 161–181. New York: Columbia University Press.

Jin, Han-jun. 2017. Interview by the author. Cheonan, South Korea, 14 February 2017.

Kang, Dongkook. 2006. "Kŭndae han'gugŭi kungmin, injong, minjok kaenyŏm [The Conceptual History of Political Identity in Modern Korea: Gukmin, Injong, Minjok]." *Han'guk Tongyang Jŏngch'isasangsa Yŏn'gu* [The Review of Korean and Asian Political Thoughts] 5: 22–24.

Kang, Jinwoong. 2019. "Han'guk minjokchuŭi yŏn'guŭi chaengjŏmgwa kwaje [The Issues and Tasks in the Studies of Korean Nationalism]." *Minjok Yŏn'gu* [Minjok Yeongu] 73: 4–25.

Kang, Jinwoong. 2022. *Minjogŭi sahoehak* [The Sociology of Nation]. Seoul: Somyong Publishing.

Kim, Eung-soo. 2020. *NGS kisurŭi pŏpkwahakchŏk chŏkyongŭl wihan kiban'gisul yŏn'gu* [The National Forensic Service Final Report on Research on the Foundational Technology for the Forensic Application of Next Generation Sequencing]. Seoul: The National Forensic Service.

Kim, Heewon. 2020. "Injonggwa injongjuŭirŭl nŏmŏ p'oyongsahoero [Anti-racism and Inclusive Society." *Yŏksa Yŏn'gu* [The Journal of History] 39: 335–350.

Kim, Kwang-Ok. 2005. *Chongjokkwa minjok: Kŭ tanilgwa pop'yŏnŭi shinhwarŭl nŏmŏsŏ* [Beyond the Myths of Ethnicity: Anthropological Approaches]. Paju: Acanet.

Kim, Kyung-yong. 2018. *Han'gukyŏn'gujaedan ch'oejongbogosŏ: Yetsaramppyŏesŏ DNA punsŏkkibŏbŭi kaebalgwa iyong* [The National Research Foundation's Final Report on the Development and Application of DNA Analysing Methods on Ancient Human Bones]. Seoul: Chung Ang University.

Kim, Yung Jin. 1988. *Han'gukkwahakchaedan ch'oejongbogosŏ: Han'gugin chiptanŭi punja yujŏnhakchŏk yŏn'gu 5. Esterase, Esterase D mit s-adenosylhomocysteine hydrolaseŭi yujŏnja pyŏnie kwanhayŏ* [The Korea Science and Engineering Foundation's Final Report on the Molecular Genetic Studies of Korean Population 5. Genic Variation of Esterase, Esterase D and S-adenosylhomocysteine Hydrolase]. Daejeon: Chungnam National University.

Kim, Yung Joo. 2005. *Han'gukkwahakchaedan ch'oejongbogosŏ: han'gugin Haplotype chŏngbogaebalsaŏp: han'gugin HapMap teit'ŏbeisŭ kaebal* [The Korea Science and Engineering Foundation's Final Report on the Development of Korean Hap-Map Database]. Daejeon: Korea Research Institute of Bioscience and Biotechnology.

Kwon, Boduerae. 2007. "Kŭndae ch'ogi 'minjok' kaenyŏmŭi pyŏnhwa: 1905~1910nyŏn taehanmaeilshinborŭl chungshimŭro [The Concept of 'Nation' in 1905–1910: Focused on Korea Daily News]." *Minjonk Munhaksa Yŏn'gu* [Journal of Korean Literary History] 33: 189–213.

Lee, Hyoseok. 2017. "Han'gugine taehan hwanginjong kaenyŏmŭi hyŏngsŏnggwa naemyŏnhwa kwajŏng yŏn'gu [A Study on the Formation of the Concept of the Yellow and Its Internalization into Modern Korea]." *Han'gunk Minjok Munhwa* [Journal of Koreanology] 64: 145–167.

Lie, John. 2014. "Introduction: Multiethnic Korea." In *Multiethnic Korea? Multiculturalism, Migration, and Peoplehood Diversity in Contemporary South Korea*, edited by John Lie, 8–15. Berkeley: Institute of East Asian Studies, University of California.

Lim, Yoo. 2019. "A 142kae chongjok yujŏnch'e DB kuch'uk… Mak'ŭrojen injongbyŏl match'umch'iryo kil yŏllyŏ [142 Asian Jongjok's Genomic Database Was Established and Injong-Specific Treatment Was Enabled by Macrogen]." *Hanguk Kyungje*, 5 December 2019.

Macrogen Inc. 2019. Macrogen Research Outcome from 'Genome Asia 100K' Covered in Nature. Accessed November 19, 2024. new.macrogen.com/en/newsroom/news/13100000039.

Na, Inho. 2022. "Injongesŏ minjokŭro: Han'gukchŏk injongjuŭiŭi hyŏngsŏnge kwanhan kaenyŏmsajŏk koch'al, 1880~1910" [From 'Race' To 'Nation/Volk': A Conceptual Historical Study on the Formation of Korean Racism, 1880–1910]." *Han'guk Tongyang Jŏngch'isasangsa Yŏn'gu* [The Review of Korean and Asian Political Thoughts] 21: 81–82.

National Academies of Sciences, Engineering, and Medicine; Health and Medicine Division. 2023. *Using Population Descriptors in Genetics and Genomics Research: A New Framework for an Evolving Field*. Washington D.C.: National Academies of Sciences, Engineering, and Medicine.

Pak, Sunyoung. 2004. "Ilche shingmint'ongch'ihaŭi chosŏn ch'ejirillyuhagi namgin hangmunjŏk kwajewa sŏgu ch'ejirillyuhaksarobut'ŏŭi kyohun [Academic Tasks Left by Physical Anthropology under Japanese Colonialism and Lessons from Experiences of Western Physical Anthropology]". *Pigyo Munhwa Yŏn'gu* [Cross-Cultural Studies] 10: 191–220.

Panofsky, Aaron, and Joan Donovan. 2019. "Genetic Ancestry Testing among White Nationalists: From Identity Repair to Citizen Science." *Social Studies of Science* 49 (5): 653–681. DOI: https://doi.org/10.1177/0306312719861434

Park, Geungap. 2011. "Yohan k'asŭp'a pŭlluncli'illi, Allgemeines Staatsrecht; kat'o hiroyuk'i,kukpŏppŏmnon [Notes on Johann Caspar Bluntschli's Allgemeines Staatsrecht and Hiryouki Kato's Kokuhō-Hanron]." *Kaenyŏmgwa Sot'ong* [Concept and Communication] 7: 243–278.

Park, Jin-bin. 2019. "Injongjuŭiŭi yŏksawa onŭrŭi han'guk [Understanding Racism in the 21st Century Korea]." *Yŏksa Bip'yŏng* [Critical Review of History] 129: 293–317. DOI: https://doi.org/10.38080/crh.2019.11.129.293

Park, Jong Hwa. 2018. Interview by the author. Ulsan, South Korea, 26 October 2018.

Park, Leeyoung. 2016. *Pok'apchirhwan palbyŏngŭi yujŏnjŏk shisŭt'em kusŏnggwa hŭnhan chirhwanbyŏni palgul* [The National Research Foundation's Final Report on Genetic Architectures of Complex Diseases and Common Disease Variants]. Seoul: Yonsei University.

Sakano, Tōru. 2005. "Jinshu. minzoku. nihonjin: Senzen Nihon no jinrui-gaku to jinshu gainen [Race, Minzoku, Japanese: Prewar Japanese Anthropology and the Idea of Race]." In *Jinshu gainen no fuhen-sei o tou: Seiyō-teki paradaimu o koete* [Is Race a Universal Idea? Transcending the Western Paradigm], edited by Yasuko Takezawa, 229–254. Kyoto: Jinbun shōin.

Seo, Hee Jin, Sohee Cho, Ji Hyun Lee, Sung Hee Lyoo, Moon-Young Kim, and Soong Deok Lee. 2017. "Forensic DNA Phenotyping: A Review in Korean Perspective." *The Korean Journal of Laboratory Medicine* 41 (2): 23–31. DOI: https://doi.org/10.7580/kjlm.2017.41.2.23

Shin, Gi-Wook. 2006. *Ethnic Nationalism in Korea: Genealogy, Politics, and Legacy*. Stanford: Stanford University Press.

Tikhonov, Vladimir. 2015. "Discourses of Race and Racism in Modern Korea, 1890s–1945." In *Race and Racism in Modern East Asia*, edited by Rotem Kowner and Walter Demel, 242–273. Leiden: Brill.

Woo, Eunjin. 2021. "Han'guginŭi kiwŏn'gwa hyŏngsŏnge kwanhan yŏn'gu kŏmt'o: Tamnonŭi hyŏngsŏnggwa chŏn'gae [A Review of Research on the Origins and Formation of the Korean People: Discourse Formation and Development]." *Han'guk Sanggosahak'oe Haksulbalp'yohoe* [Proceedings of the Korean Ancient Historical Society] 53: 3–14.

Zhang, Ai Hua, Hye Young Lee, Seung Bum Seo, et al. 2012. "Han'guginesŏŭi Y hap'ŭllogŭrubŭi punp'o yangsang mit tarŭn minjokkwaŭi pigyo [Y Haplogroup Distribution in Korean and Other Populations]." *Taehan Bŏbŭihak Hoeji* [Korean Journal of Legal Medicine] 36 (1): 34–44. DOI: https://doi.org/10.7580/KoreanJLegMed.2012.36.1.34

Reconstructing Narratives

Exploring the Significance of Descent and Ancestry for People of African Descent

Nicole Sommer

Abstract *This article examines the relationship between "descent" and "ancestry" within the context of people of African descent. It explores the emergence of the term "people of African descent" as a significant political category and its recognition by international political institutions. By analyzing the meanings and implications of descent and ancestry from a sociological perspective, the article asks how these terms intersect or diverge to shape the meaning and narrative of people of African descent. Additionally, the article discusses the usage of ancestry DNA testing and its role in validating historical narratives. This analysis uncovers the connections between these terms and their influence on understanding collective identity among people of African descent.*

Introduction

Alex Haley's groundbreaking book *Roots*, published in 1976, marked the beginning of a new trend in exploring individual ancestry. In his captivating narrative, Haley traced his African ancestry back through numerous generations to the African continent. The success of *Roots* made him a pioneer of literature genre dealing with the search for identity. Haley's influence inspired a wave of subsequent literature. This was followed by novels such as *The Twelve Tribes for Hattie* or *Americanah*, which deal with the search for an African identity. The increasing popularity of these works underscoring the common desire to not only understand personal identity, but also to connect to the past and cultivate a sense of belonging. For many people of African descent, however, this journey is clouded by a difficult reality—the virtual absence of knowledge about their personal ancestral history. The lingering effects of enslavement forced migration, and in particular the transatlantic slave trade have severed connections to origins and obscured family histories, leaving a void that needs to be explored.

While the term "African descent" emerged as a population descriptor only in the early twenty-first century as a comprehensive designation encompassing a diverse range of individuals in the African Diaspora worldwide and predominantly referring to the descendants of enslaved Africans, the term has also come to encompasses recent migratory movements.[1] Recognized as a political designation in international politics, mainly through the efforts of the United Nations, it aims to address the global community of people of African descent and advocate for their rights.

In the realm of lineage and genealogy, descent derives from the French and Latin *descendre* which combines *de* (down) and *scandere* (to climb). Indicating a progression or movement from a higher position to a lower one, it specifically refers to the successive generations that follow an individual, representing a downward movement from ancestors to their descendants. In contrast, the term *ancestry*, commonly used in genetics, focuses on the biogeographical origin of individuals, and involves tracing back lineage from the current individual to their ancestors. This process entails a retrospective movement from the present to the past, establishing connections between individuals and their ancestors, which can be seen as an upward movement. Both concepts, *descent* and *ancestry*, connect present individuals to their ancestors, albeit in different directions. In doing so, they represent a flow of time, although in opposite directions (Abel and Schroeder 2020, 202; Zerubavel 2000, 363). In the context of people of African descent, the separation from their ancestral roots due to historical circumstances has fueled a growing desire to explore and trace one's ancestry. Commercial companies such as African Ancestry, Afroroots, or AfricanDNA that offer consumer DNA tests have identified people of African descent as a distinct target audience, offering them the possibility to gain insights into their genetic, geographic, and even ethnic origins.

In the current era, DNA testing spills over well beyond the medical domain, as it intertwines with the politics of identity at an individual level, contributing to the politics of belonging and supporting claims related to political rights (Brubaker 2015, 74). Consequently, reliance on these tests raises questions about the social meanings and implications of intertwined notions of ancestry and descent. The intertwined concepts of ancestry and descent serve as a foundation for this article to explore the relationship between these terms. The article examines the influence of political institutions on the terminology used to refer to people of African descent collectively at a global level. In particular, it asks why the United Nations uses the term "descent" and what meaning it has for this group. At the same time, the article examines the role of individual ancestry, as evidenced by DNA testing, in validating the political narrative associated with the collective identity of African descent. It explores how the concept of descent intersects with or diverges from the concept of ancestry, and it deciphers the interplay between political discourse and individual identity in this context.

To that end, this article is divided into three parts. The first part discusses the emergence of the term "people of African descent" as a political category and a category of persons. Special attention is placed on obtaining insights about the various meanings of the term within international political organizations, especially the United Nations, and the specific range of meanings associated with the idea of descent. The second part provides a more nuanced analysis of the relationship between ancestry and descent. This analysis

---

1 For example, in the background paper published by the Commission on Human Rights in 2003, entitled "Identification and Definition of People of African Descent and how Racial Discrimination is Manifested in Various Regions" (UN Doc. E/CN.4/2003/WG.20//WP.3).

focuses on consumer genetic ancestry testing, emphasizing its significance in relation to people of African descent. The third and final part summarizes these findings. This examination explores the possible reasons behind the employment of the term "descent" by political institutions such as the United Nations. It highlights the implications for the term of African descent and how the concepts of descent and ancestry are intertwined.[2]

The article is anchored in a qualitative content analysis of various text documents from different UN bodies (Human Rights Council, General Assembly, Economic and Social Council, Committee on the Elimination of Racial Discrimination, OHCHR), including resolutions, decisions, declarations, recommendations, reports, minutes of debates and background papers. In a first step, I identified all documents dealing with people of African descent or the term "descent" and its use, which comprised about 297 documents. In a second step, I drew a sample from the collected corpus according to predefined criteria related to the research question. The documents were then coded and subjected to a structured content analysis.

People of African Descent as a Population Descriptor in the Political Context and Meaning of Descent

It is essential to understand the underlying principles governing categorization in order to delve into the implications of what it means to be of African descent or who qualifies as falling within the category. In this context, two central processes, lumping and splitting (Zerubavel 1996), play a crucial role. Lumping involves grouping individuals based on a shared characteristic while downplaying differences and thus simplifying complexity (Starr 1992, 271). For instance, a person is considered old when they reach a specific numerical age, say eighty numerical years. In this case, an old person is defined by the common characteristic of being old regarding a specific age, regardless of the person's gender, whether male, female, or another gender identity. Consequently, gender identity becomes less important in the context of age. Here, the focus is on highlighting the distinctions from other categories. Using the previous example implies that an individual categorized as old is inherently distinct from—and cannot simultaneously belong to—the category young. Consequently, the emphasis is placed on underscoring the disparity or contrast between the young and old classifications. Therefore, lumping and splitting represent two opposing yet complementary procedures within the categorization process (Zerubavel 1996, 421).

Considering this understanding, let us focus on the categorization concerning individuals who fall under the designation of people of African descent by reconstructing the lumping concerning the category. The term "people of African descent" is relatively new[3]

2 Since my interest lies in the social subfield of international politics and international law, I focus specifically on the naming practices in political documents and less on their use in (locally) situated everyday practices or individual self-understandings.

3 It is crucial to note that the term African descent was coined by Pan-Africanists such as W.E.B. Du Bois and George Padmore in the early twentieth century. The usage of the terminology "Africans and People of African descent" in the Final Declaration of the conference reflects this connection.

and has only recently emerged as an important designation in international politics. It goes back to the UN World Conference against Racism, Racial Discrimination, Xenophobia and Related Intolerance held in Durban, South Africa, in 2001.[4] At the Durban Conference, the United Nations officially adopted the concept, which generally refers to individuals in the African Diaspora. The conference's primary objective was to identify specific groups that experience disproportionate levels of racism and racial discrimination. Although individuals of African descent in every part of the world have faced racism and discrimination for centuries (Jean Emigh, Riley, and Ahmed 2015; Magubane 2017; Mills 1997; Rodney 2000; Schramm 2016; Turner 2002), historically, categorizations tended to be fragmented along local or national lines, such as African American, Afro-Columbian, Afro-Latino, or labels based on status of citizenship (e.g., refugee, migrant, migration background). One notable characteristic of the term African descent is its global scope. By incorporating the notion of descent, it encompasses individuals from diverse geographical and cultural backgrounds, albeit with African lineage.

The Durban World Conference and its final Declaration introduced the term people of African descent with a specific framing that is constitutive for the category. This framing emphasizes the shared characteristics of those it designates, namely, individuals of African descent who have historically faced particular vulnerability to racism and discrimination

One shared characteristic among individuals within the category is their historical experience as victims of (mainly transatlantic) slavery and its consequences. In the spirit of the Durban Conference, the term "African descent ... refers[s] to Africans living in the diaspora, particularly those who, because of the slave trade, were enslaved and taken to various parts of the world, mainly the Americas" (UN HR/MEX/SEM.1/2002/BP.2, 17). Further, the Durban Declaration and program of action acknowledge that:

> Slavery and the slave trade, including the transatlantic slave trade, were appalling tragedies in the history of humanity ... especially their negation of the essence of the victims, and further acknowledge that slavery and the slave trade are a crime against humanity and should always have been so, especially the transatlantic slave trade and are among the major sources and manifestations of racism, racial discrimination, xenophobia, and related intolerance and that Africans and people of African descent ... were victims of these acts and continue to be victims of their consequences. (UN A/CONF.189/12, para 13)

---

4 The term "African descent" is derived from the Declaration of the Preparatory Conference of the American States held in Santiago de Chile (UN Doc. A/CONF.189/PC.2/7). Here, mainly the Latin American and Caribbean countries pushed for the recognition of the term "people of African descent" (Afrodescendiente). Latin America and the Caribbean have the highest population density of people of African descent since most of the enslaved Africans were brought to Latin American countries and the Caribbean during the transatlantic slave trade. Unlike in the United States or even today in many parts of Europe, the term Black has long not been established as a term of self-identification and empowerment. The term "African descent" was intended as a political term to replace the naming practices [terms of Blackness] that had negative connotations in many of these countries (Paschel 2013; Reiter and Simmons 2012; Telles and Paschel 2014; Tsri 2016; Torres and Whitten, Jr. 1998).

The designation of people of African descent involves here a temporal universalization of the shared characteristic of the historical experience of enslavement due to its ongoing consequences into the present; it also underscores the notion of forced migration. Individuals of African descent live within a diasporic context because of the historical events of slavery and the transatlantic slave trade. Emphasizing the diasporic context highlights the global nature of the category, as individuals are dispersed across the world as a consequence of the slave trade and its repercussions. In other words, implementing the category of African descent aims to make visible—and therefore politically addressable—the historical events of slavery, colonialism, and their enduring consequences, particularly the past and present racism and discrimination faced by individuals of African descent.

Examining the formation of the category through the lens of lumping reveals how the concept of descent operates to unite individuals from diverse backgrounds under a collective identity. In this sense, the term is intended to promote a sense of cohesion and common belonging between these people. The concept of descent therefore recognizes and accommodates the diverse experiences and identities within this category. It is important to note that descent had already been recognized within international law as grounds for discrimination prior to the recognition of people of African descent in the 1960s (Keane 2005; Keane and Waughray 2017). Moreover, it is legally protected as a ground of discrimination through international human rights treaties. As a discrimination criterion, descent is identified in the International Convention on the Elimination of All Forms of Racial Discrimination (ICERD), the first binding human rights instrument to address racial discrimination. As Article 1 of the ICERD states, racial discrimination

> shall mean any distinction, exclusion, restriction, or preference based on race, colour, descent, or national or ethnic origin which has the purpose or effect of nullifying or impairing the recognition, enjoyment, or exercise, on an equal footing, of human rights and fundamental freedoms in the political, economic, social, cultural or any other field of public life. (UN A/RES/2106[XX], 2)

Strikingly, the term "descent" was not initially included in the draft of the Convention but was subsequently added to clarify certain ambiguities arising from terms such as *place of origin* and *national origin* (UN CERD/C/SR.1531, 5). It is important to note that descent has its own meaning and should not be confused with those concepts (UN CERD/C58/Misc.17/Rev.3, 3). While descent can be understood in the realm of social inheritance, the Committee on the Elimination of Racial Discrimination (CERD) explicitly clarifies in its statement that the term should not be interpreted in a biological sense. Politically, descent refers to the social implications of belonging to a group experiencing discrimination. Descent serves to establish affiliations between individuals, not through genetic connections or biological lineage, but rather through the shared experiences of discrimination and marginalization (UN CERD/C/SR.1531, 4). As such, descent is predominantly influenced by social and cultural factors rather than geographic-biological determinants.

In summary, in the political context[5], the term "descent" points to social implications, to the fact of explicitly belonging to a group of individuals who are subject to discrimination or its consequences. The notion of descent becomes relevant in political discourse when associated with experiences of discrimination, serving as a defining criterion for the categorization of marginalized groups and the discrimination associated with them. The ascription of an individual to a marginalized group arises out of descent (UN E/CN.4/Sub.2/2003/24, para 45).[6]

The term "descent" serves a political purpose and is crucial in constructing collective identities by establishing a temporal continuity. From a genealogical standpoint, descent signifies the connection between individuals and their various ancestors (Zerubavel 2000, 362). It functions as a means of identifying a common lineage. The focus lies in individuals positioned within a temporal line with their ancestors, accentuating their ties to the past. Consequently, it is not only that this lineage connects individuals to their historical roots, but also that a shared descent creates a deep sense of belonging and collective identity among the individuals in the present (ibid.). Regarding the political discourse, the term "descent" is intricately tied to Max Weber's social constructivist interpretation of how social groups relate to ethnicity. Weber describes ethnicity as a group of people founded on

> [t]he belief in group affinity, regardless of whether it has any objective foundation, [which] can have important consequences especially for the formation of a political community. We shall call "ethnic groups" those human groups that entertain a subjective belief in their common descent because of similarities of physical type ... or because of memories of colonization and migration; this belief must be important for the propagation of group formation; conversely, it does not matter whether or not an objective blood relationship exists. (Weber 1978, 389)

Following Weber, commonality and membership in a group is primarily based on the subjective belief in a common ancestry based on racial and cultural differences. Thus, it is primarily a matter of self-identification regarding a shared ancestry: "persons who identify themselves in terms of who they believe their ancestors to be, whether they act on this basis or not" (Jackson 1982, 5). The decisive criterion for belonging to a social group in this sense, then, is the belief in a common ancestry, and not an actual common ancestry in the sense of kinship, but "in the sense of a presumed common past" (ibid.).

By incorporating the notion of descent into the framework of the term "African descent" one establishes a profound connection to ancestral roots, mainly enslaved Africans, and emphasizes the historical narrative of slavery and enslavement. Consequently, descent is a conduit that connects historical events with contemporary

---

5 Between 2001 and 2004, a comprehensive discussion on the meaning of descent evolved in international political debates. This debate led to various outcome documents, including the formulation of General Recommendation XXIX on Article 1(1) of the Convention (on descent) (see also Keane 2005; Keane and Waughray 2017; UN Doc. CERD, General Recommendation XXIX).

6 Descent is acquired by birth, but only in the sense that the individual is born into a certain social group from which they cannot step out because the marginalized status is bound to this group (UN E/CN.4/Sub.2/2003/24, para 45).

manifestations of racism and racial discrimination. The social implications of descent encompass the development of a shared present that fosters a profound sense of interconnectedness (Zerubavel 2000, 362). For example, people of African descent in both the United States and the United Kingdom face racism despite their diverse ancestral backgrounds from different geographic regions. Nevertheless, they share a collective sense of belonging. Thus, the historical context becomes a powerful basis for developing a sense of commonality among people of African descent across the globe.[7]

Consequently, the recognition of a shared past provides individuals with a collective present, thereby fostering a sense of unity (Zerubavel 2000, 363). Furthermore, it creates the notion of social solidarity (Durkheim 1933) among individuals of African descent and, in this sense, emphasizes the importance of shared consciousness and connectedness, evoking or enabling collective action. Within the categorical framework of African descent, which encompasses a collective history of slavery and its descendants, a deep sense of unity can foster a stronger collective identity. It contributes to the stability of the category globally. In the political sphere, the term "descent" also reflects the tendency to emphasize cultural aspects more strongly. Culturalization emphasizes social and cultural characteristics in general (Geertz 1973) and in specific categorical descriptions of people (Bennani and Müller 2018). It is worth noting that this increasing emphasis on cultural aspects is also due to the participation of social scientists in these discourses and debates. This participation shapes the meaning of concepts and terminology used in political discourses.

Unraveling "African descent" in Politics: Understanding the Meaning and Usage of Ancestry Tests

To better understand why the term "descent" is used by political institutions such as the United Nations and how it is interwoven with the concept of ancestry, it is essential to examine the meaning and use of the term ancestry. While descent places considerable emphasis on the centrality of the present individual, the term ancestry in contrast, is commonly used in life science and refers to a genealogical link to the past transmitted through various generations. Although it is often not well defined, in the biomedical context it refers to a fixed characteristic within an individual genome (Byeon et al. 2021, 2215; Dauda et al. 2023, 2). It is frequently used in scientific research and literature to describe genetic lineage and biological differences among human populations and thus to categorize and describe populations. A shared ancestry encompasses the connection between individuals based on factors such as DNA, population affinity, and geographic origin, establishing a genealogical link which traces their familial or ancestral relationships (Dauda et al. 2023, 6). In this regard, individuals with the same ancestry have a common genetic heritage, indicating their affiliation with others and their families.

7 It is important to note that the category of people of African descent, as a global category, does not replace national or local categories and concepts or self-identifications, but rather exists in parallel.

Due to the curiosity of individuals about their family histories and ancestral linkages, consumer DNA tests have emerged as a popular avenue for people who are interested in exploring their (personal) family histories and ancestral linkages, complementing the scientific and medical applications of genetics. Companies who offer these tests advertise with the promise of enabling individuals to dive into their familial history in terms of geography and even infer their ethnic and racial origin (Nelson 2008, 761). These tests seem to allow individuals to set out on a personal genetic quest, seeking a deeper understanding of their roots. Advancements in genetics and the increasing prevalence of genetic technology have contributed to the high demand for these tests (Abel 2018, 6; Brubaker 2015, 69–73; Nelson 2008, 759). The African Diaspora has been identified as a specific target market for companies serving this demand. This can be ascribed to people of African descent seeking knowledge of their preslavery or precolonial origins and a desire to reconnect with their lost heritage as well as to fill in the gaps in their identities (Abel 2018, 23; Baylis 2003, 143; Hacking 2006, 87; Jackson and Borgelin 2010, 76; Nelson 2008, 764). These circumstances arise since individuals seeking to trace their African ancestry beyond the nineteenth century present significant challenges due to the lack of written documents and records from the era of the slave trade (Nelson 2008, 764). However, with the promises made by consumer DNA testing companies to unveil the "missing pieces of one's ancestors puzzle,"[8] individuals are offered an opportunity "to find the missing pieces of your identity … [and] reconnect with Africa [to] find their lost heritage."[9]

The procedure for consumer DNA testing typically involves several steps. The test taker orders a test kit from the company, which is then delivered to their home. The individual then takes the test by collecting a sample, for example through a swab taken from the mouth and throat region.[10] Subsequently, the collected sample is returned to the testing company for analysis. The company compares and examines the DNA sample with the existing DNA profiles in their database. DNA testing companies use various methods to determine the ancestry of their customers. I will highlight three types, distinguished by the information they provide and their social significance, particularly regarding African ancestry as outlined by Nelson (2008, 765). Ethnic lineage testing utilizes the Y-chromosome, passed down from the paternal side to males, and the mitochondrial DNA (mtDNA), passed down maternally to both male and female offspring and exhibiting a higher level of variation. By analyzing the Y-DNA and mtDNA, the testing company can make inferences about nation-states or ethnic groups. Suppose an individual's submitted sample matches a minimum number of genetic markers in the database: in this case, it can be stated that the person shares distant maternal or paternal ancestors with the individual who provided the reference sample for the database. This process enables the identification of genetic similarities between individuals, such as determining whether the two sample providers share, for instance, genetic ancestry with the Yola

8 https://www.afrorootsdna.com/lets-stay-in-touch.

9 https://africanancestry.com.

10 Some companies also use saliva samples for DNA testing that is done in a test tube (see, for example, https://support.ancestry.com/s/article/Collecting-a-Saliva-Sample.

People residing in Sierra Leone. Spatiotemporal testing, by contrast, can identify an individual's ancestral and geographical roots and the timeframe in which they exist. For example, the haplogroup obtained from mtDNA can indicate that their ancestors lived in Africa approximately 60,000 years ago. Thus, it determines the depth of time to which the lineage's ancestry can be traced back. The third method that comes into play can be referred to as racioethnic compositive. This is performed by analyzing autosomal DNA, which is unique to each individual and determines their genetic makeup. It provides information about an individual's ancestries in terms of admixtures. There are four standard categories: African, Native American, Asian, and European. Through this testing, a test taker can learn, for example, that they are 80 percent African and 20 percent Native American (Brubaker 2015, 72; Nelson 2008, 765; Skinner 2006, 460; Via, Ziv and Burchard 2009, 4).

These three types of ancestry tests generate distinct insights concerning ancestral origins. Despite their differences, they all share a common objective of tracing ancestral lines into the past, albeit from varying perspectives. This process establishes a connection between present-day individuals and their historical roots. Notably, this has significant implications for individuals of African descent. The validation of African descent through ancestry testing encompasses social dimensions. Zerubavel (2000) emphasizes that the validation of descent, exemplified by these tests, can foster a sense of identity. Because

> [w]hen such "descent" can be articulated in biological terms, especially in terms of blood. In other words, social reproduction is even more symbolically effective when it involves elements of actual biological reproduction. The symbolic significance of blood … establish historical continuity is most pronouncedly evident in essentialist narratives that try to portray the contact between past and present objects as more "real." (Zerubavel 2000, 361)

For individuals of African descent, knowledge of ancestry plays a crucial role in this regard, going beyond a narrative that addresses issues of personal identity, such as *who they are* and *where they come from* (Brubaker 2015, 69; Hacking 2006, 88). It also fosters a connection to the historical narrative of slavery, enslavement, and forced migration of Africans from the African continent. Moreover, ancestry tests have the potential to destabilize "taken-for-granted identities" (Brubaker 2015, 69). For instance, when an African American person learns about their African roots belonging to a specific ethnic group, they may adopt and integrate aspects of that culture or even change their name to claim their African identity (Brubaker 2015, 74; Schwartz-Marín and Wade 2015, 897; Skinner 2006, 460). In this context, ancestry is socially constructed as individuals actively choose which ancestral line to explore or express (Roth and Ivemark 2018). For instance, individuals who discover that they possess 30 percent Asian, 10 percent American, and 60 percent African ancestry, could also equally choose to identify themselves primarily as Asian. Thus, unlike in the scientific or medical fields, these tests provide room for interpretation and personal choice, granting them a more subjective character (Nelson 2008, 69). In addition, ancestry testing has political implications, particularly in relation to claims for citizenship or other political rights.[11] In the case of people of African descent, in particular, proof of being the descendants of enslaved people could be a potential tool in legal actions to seek reparations. The growing prominence of reparations movements advocating for legal reparations based on slave-descendant evidence may underscore the political importance of these tests in validating claims for reparatory justice.

The legal and political relevance of DNA ancestry tests for consumers has become increasingly clear since they came onto the market. Plaintiffs and activists have used these tests to prove their ancestry from enslaved Africans, e.g., in *Paellmann v. FleetBoston*. This landmark slavery trial took place in 2002 against FleetBoston, the seventh largest bank in the United States, which played a significant role in the institution of slavery. Deadria Farmer-Paellmann sued the company on behalf of her ancestors, who were forced into the slavery from which the company profited, for conspiring with slave traders and knowingly committing a crime against humanity (*Farmer-Paellmann vs. FleetBoston Financial Corp.*, 2002).

This case represents a milestone in using genetic ancestry testing as evidence. The plaintiff, Farmer-Paellmann, turned to the company African Ancestry and used their test results as evidence in court. The DNA results provided by African Ancestry indicated a direct link: "The plaintiffs' filing declared that 'DNA testing proven beyond a doubt' that there was a fiduciary relationship between the plaintiffs' ancestors and the defendants' companies" (Nelson 2016, 131). Moreover, this case is part of a broader historical-political trend: the plaintiffs were able to rely on the international human rights laws which were available after the formation of the United Nations. This legal framework, which declared slavery as a crime against humanity, was central to the Farmer-Paellmann case (Nelson 2016, 108; Loth 2020, 197).

In 2004, another significant case emerged when the descendants of enslaved Africans filed a lawsuit amounting to one billion dollars against the British company Lloyds of London. The company insured enslaved Africans as part of the ship's cargo. In addition, much of Lloyd's business also included grain such as wheat and rice, industrial crops such as cotton or rubber, luxury foods such as sugar, tea, coffee, and tobacco, and minerals such as copper and gold, produced by enslaved and unfree people.[12] Black leaders and activists who were involved in the case used DNA technology to link themselves to the recorded slave ships.[13]

Another example is the case of Georgetown University, which sold 272 enslaved persons to plantations in the Deep South. Georgetown, which was run by Jesuit priests, held hundreds of enslaved people. The Harvard Slavery Remembrance Program and the

---

11 The relevance of DNA testing goes beyond African descent and includes various other groups. These include Native Americans in the United States, where such tests can be important in determining tribal affiliation. For the concept of indigeneity, moreover, DNA testing transcends political claims. It includes attempts to reconstruct prehistoric migrations and settlement patterns. In India, identity claims are closely intertwined with discussions of origin and antiquity, which is further complicated by the complexity of caste systems (Brubaker 2015, 78–79).

12 https://www.lloyds.com/about-lloyds/history/the-trans-atlantic-slave-trade/lloyds-marine-insurance-and-slavery.

13 https://www.theguardian.com/money/2004/mar/28/insurance.usnews.

Georgetown Memory Project identified more than 10,000 descendants of enslaved people related to Georgetown University.[14] The project used two methods to identify the descendants of enslaved Africans from Georgetown. The first method used genealogy research in which family trees were built based on birth certificates, death records, and other documents. The second method employed DNA ancestry testing, where potential descendants underwent a DNA ancestry test and discovered that they may be genetically related to other descendants of the Georgetown enslaved people.[15] Consequently, Georgetown University established a Reconciliation Fund, which awards 400,000 dollars annually to community-based projects that "directly impact Descendant communities whose ancestors were enslaved on the Jesuit plantations."[16] The use of ancestry testing for legal claims is, nevertheless, only one possible motivation. The primary motivation for individuals who take these tests is often to explore and gain insight into their family history.

Conclusion

This article examined the term "people of African descent" and explored its meaning within the context of political debates. Additionally, it analyzed the specific discourse surrounding the concept of descent and its relationship to ancestry, demonstrating that "people of African descent" emerged as an international population descriptor only at the beginning of the twenty-first century to characterize individuals in the African Diaspora. The unifying characteristic lies in the dispersion of people of African descent across the world due to historical circumstances such as enslavement, colonization, and the slave trade. Furthermore, individuals of African descent worldwide share the common experience of racism and racial discrimination.

Despite variations in ancestry, geographical origins, and cultural diversity among people of African descent, their shared historical experience of enslavement and forced migration fosters a sense of belonging that binds them together as an imagined community (Anderson 2006; Kent, Ventura, and Wade 2014). This connection is established based on individuals' subjective belief in a shared past, aligning with Weber's characterization of a social group. Using the concept of descent by political institutions such as the United Nations serves a specific purpose. It highlights the historical circumstances of forced migration and disconnection from a specific region or homeland. Furthermore, the term "descent" has been a long-standing element of political discourses, employed to address and describe social groups vulnerable to discrimination and marginalization.

This remains relevant because institutions such as the United Nations are "specified agencies" (Strang and Meyer 1993, 499) that provide, in their fight against racism and the associated semantic structure, concrete behavioral scripts embedded in a specific language. These scripts serve as orientation patterns for the global community. In this

14 https://ethics.harvard.edu/people/richard-j-cellini.

15 https://www.washingtonpost.com/local/social-issues/they-thought-georgetowns-missing-slaves-were-lost-the-truth.

16 https://www.georgetown.edu/slavery.

way, they not only reproduce certain terminologies and definitions, but also constitute them. The negotiation process from which the term emerges is anchored in linguistic discourses, legal texts, and political documents and is also disseminated through them. Furthermore, social scientists have played an important role in the debates from which the term "descent" has emerged, and it is particularly relevant in political discussions, especially when it comes to people of African descent. Descent underscores the collective historical events of slavery, enslavement, forced migration, and the shared experiences with racism and racial discrimination that have profoundly shaped the lives of individuals of African descent.

While descent emphasizes the historical circumstances, ancestry represents a more individually centered concept. Ancestry is relevant among people of African descent in the sense that DNA ancestry testing can help to clarify the narrative and history of their ancestors using genetic analysis (Abel 2018). Furthermore, it is also a tool not primarily aimed to determining geographic origin, as many test takers already have assumptions about or knowledge of their roots on the African continent (Nelson 2008, 771–775). Instead, it validates narratives of enslavement, slavery, and forced migration. The concepts of ancestry are intermingled with the category of African descent with regards to identity exploration. However, the determining factor for belonging is not necessarily actual ancestry: that these tests are so popular suggests that what is being sought is not actual (genetic) unquestionable ancestry but more a confirmation of a shared history, and thus, DNA tests can function to support social affiliation.

References

Abel, Sarah. 2018. "Of African Descent? Blackness and the Concept of Origins in Cultural Perspective." *Genealogy* 2 (1): 1–24. DOI: https://doi.org/10.3390/genealogy2010011.

Abel, Sarah, and Hannes Schroeder. 2020. "From Country Marks to DNA Markers: The Genomic Turn in the Reconstruction of African identities." *Current Anthropology* 61 (S22): 198–209. DOI: https://doi.org/10.1086/709550.

African Ancestry. Discover Your African Roots: By Black People, for Black People. Accessed May 12, 2024. https://africanancestry.com.

AfroRoots DNA. Let's Stay in Touch. Accessed May 12, 2024. https://www.afrorootsdna.com/lets-stay-in-touch.

Anderson, Benedict. 2006. *Imagined Communities: Reflections on the Origin and Spread of Nationalism*. London: Verso.

Ancestry. Collecting a Saliva Sample, Accessed May 12, 2024. https://support.ancestry.com/s/article/Collecting-a-Saliva-Sample?language=en_US.

Baylis, Francois. 2003. "Black as Me: Narrative Identity." Developing World *Bioethics* 3 (2): 142–150. DOI: https://doi.org/10.1046/j.1471-8731.2003.00070.x.

Bennani, Hannah, and Marion Müller. 2018. "'Making up People' Globally: Institutionalizing 'Indigenous Peoples' and 'Persons with Disabilities' as Global Categories." *Zeitschrift für Soziologie* 47 (5): 306–331. DOI: https://doi.org/10.1515/zfsoz-2018-0120.

Brubaker, Roger. 2015. *Grounds for Difference*. Cambridge, MA: Harvard University Press.

Byeon, Yen Ji Julia, Rezarta Islamaj, Lana Yeganova, et al. 2021. "Evolving Use of Ancestry, Ethnicity, and Race in Genetics Research: A survey Spanning Seven Decades." *The American Journal of Human Genetics* 108 (12): 2215–2223. DOI: https://doi.org/10.1016/j.ajhg.2021.10.008.

Dauda, Bege, Santiago J. Molina, Danielle S. Allen, et al. 2023. "Ancestry: How Researchers Use It and What They Mean by It." *Frontiers in Genetics* 14 (Jan 2023): 1–13. DOI: https://doi.org/10.3389/fgene.2023.1044555.

Draper, Nicolas. Lloyd's, Marine Insurance and Slavery, Accessed August 12, 2024. https://www.lloyds.com/about-lloyds/history/the-trans-atlantic-slave-trade/lloyds-marine-insurance-and-slavery.

Durkheim, Emil. 1933. *The Division of Labor in Society*. Translated by George Simpson. Glencoe, IL: Free Press.

Emigh, Rebecca Jean, Dylan Riley, and Patricia Ahmed. 2015. "The Racialization of Legal Categories in the First U.S. Census." *Social Science History* 39 (4): 485–519. DOI: https://doi.org/10.1017/ssh.2015.69.

Farmer-Paellmann v. FleetBoston Financial Corp., CIVIL ACTION # CV 02 1862, CLASS ACTION. Accessed May 16, 2024. https://casetext.com/case/farmer-paellmann-v-fleetboston-financial-corp.

Geertz, Clifford. 1973. *The Interpretation of Cultures*. New York: Basic Books.

Georgetown University. Georgetown Reflects on Slavery, Memory, and Reconciliation. Accessed July 19, 2024. https://www.georgetown.edu/slavery.

Harvard University. Richard J. Cellini. Accessed July 25, 2024. https://ethics.harvard.edu/people/richard-j-cellini.

Hacking, Ian. 2006. "Genetics, Biosocial Groups & the Future of Identity." *Daedalus* 135 (4): 81–95. DOI: https://doi.org/10.1162/daed.2006.135.4.81.

Jackson, Fatimah L. C., and Latifa F. J. Borgelin. 2010. "How Genetics Can Provide Detail to the Transatlantic African Diaspora." In *The African Diaspora and the Disciplines*, edited by Tejumola Olaniyan and James H. Sweet, 75–100. Bloomington: Indiana University Press.

Jackson, Maurice. 1982. "An Analysis of Max Weber's Theory of Ethnicity." *Humboldt Journal of Social Relations* 10 (1): 4–18, https://www.jstor.org/stable/23261855.

Keane, David. 2005. "Descent-based Discrimination in International Law: A Legal History." *International Journal on Minority and Group Rights* 12 (1): 93–116, https://www.jstor.org/stable/24675289.

Keane, David, and Annapurna Waughray, eds. 2017. *Fifty Years of the International Convention on the Elimination of All Forms of Racial Discrimination: A Living Instrument*. Manchester: Manchester University Press.

Kent, Michael, Ricardo Ventura Santos, and Peter Wade. 2014. "Negotiating Imagined Genetic Communities: Unity and Diversity in Brazilian Science and Society." *American Anthropologist* 116 (4): 736–748. DOI: https://doi.org/10.1111/aman.12142.

Loth, Marc. 2020. "How Does Tort Law Deal with Historical Injustice? On Slavery Reparations, Post-Colonial Redress, and the Legitimations of Tort Law." *Journal of European Tort Law* 11 (3): 181–207. DOI: https://doi.org/10.1515/jetl-2020-0141.

Magubane, Zine. 2017. “Following ‘the Deeds of Men’: Race, ‘the Global,’ and International Relations.” In *Global Historical Sociology*, edited by Julian Go and George Lawson, 101–123. Cambridge: Cambridge University Press.

McCoy, Terrence. 2018. “They Thought Georgetown University’s Missing Slaves Were ‘lost.’ The Truth Was Closer to Home Than Anyone Knew.” *The Washington Post*, 28 April 2018. Accessed August 21, 2024. https://www.washingtonpost.com/local/social-issues/they-thought-georgetowns-missing-slaves-were-lost-the-truth-was-closer-to-home-than-anyone-knew/2018/04/28/074beb66-3e65-11e8-a7d1-e4efec6389f0_story.html.

Mills, Charles W. 1997. *The Racial Contract*. Ithaca: Cornell University Press.

Nelson, Alondra. 2008. “Bio Science: Genetic Genealogy Testing and the Pursuit of African Ancestry.” *Social Studies of Science* 38 (5): 759–783, https://www.jstor.org/stable/25474607.

Nelson, Alondra. 2016. *The Social Life of DNA: Race, Reparations, and Reconciliation after the Genome*. Boston: Beacon Press.

Paschel, Tianna. 2013. “‘The Beautiful Faces of My Black People’: Race, Ethnicity and the Politics of Colombia’s 2005 Census.” *Ethnic and Racial Studies* 36 (10): 1544–1563. DOI: https://doi.org/10.1080/01419870.2013.791398.

Reiter, Bernd, and Kimberly Eison Simmons, eds. 2012. *Afro-Descendants, Identity, and the Struggle for Development in the Americas*. East Lansing: Michigan State University Press.

Rodney, Walter. 2000. “How Europe Became the Dominant Section of a World-wide Trade System.” In *Caribbean Slavery in the Atlantic World: A Student Reader*, edited by Verene Sheperd and Hilary Beckles, 2–11. Kingston, Jamaica: Ian Randle Publishers.

Roth, Wendy D., and Biorn Ivemark. 2018. “Genetic Options: The Impact of Genetic Ancestry Testing on Consumers’ Racial and Ethnic Identities.” *American Journal of Sociology* 124 (1): 150–184. DOI: https://doi.org/10.1086/697487.

Schramm, Katharina. 2016. *African Homecoming: Pan-African Ideology and Contested Heritage*. London: Routledge.

Schwartz-Marín, Ernesto, and Peter Wade. 2015. „Explaining the Visible and the Invisible: Public Knowledge of Genetics, Ancestry, Physical Appearance and Race in Colombia.” *Social Studies of Science* 45 (6): 886–906. DOI: https://doi.org/10.1177/0306312715621182.

Skinner, David. 2006. “Racialized Futures: Biologism and the Changing Politics of Identity.” *Social Studies of Science* 36 (3): 459–488, https://www.jstor.org/stable/25474453.

Starr, Paul. 1992. “Social Categories and Claims in the Liberal State.” *Social Research* 59 (2): 263–295, https://www.jstor.org/stable/40970693.

Strang, David, and John W. Meyer. 1993. “Institutional conditions for diffusion.” *Theory and Society* 22 (4): 487–511, https://doi.org/10.1007/BF00993595.

Telles, Edward, and Tianna Paschel. 2014. “Who is Black, White, or Mixed race? How Skin Color, Status, and Nation Shape Racial Classification in Latin America.” *American Journal of Sociology* 120 (3): 864–907. DOI: https://doi.org/10.1086/679252.

Tsri, Kwesi. 2016. “Africans are Not Black: Why the Use of the Term ‘Black’ for Africans Should Be Abandoned.” *African Identities* 14 (2): 147–160. DOI: https://doi.org/10.1080/14725843.2015.1113120.

Turner, Michael (2002): "The Road to Durban—And Back." *NACLA Report on the Americas* 35 (6): 31–35. DOI: https://doi.org/10.1080/10714839.2002.11722524.

UN Commission of Human Rights. Identification and Definition of "People of African Descent" and How Racial Discrimination against Them Is Manifested in Various Regions: Working Paper Prepared by Ambassador P. L. Kasanda. E/CN.4/2003/WG.20/WP.3 (28 January 2003).

UN Committee on the Elimination of Racial Discrimination (CERD). Concluding Observations, 58th session, 6–23 March 2001. CERD/C58/Misc.17/Rev.3.

UN Committee on the Elimination of Racial Discrimination (CERD). Summary Record of the 1531st Meeting, 61st Session, 9 August 2002. CERD/C/SR.1531 (16 August 2002).

UN Committee on the Elimination of Racial Discrimination (CERD). CERD General Recommendation XXIX on Article 1, Paragraph 1, of the Convention (Descent). 1 November 2002.

UN Economic and Social Council. Prevention of Discrimination: Discrimination Based on Work and Descent. E/CN.4/Sub.2/2003/24 (26 June 2003).

UN General Assembly. International Convention on the Elimination of All Forms of Racial Discrimination. A/RES/2106(XX) (21 December 1965).

UN General Assembly. Reports of Preparatory Meetings and Activities at the International, Regional and National Levels: Report of the Regional Conference of the Americas, Santiago, Chile, 5–7 December 2000. A/CONF.189/PC.2/7 (24 April 2001).

UN Latin America-Caribbean Regional Seminar. Implementation of Provisions of the Durban Programme of Action for Specific Groups of Victims: Africans and People of African Descent: Background Paper Prepared by Gustavo Makanaky Córdoba. HR/MEX/SEM.1/2002/BP.2 (4 June 2002).

UN World Conference. Report of the World Conference against Racism, Racial Discrimination, Xenophobia and Related Intolerance, Durban, 31 August–8 September 2001. A/CONF.189/12.

Via, Marc, Elad Ziv, and Esteban González Burchard. 2009. "Recent Advances of Genetic Ancestry Testing in Biomedical Research and Direct to Consumer Testing." *Clinical Genetics* 76 (3): 225–235. DOI: https://doi.org/10.1111/j.1399-0004.2009.01263.x.

Walsh, Conal. 2004. "Slave Descendants Sue Lloyd's for Billions." *The Guardian*, 28 March 2004: Accessed August 21, 2024. https://www.theguardian.com/money/2004/mar/28/insurance.usnews.

Weber, Max. 1978. *Economy and Society: An Outline of Interpretative Sociology*. Translated by Guenther Roth and Claus Wittich. Berkeley: University of California Press.

Whitten, Norman E., and Arlene Torres, eds. 1998. *Blackness in Latin America and the Caribbean: Social Dynamics and Cultural Transformations: Eastern South America and the Caribbean*, Vol. 2. Bloomington: Indiana University Press.

Zerubavel, Eviatar (1996): "Lumping and Splitting: Notes on Social Classification." *Sociological Forum* 11 (3): 421–433. DOI: https://doi.org/10.1007/BF02408386.

Zerubavel, Eviatar. 2000. "Social Identity and Social Descent: Some Sociological Reflections on Population Genetics." *Sociological Forum* 15 (2): 361–366. DOI: https://doi.org/10.1023/A:1007581827058.

Mapping Genomes through History[1]

Using Global Ancestry in a Contemporary Population Genetics Approach to Dengue

Ricardo Gomes Moreira

Abstract *In this chapter I examine the operational value of the concept of ancestry for a group of population geneticists from a Portuguese laboratory working closely with a Cuban team of virologists. Developed as an ethnomethodological inquiry about scientific concepts sustaining the work of these geneticists, it attempts to understand the rationale behind the use of broad classification categories for analyzing the genetic structure of populations and identifying potential genotype-phenotype associations of biomedical value. Arguing that there is an evolutionary dimension to the concept of ancestry that is consistently present in geneticists' naming practices, I propose that it is precisely this link of ancestry with evolutionary theory that enables geneticists to follow the historical trajectories of populations in order to gain an understanding of the selective processes that may underpin the role of specific genetic traits in disease and its clinical outcomes.*

Introduction

With the rise of biocomputational power and new automated (high-throughput) DNA sequencing technologies, the notion of ancestry has in recent decades become a central concept in human genetics. Particularly useful for population geneticists as a category for addressing biological diversity, "ancestry" enables laboratory practitioners to distinguish between genetic profiles of individuals and populations. This study describes and analyzes the practical and symbolic dimensions of this concept in a medical genetics context focused on dengue research, grounding the analysis in ethnographic laboratory fieldwork and qualitative analysis of biomedical and population genetics literature.

1 The research and writing of this chapter were funded by the European Research Council (ERC) under the European Union's Horizon 2020 research and innovation programme, grant agreement No. AdG 695573, "The Colour of Labour," coordinated by Cristiana Bastos; and by National Funds through the Portuguese funding agency FCT, project "GeneAntropMag" (CIRCNA/CIS/0142/2019), coordinated by Luísa Pereira.

Aiming to clarify how geneticists understand "ancestry" as both a methodological and interpretative tool to account for certain dimensions of human diversity, it seeks a critical understanding of the logics and rationale underlying its use.

In recent decades, the idea of biogeographic ancestry has consolidated in population genetics as a core conceptual tool linking the historical and geographical origins of one's ancestors. It has become particularly useful for tracing remote forebears or mapping ancient migratory routes beyond memory. Several STS studies have examined how this concept is applied in geneticists' work. Some scholars warn of the risks of using "ancestry" in biomedical and genetics contexts in ways that may be interpreted as racial.[2] Others show that its conceptual flexibility and variable definition depend on methodological configurations and research designs, and that its referents and categorical distinctions hinge on related concepts such as "population" (Fujimura and Rajagopalan 2011; Panofsky and Bliss 2017; Lipphardt 2017; Byeon et al. 2021). Fujimura and Rajagopalan (2011) have persuasively demonstrated the analytical value of an ethnographic approach that considers this concept from the perspectives of scientists engaged in laboratory work. Examining how classificatory concepts such as ancestry operate in laboratory knowledge production reveals their contingent nature, shaped by technology, methods, research design, and—not least—theory.

Some STS scholars engaged in sustained critical examination of genetically applied notions of ancestry have recently produced scholarship focused on the North American context. They highlight the entwining of genetic ancestry classificatory concepts with US sociopolitical notions of race and their propensity to slip from laboratory into identitarian and politically charged contexts, where dynamics of power and relatedness may be intentionally or unintentionally reinforced by scientific practice (e.g., Nash 2015; Nelson 2016; Fullwiley 2024).

In this chapter, I propose to explore how the concept of biogeographic ancestry is used in a population genetics laboratory according to its practical applicability to the scientific problem of dengue, its objects of study, and various aspects of scientific work. My analysis draws on ethnographic methodology, consisting mainly of interviews with geneticists, direct observations, documentation of their views and commentaries on their own work, and analysis of publications by a population genetics team from Portugal. An ethnomethodological approach was adopted to yield insights into both the rationale for using a classificatory form of genetic ancestry and its possibilities across different stages of research: from planning, through collection building and statistical analysis, to comparison and writing of results.

The chapter examines conceptual and practical configurations of genetic ancestry, understood as procedurally formed by two mutually constitutive aspects—molecular and sociopolitical—combined to produce a scientific concept of analytical value and identitarian character.

The first section describes the configuration of a medical genetics project on dengue, in which this concept of ancestry was modeled and applied to name specific popula-

2 Cogent discussions on this topic include Fullwiley (2008, 2014), Gannett (2014), Wade et al. (2014), and Lewis et al. (2022). An earlier, important contribution to the debate was the edited volume by Koenig, Soo-Jin Lee, and Richardson (2008).

tion groups. The second explores how ancestry became useful as a research tool, associated with traits of immunological resistance and susceptibility, and gained significance through translation from epidemiological history and virology into population genetics.[3] The third identifies the importance of ancestry in linking laboratory research to evolutionary theory, enabling theoretical interpretation of results and connecting analyses to other projects that may lead to broader generalizations and hypothesis formulation.

Cuba Meets Europe: The Making of a Medical Genetics Approach to Dengue

Dengue virus (DENV) is recognized by international organizations as one of the most significant arthropod-borne diseases and a major global public health concern, affecting nearly all tropical regions. In the 2000s, about 2.5 billion people in over one hundred countries were considered at risk, while incidence patterns shifted as the disease became endemic in several regions and rates rose alarmingly (Guzmán et al. 2010, 57; see also Jaenisch et al. 2013).

In 2012, the European Commission funded a major dengue research project, organized as three independent consortia involving multiple scientific institutions. One consortium, DENFREE (Dengue Research Framework for Resisting Epidemics in Europe), aimed to provide "a new strategy for surveillance," "better control of DENV transmission," and determine epidemic factors and trajectories, as well as the underlying factors of "pathogenesis with respect to viral and host factors that can predict disease severity and prepare for further development of new vaccines [and] antiviral compounds" (Jaenisch et al. 2013, 2). In short, DENFREE sought to explore the genetic basis of resistance and susceptibility to dengue and to identify genetic configurations causal to the immunological determinants of infection outcome (Jaenisch et al. 2013, 2).

Led by a French team from the Institute Pasteur, the consortium included the Pedro Kourí Institute of Tropical Medicine (PKI) in Cuba and a team of population geneticists from IPATIMUP, Portugal, invited for their expertise in applying populational methods to medical genetics. Their task was to investigate possible genetic causes for different immunological responses observed in Cuban epidemiological records, which had been previously studied by Cuban biomedical scientists and virologists.

Cuba was selected as a case study partly because of its epidemiological history. Dengue has long been present in the Americas and Caribbean, with reports as early as 1699 in Panama (Guzmán 2012). The year 1827 marked the start of regular epidemic records. In Cuba, outbreaks were reported in 1906 (Sierra et al. 2017) and 1944–45, but the large-scale epidemic of 1977—with more than 400,000 cases reported—is regarded as the beginning of the modern presence of DENV on the island (Guzmán 2012).

Over the last five decades, Cuba has built substantial clinical, epidemiological, and virological knowledge on dengue, most of it developed at the PKI in Havana. Following the epidemics of 1981 and 1997, Cuban researchers documented distinctive aspects of dengue incidence, promoting the country as a relevant case study (e.g., Guzmán 2012).

3 I use the notion of translation in the sense proposed by Michel Callon and John Law, as part of the scientific work of connecting different spheres of interest (Callon 1984; Callon and Law 1982).

Internationally, they highlighted successful public health interventions that prevented endemic dengue, eliminated transmission chains, and disseminated research results through organizations such as the Pan American Health Organization, the Special Programme for Research and Training in Tropical Diseases, and the Dengue Vaccine Initiative (e.g., Guzmán et al. 1999; Guzmán 2012).

Some of these results were especially relevant to population geneticists, notably factors and patterns of risk regarding the Cuban population. In 1999, in the *WHO Dengue Bulletin*, Maria Guzmán and her team put forward what they called "some interesting observations" about the dengue hemorrhagic fever (DHF) epidemics of 1981 and 1997 (Guzmán et al. 1999), arguing that the "Cuban experience is probably unique." Among these were empirical demonstrations of population behavior under epidemic contexts, the crucial role of "secondary infections" in disease outcome, and the relevance of "race" as an indicator of potential severity. Based on epidemiological data, they hypothesized that "black people were at a reduced risk to manifestations of [dengue shock syndrome (DSS)] as compared to white people,"[4] an argument traceable to early twentieth-century outbreak reports.

Cuban virologists further explored these observations in a 2007 paper in the *Archives of Virology*, specifically focusing "race" as "a risk factor for dengue hemorrhagic fever." They state that:

> Cuban DHF/ DSS outbreaks have provided evidence of a reduced risk of people of Negroid race for DHF/ DSS compared to those of Caucasoid race. These observations from Cuban dengue outbreaks have significant epidemiological interest, as the differences in susceptibility to DHF/ DSS among racial groups in Cuba coincide with that reported in African and Black Caribbean populations. (Sierra et al. 2007)

As we will see, these conjectures became the touchstone of geneticists' approach to the DENFREE scientific problem. It is precisely the translation of these epidemiological observations into a population genetics framework that this chapter will explore. The fact that these observations, concerning population groups with different "racial" profiles, hold valuable hints for contemporary geneticists raises considerable questions: Why would differential skin pigmentation or phenotypic constitutions, considered in racial terms like "Caucasoid" or "Negroid," serve as an empirical basis for population genetics? What are geneticists seeking in associating "racial" profiles with differences in clinical outcomes and dengue severity?

Such questions, which require understanding how population genetics practices are applied in biomedical research, speak to the epidemiological and medical approach of the DENFREE consortium and become relevant within the European dengue research agenda. The Portuguese team's background shows important research lines in demographic history and medical genetics. Since the 2000s, this lab has extensively used mitochondrial DNA and Y-chromosome analyses to study genetic structure and variability in Portuguese and Iberian contexts, as well as in African Portuguese-speaking countries,

4 DSS is the most severe, life-threatening form of dengue, which follows as a series of complications of Dengue hemorrhagic fever.

producing narratives on demographic history, migratory routes, and past population encounters across continental regions. In the following decade, whole-genome technologies expanded the practical scope of genomic analysis. Genome-wide association studies (GWAS), for instance, were devised to identify DNA regions linked to phenotypic variations. The lab's publishing history also shows methodological strength in analyzing genetic structures in diverse contexts. Since the early 2010s, several projects targeted highly diverse populations, consolidating genetic ancestry as a central concept.

The integration of IPATIMUP's population genetics laboratory into DENFREE fit the scientific interests of both sides. The team's expertise in population structure met the needs of the consortium, while the research problem—rooted in epidemiological and virological observations at the population scale—aligned with the geneticists' preferred approaches. Another factor in the success of this partnership lies in the technological and methodological capacity of contemporary genomics, which can statistically link genotypical and phenotypical dimensions at population scales. With computational and statistical tools, geneticists can propose relations between specific DNA loci and biological traits selected during sample collection. In this sense, Cuban epidemiologists and virologists provided the starting conditions for a research design in which population geneticists could examine Cuban epidemiological and virological findings genetically.

In sum, the DENFREE project analyzed here was built on close collaboration between a Cuban team of virologists and a Portuguese team of population geneticists, jointly seeking answers to scientific problems posed by dengue.

Genetic Ancestry: Reframing Natural History in Dengue Research

As argued above, approaching dengue as a study of its "pathogenesis with respect to viral and host factors," in order to "predict disease severity," and aiming for the "development of new vaccines [and] antiviral compounds," meant directing research toward understanding genetic causes of different immunological responses, specifically the resistance and susceptibility patterns reported in Cuban epidemiological records and studied by PKI virologists.

This research did not begin with DENFREE, however. Over the preceding three decades, Cuban scientists had produced extensive work, summarized in a 2012 article listing major results since the first epidemic of 1981 (Guzmán 2012). These included populational and genetic configurations relevant to DENFREE's design, such as the observation that DHF risk factors included "white skin color," or that "higher association with DHF was found for persons of European descent, with stronger memory T-cell response and cross-reactivity in white compared with black dengue-immune persons." Other results identified biomarkers and immunologically relevant genes (ibid.).

As argued above, Cuba's selection as a case study was due less to public health impact (other serious candidates for that matter existed) than to the prior biomedical work of its physicians and virologists. The phenotypic diversity of the Cuban population, it can well be argued, was equally crucial. This diversity, long part of Cuban clinical and epidemiological records, was described in DENFREE as "admixed composition" (Oliveira et al. 2018). Geneticists considered such "admixture" advantageous, as they understood it to be directly translatable into a genetic structure that could be empirically detected and then analyzed; this meant, in other words, that different populations with distinct genealogical histories and genetic profiles could be discerned within the contemporary Cuban population. Within DENFREE, an analytical framework was thus devised to relate population genetic structure (and ancestry categories) to observed clinical responses.

Thus, the DENFREE geneticists, using the classificatory concept of genetic ancestry, approached the problem as one of "ancestry-conferred DHF susceptibility/protection."[5] As they write: "admixed populations are a great advantage in cases of differential ancestry-conferred susceptibility/resistance to a disease through the use of admixture mapping" (Sierra et al. 2017, 3)—a methodological position that will be further considered here.

In molecular anthropology and human population genetics, genetic ancestry is often applied to statistically detect patterns of human biogenetic diversity, grounded in hereditary transmission and ascendancy in comparisons between population groups.[6] Despite being a concept developed and applied within the field of population genetics, the notion thus combines historical, sociopolitical, and molecular dimensions (Fujimura et al. 2010; Rajagopalan and Fujimura 2012).

I should note, however, that this dengue study illustrates the contingent character of genetic ancestry as it adapts to specific research goals. Unlike studies explicitly comparing groups defined along racial,[7] ethnic,[8] or national lines,[9] in which the idea of ancestry is thus superimposed on such sociopolitical labels in ways that "biologize" race or other identities (El-Haj 2007; Fullwiley 2007, 2008; Lipphardt 2017), DENFREE examined groups defined by clinical outcomes and probable hereditary variations explaining those outcomes. Here, researchers employed a notion of ancestry that linked "genes" to hypothetical past origins inferred from physicians' observations in order to identify and explain the relevance of these hereditary traits. Race was not the object of study but a shortcut toward the evolutionary theory of population genetics.

The link between ancestry categories and genetic material, traced through a populational approach, serves as a connection between specific ecogeographic environments and those molecular configurations that—because they constitute cases of variation—were likely positively selected over time under the influence of such environments. In seeking to understand the research process and its methodologies, I argue that, from the geneticists' perspective, identifying populations through ancestry categories is necessary to associate particular geographic and ecological origins with specific genetic features and to understand why such configurations have been transmitted and preserved across generations. Following the geneticists' theoretical position, this intermediate step—between identifying molecular (DNA) configurations related to clinical

5 See Sierra et al. (2017, 3). DHF and DSS are used as abbreviations for dengue hemorrhagic fever and dengue shock syndrome, respectively.

6 For an informative and detailed historical account about how these genetic variations have been studied by population geneticists and their sociotechnological aspects, see the study by Rajagopalan and Fujimura (2018).

7 See Gannett (2014) for a critique of the concept of global biogeographic ancestry.

8 See Lipphardt et al. (2021) for a critique of studies on vulnerable populations.

9 See, e.g., Novembre 2008.

outcomes and understanding the biological processes behind the genotype–phenotype nexus—makes the application of ancestry labels productive. These labels allow for the tracing of population histories and the formulation of evolutionary explanations for the causes, effects, and probable molecular pathways of certain genetic configurations and their phenotypic consequences. This rationale for ancestry categories, as a foundation for evolutionary explanation, is exemplified by the dengue research project discussed above.

In this case, Cuban samples were collected "during the 2006 dengue outbreak" and included "67 subjects from Havana ... and 70 subjects from Guantanamo" with clinical symptoms, plus 32 asymptomatic individuals from Havana and 16 from Guantanamo, for a cohort of 185 individuals infected with one of the dengue virus strains (Sierra et al. 2017, 14). These samples, from hospital institutions, were classified according to the reported clinical outcomes of patients, which in this project is the fundamental information for the collection, with the additional stipulation that *all* samples came from patients infected with DENV. The primacy of clinical outcomes as the first classificatory criterion, and the most fundamental aspect in forming the object of study, is reflected in Figure 1, where genetic diversity of samples classified by national labels and analyzed with ancestry components is shown next to groups classified by dengue outcome severity.

In addition to DENV patient samples, the historical and scientific background of dengue epidemiological and medical research in Cuba provided DENFREE with essential empirical foundations. On that basis, geneticists analyzed variants and evaluated the relative weight of two ancestries—African and European—in the target Cuban population's genetic structure. Drawing on historical medical sources and Cuban research, they tested the hypothesis that genetic differences might exist between groups with different proportions of these ancestries, and that these differences might influence susceptibility and resistance to dengue. The hypothesis to be tested was made clear by quoting an early medical report:

> As early as in 1906, it was reported that Cuban *dark-skinned* individuals showed a remarkable resistance against dengue disease compared with *light-skinned* individuals. This early observation was confirmed during the 1981 Cuban DHF/DSS epidemic of DENV-2 when *ethnicity* was recognized for the first time as a possible host risk factor and confirmed afterwards in several other dengue Cuban outbreaks. The low occurrence of dengue disease in *Haitians* and in *African populations* adds further support for this *ancestry* influence. (Sierra et al. 2017, 2–3, emphasis added)

This historical account and its implications for research design appear to have posed a terminological challenge to researchers, who in their published findings sometimes slip from the notion of ancestry toward ethnicity, skin color, and population designations. This is likely because historical records refer to what would be considered racial groups among Cuban populations in the early twentieth century. It is not clear why the notion of ethnicity emerged as an acceptable choice for geneticists, nor whether it was meant to identify broadly the physical characteristics or the genomically determined ancestry of Cubans that earlier scientific works would have described in racial terms.[10] However, as another DENFREE publication shows, the idea of "genetic diversity" guiding the research design is entirely at odds with the notion of ethnicity. In published articles, when presenting and discussing results, the notion of ethnicity yields to an idea of diversity defined explicitly around genetic ancestry, in the form of *global ancestry*.

In essence, global ancestry is conceptualized as a partition of contemporary human populations into four major continental categories of genetic ancestry: African, European, East Asian, and Native American (Figure 1). The question that arises is whether these continental categories diverge fundamentally from the nominally similar racial classifications used by physical anthropologists in past centuries, or whether they represent a continuation of that history of racial classifications.

Figure 1: Representation of four different categories of biogeographic ancestry in different populations. Cuban samples are shown on the left side of the graph under clinical labels corresponding to different forms of dengue severity and sampling locations (Havana: HC, HA, HH; Guantánamo: GC, GA, GH, GF). Each vertical line represents an individual genomic analysis: African ancestry in blue, European in red, Native American in yellow, and East Asian in orange.

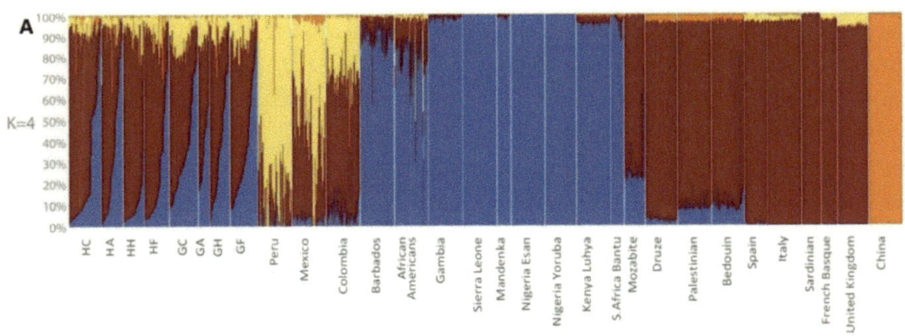

Source: Sierra et al. 2017.

I believe these are distinct types of classification. To answer this question, we must first understand the *nature of the "objects" to which these categories refer and the relevance of using such classifications, given the research questions and objects of study*. What do geneticists (as in Sierra et al.) mean by global ancestry, and why have they adopted a fourfold continental geographic division?

A crucial point in geneticists' work is that these categories are not meant to represent real populations but rather segments of DNA; and if they do not represent populations,

10 Terminological uncertainty sometimes arises and both terms are used interchangeably, in DENFREE's published works, as if they were equivalent, which can be confusing: "If an *ancestry* is associated with increased susceptibility, the disease locus will present a significant higher proportion of that ancestry in cases versus controls; when an *ethnicity* is protective, the disease locus will present a significantly lower proportion of that ancestry in cases versus controls" (Oliveira 2018, 3; emphasis added).

perhaps they should not be mistaken for racial labels. One reason to avoid such confusion is that, unlike the twentieth-century anthropological concept of "race," which defined broad population groups with fixed and unmixed hereditary biological traits (essentially a biological equivalent of "subspecies"), global ancestry categories are assigned to certain individual genomic features derived from presumed ancestral populations. In contrast to the physical and visual basis of racial classifications, geneticists present global ancestry probabilistically—as statistically inferred geographic origins for segments of DNA in present-day individuals, based on similarities with analogous segments found in populations from those regions.

While endorsing population-based methods and broad ancestry classifications, one interviewed researcher argues that, in studying genetic configurations—especially those with medical relevance—the use of labels may reflect population identity, but the use of ancestry is justified by its heuristic utility. That is, ancestry categories help differentiate historical trajectories of genes, which is essential for situating genetic configurations in time and geography. In this sense, populational approaches have adopted methodological designs that combine genomic sequencing technologies with historical and sociographic data to generate new biological knowledge with medical value. As one geneticist argued:

> Finally, we are able to say something about the selection effects. In early days we didn't have selection—a big concept that was introduced in biology and after became highly important in genetics; but we didn't have the necessary tools to study selection. Now, the chip gave us the power to study the entire genome in a random fashion. […] I don't think we are already at the level of personalized medicine. I think there is an intermediary step that implies studying genetics with biomedical implications at the level of population groups. (interview with DENFREE geneticist)

These researchers argue that although admixture has often been viewed as a "major confounding factor" (Sierra et al. 2017, 3; see also Price et al. 2006; Fujimura and Rajagopalan 2011), admixed populations may now offer a valuable research opportunity if admixture mapping is used as an auxiliary methodological tool. Rather than relying solely on principal component analysis (PCA) to "correct for stratification in the cohort" and adjust samples' "genotypes and phenotypes by amounts attributable to the ancestry" (Sierra et al. 2017, 2)—a procedure that makes cohort samples more comparable to controls—*ancestry mapping* is employed to exploit stratification itself (i.e., the internal variation of cohort samples), particularly when researchers suspect that phenotype and gene variability will show a strong link with ancestry.

When geneticists refer to the "great advantage" of admixed populations in this type of analysis, they mean that the technique of "admixture mapping" can yield promising results when applied to populations descended from two or more ancestral groups of distinct geographic origins. Given that different medical outcomes reflect phenotypic variations—such as susceptibility or resistance factors linked to genetic constitutions—admixture mapping is considered valuable for associating these phenotypic traits with specific ancestries. It is especially useful when the target populations exhibit a high degree of admixture. In short, admixture mapping is presented as one of the most efficient methods for identifying which ancestries confer resistance and which confer greater susceptibility to the pathogen. As Sierra at al. (2017, 3) note: "It has been shown that this test is statistically more powerful than traditional GWAS: around 250 samples can provide a 60% power to detect a two-fold risk due to ancestry, compared to the thousands of samples required in GWAS."

However, not long before this project began, admixed populations were often seen as obstacles to genetic analysis, particularly in genome-wide association studies (GWAS).[11] GWAS, used to identify genes of clinical or epidemiological relevance, typically required prior analysis and careful selection of samples so that cohorts and controls shared comparable genetic structures. Without such precautions, DNA regions or loci identified as statistically significant for disease outcomes could reflect structural genetic differences between those two sets of samples rather than meaningful trait variation.

Fujimura and Rajagopalan (2011) emphasized the importance of calibration tools in GWAS to address population substructure associated with different ancestries. This was the function of the software Eigenstrat, which geneticists applied to adjust data for ancestry. The adjustment allowed aggregation of samples sharing the same ancestry—and thus the same "population substructure"—ensuring that GWAS could identify genome regions genuinely associated with disease, rather than those reflecting different ancestry substructures of the sampled populations and controls. The central premise of Eigenstrat's statistical algorithm, as developed by Price et al. (2006) and discussed by Fujimura and Rajagopalan (2011, 12–15), is that distinct ancestry structures correspond to distinct genetic substructures (Price et al. 2006; Fujimura and Rajagopalan 2011, 12–15). Once principal component analysis (PCA) and Eigenstrat remove ancestry-related stratification within population samples, it becomes possible to disregard population labels, ancestry information, or identity categories in direct cohort-control comparisons (Fujimura and Rajagopalan 2011, 16).

But whether retaining or removing ancestry information, the relationship between population genetic structure and ancestry remains central to population genetics. With the growing volume and diversity of reference population data, it has become increasingly feasible to infer individual ancestry through genomic analysis. As shown in ancestry mapping, the transaction between ancestry and genomic information—what Fujimura and Rajagopalan call classificatory "slippage" (2011, 14–16)—moves in both directions: from molecular data to population ancestry, and from population ancestry back to DNA as molecular ancestry. This dynamic has significant epistemic implications for understanding how "genetic ancestry" functions as a classificatory tool.

Mapping Ancestry: Tracing the Historical Temporalities and Geographies of Dengue Immunity

A brief general description of how ancestry mapping operates within the framework of populational research helps clarify how ancestries are attributed to specific portions of DNA, producing a chromosomal map. In laboratory practice, "admixture mapping"

11 GWAS is a common technique used in medical genetics to identify loci or genes with potential clinical significance.

is conducted after the standard sequencing of somatic DNA. Once genomic data are obtained for each individual sample, geneticists use computational analysis to identify chromosomal sections based on comparative differences, assigning each section an ancestry label from a predefined set relevant to the research question. These labels are applied using ancestry informative markers (AIMs)—previously identified genetic (molecular) locations that statistically vary across regional populations and are considered significantly more prevalent in certain populations.

Figure 2: Schematic representation of "admixture mapping" in chromosome 9, "comparing patient and control groups." Sections colored in orange (here: dark grey) indicate sections of the chromosome associated with African ancestry. The graphic on the right displays a line graph measuring statistical standard deviation in African ancestry (Y-axis) along chromosome 9 (X-axis), represented as an ordered sequence of hundreds of millions of base pairs. It shows significant deviation values—exceeding 3 standard deviations (3SD)—near the end of the chromosome, around position 1.4, corresponding to base pairs close to the 140,000,000th position.

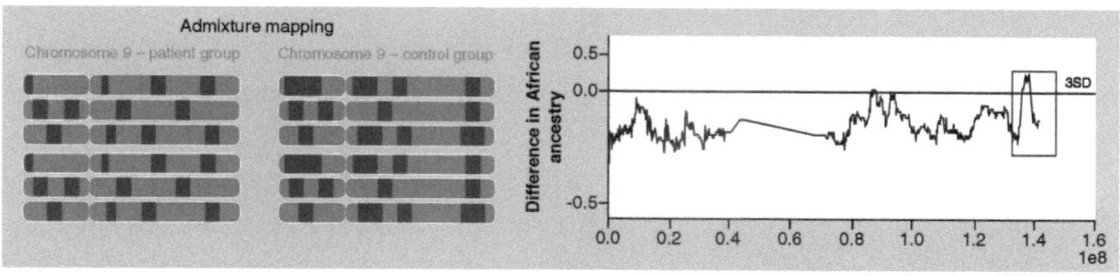

Source: Oliveira, 2018.

As noted above, in studying the symptomatology of dengue epidemics, geneticists have associated ancestry categories with continental geographic regions. The global ancestry classification system gains relevance from the idea that phenotypic profiles of resistance or susceptibility to dengue reflect evolutionary dynamics—specifically, the selection of genetic traits shaped by the ecological contexts in which populations were, or were not, historically exposed to similar pathogens.

The role of evolutionary processes in selecting protective traits—or in the loss of inherited protection—is not only cited in published studies but was also explicitly addressed by geneticists during interviews. Within this project, such discussions frequently concern genes related to lipid metabolism. One interviewed researcher expressed a particular interest in evolutionary questions, especially in understanding how "African populations developed these variants, related to lipid metabolism, which protects them against infections." She noted that while two genes linked to lipid metabolism may confer dengue protection, many others are likely involved. She thus shifted her focus from studying migration and diversity alone to examining how genetic diversity impacts disease susceptibility.

By identifying chromosomal segments with AIMs corresponding to reference ancestral "origins," researchers can map the distinct ancestral genetic contributions inherited from previous generations, which can aid in comparing the sampled populations. As outlined in the DENFREE research design, mapping ancestries along the chromosomes of individuals with resistance or susceptibility to dengue allows geneticists to statistically correlate ancestry configurations in specific genome sections with variations in dengue severity as presented in individual and population records (Figure 2).

In this sense, admixture mapping is a method for extracting relevant information from genetic analysis, based on the idea that "ancestry blocks" can be identified across individual genomes, facilitating the tracking of relevant polymorphisms. These blocks—segments of DNA inherited from ancestral populations—form part of the chromosomal material of contemporary individuals and populations. When searching for a clinical phenotype potentially associated with a particular ancestry, some of these blocks may be statistically more prevalent in groups carrying the variants linked to that phenotype, reflecting higher proportions of that ancestry. If dengue protection were associated with African ancestry, as hypothesized by the Cuban researchers in the consortium, genomic analysis would reveal a higher proportion of African-ancestry DNA blocks in individuals with more favorable clinical outcomes. Protective genes would thus be expected within those same blocks. As Sierra et al. (2017, 3) note: "[A]ncestry blocks will be distributed at random across the genome, reflecting the admixture proportions of the parental ancestries, except in candidate gene locations where statistically significantly different proportions for the ancestry with higher disease levels will be observed in cases versus controls."

The central premise here is that, in order to determine how past populations have contributed differently to the biomolecular composition of present-day humans, geneticists must first conceptualize those ancestral populations in biomolecular terms. This involves constructing categories for ancestral groups and presuming knowledge of their "genetic composition," at least in terms of ancestry informative markers (AIMs) now found in contemporary individual genomes. Using admixture mapping, geneticists claim to identify the contributions of these ancestral populations within individual genomes. In theory, a population's genetic structure is the aggregate of its "composite" (admixed) individual genomes when considered together in a populational set.

A core principle of contemporary population genetics—criticized by STS scholars—is the linking of chromosomal segments to ancestry labels. This entails applying classification systems that carry social or identity categories (such as ethnic, racial, or national labels) into the domain of biological research. Critics argue that such research practices reproduce biological differences grounded in historical and ongoing processes of social differentiation (Duster 2006; El-Haj 2007; Fullwiley 2007, 2014, 2024; Nelson 2008; Reardon 2005; Schramm et al. 2012; TallBear 2013; Wade et al. 2014; Wailoo et al. 2012).

An examination of the methodological dimensions of genetics research in the DENFREE consortium—as described by geneticists in conversations, interviews, or published articles—reveals what may initially appear to be a simple reframing of racial categories (previously used by Cuban researchers) into global ancestries. This shift, however, constitutes a deeper methodological choice, not always made explicit in publications, but one that can be elucidated through the theoretical reasoning underpinning the evolutionary framework of population genetics. The first critical move by geneticists, allowing them to justify the link between genetic data and ancestry labels, is the appropriation of historical narratives that identify and establish the original populations serving as references for ancestry categories. These narratives—transcending the specifics of individual research projects and established across the discipline as foundational historical frameworks—function as "origin myths" for population genetics.[12] Broadly, they recount innumerable human movements and encounters over centuries or millennia, shaped by different temporal frameworks and addressing specific research questions concerning prehistoric periods, historical epochs, or the present. In these narratives, historical contexts are proposed as setting the stage for major past migration movements, generating encounters and admixtures among ancestral populations, which are then used to explain the formation of later composite generations. Geneticists argue they can trace these events molecularly in present-day populations, as DNA analyses reveal two or more *historically inferred* ancestral genetic contributions within individual samples.

In the context of the contemporary Americas or the New World (to which Cuba belongs), a key historical moment frequently invoked in the analysis of living populations is the onset of the modern age, marked by the early globalization that followed European expansion and the establishment of intercontinental oceanic routes. Seen as one of the historical conditions that led to the emergence of the "modern melting pot," this moment is recurrently memorialized in the oceanic voyages of Columbus or Magellan and finds its historical relevance in the development of oceanic traffic, which inaugurated a new phase in the circulation of people and things across continents and latitudes. This historical moment is frequently employed as a temporal boundary for defining ancestry references, grounded in a distribution of biological (or genetic) diversity attributed to the pre-Columbian world—a proposed 500-year-old distribution that enables the production of a meaningful, globalized analysis of genetic configurations across populations, accounting for evolutionary dynamics.[13]

Historians of science and STS scholars have frequently critiqued this framework. Their central concern is that this historical moment, while serving as a significant reference point for comparative analysis in population genetics, also marks the beginning of European colonialism and its associated violence, including the racialized exploitation of labor. Acknowledging the capacity of historical narratives to reproduce distinct worldviews and sociopolitical interpretations, critics have shown that global approaches

---

12 Some of these relatively well-known historical narratives include, for example, the "Mitochondrial Eve" theory, which refers to the 50,000-year-old movement of human populations originating in Africa, developed by geneticists in the 1980s (e.g., Cann et al. 1987); the multiple waves of migration from eastern into western Eurasia, known to have occurred throughout the Neolithic period and into the Middle Ages (e.g., Reich 2018, chap. 5); and, particularly relevant for the genetics of contemporary populations, the global population movements that followed the rise of capitalism and its expansion in the sixteenth century. When applied to specific research objectives, these narratives establish the key temporal and spatial contexts in which populations with distinct historical backgrounds encountered one another and produced "admixed" descendant generations.

13 See, e.g., Winkler et al. (2010). Other temporal frameworks in genetic analysis that focus on past populations have, however, enabled the tracking of older, meaningful historical moments of encounter and admixture, as in the case of ancient DNA analysis using biological material extracted from preserved ancient bones (Pääbo 2014).

to human biodiversity can, often inadvertently, perpetuate colonial or racial imaginaries (e.g., Reardon 2005; M'charek 2005).

As was evident to the DENFREE researchers from the beginning, the Cuban samples revealed a combination of two genetic ancestries inherited from African and European origins. Any other possible admixtures that may have occurred earlier in history—before African and European populations came together in the New World—were disregarded in the analysis. For the purposes of this dengue research, such prior encounters and admixtures were deemed either unknown or irrelevant. By taking the rise of the post-Columbian Americas as the "original" moment in the history of the present Cuban population, the European and African populations that mixed in Cuba over the past five centuries were hardly seen as the products of prior ancient admixtures. These populations were conceived as homogenous; earlier admixtures were effectively treated as part of European or African genetic heritage, since the phenotype analyzed for dengue resistance appeared from the beginning to vary distinctly between African and European ancestries, and not within each group.[14]

We should keep in mind that this understanding of global ancestry is explicitly molecular. As noted earlier, with the aim of identifying genes with phenotypic outcomes of medical or clinical relevance, DENFREE geneticists labeled certain DNA segments with global ancestry categories, indicating geographic origin and thus pointing to historical trajectories and circulations of genetic information. To understand global ancestry as used in biology and population genetics, it is crucial to recognize that these origins are attributed to "blocks" of DNA—not to individuals or populations. At this stage of genetic research, what is being classified is not people, but portions of the DNA molecule within individuals.

The process of tracing ancestries in genomes to determine the history of specific molecular configurations should not lead us to conflate molecular "identities" with the social or historical identities of people. Ancestry labels, as assigned through genomic analysis, refer solely to molecular configurations. Even when established through a populational approach, "genetic ancestry" concerns DNA and its variability. Therefore, in a deeper sense, we can ask: What is involved in the idea of ancestry? What do geneticists mean by genetic ancestry and its relation to DNA segments, and why is it so often misinterpreted—as much in scientific literature as in public reception—as a statement about social identity?

Because geneticists speak of an evolutionary path that led to the selection of particular DNA configurations over others, their concept of ancestry refers to an original population from which these DNA stretches disseminated—typically associated with specific geographic locations or sociocultural geographies. Although ancestry labels encode a geographic dimension—hence the term *biogeographic ancestry*—what emerges from the interviews, geneticists' publications, and other relevant works in population genetics is that evolutionary processes leading to differences in DNA molecular configurations are taken to justify the classificatory logic employed by the dengue researchers.

---

14 The *1000 Genomes* database was used as the reference for ancestry testing, with the European and African components represented by fifty samples each of Southern European (Italian) and Yoruba origin, used for genomic analysis and ancestry mapping (Sierra et al. 2017, 15).

These evolutionary processes, moreover, encompass dynamics such as mutation, selection, reproduction, and transmission, as well as environmental influences such as diet and pathogen exposure—all of which are thought to leave molecular traces in population-level genetic profiles.[15] The hypothesis I am advancing here is that ancestry is intended to identify population-geographic origins for particular DNA segments—not by ignoring evolutionary complexity, but by engaging it directly. While sociocultural classificatory labels such as ethnolinguistic categories may also be used, the global ancestry identities attributed to molecular sequences ultimately refer to hypothetical ancestral populations located in a specific geographic space at a defined historical time—populations understood as the "origins" of the DNA segments under study.

It is this complex technomethodological set of scientific norms and procedures—interrelating populations, molecules, historical narratives, and geographic and temporal locations—that must be disentangled in exploring what may lie within the concept of genetic ancestry.

Evolutionary Theory and Naming Populations in Genetics

The Theoretical Framework: Situating Change in Time and Geography

Clearly, the notion of global biogeographic ancestry was well-suited to the DENFREE project as a classificatory baseline for the molecular data analyzed by the project's geneticists. As demonstrated in the previous section, the DENFREE research proposal led geneticists to methodologically operationalize global ancestry categories deemed significant and to apply an ancestry mapping approach that, they argued, would enhance the statistical efficiency of GWAS. As shown in the team's published articles, GWAS was applied following ancestry mapping to identify genomic regions responsible for differences in dengue clinical outcomes (e.g., Sierra et al. 2017). Their conclusions pointed to the identification of loci believed to confer resistance to DENV infection, inherited as part of genome sections labeled as African ancestry blocks.

I interpret the use of four global ancestry categories by the research group as linked to the need to integrate a conventional medical genetics approach—centered on GWAS—with the evolutionary dimensions of genetic change and variation that, by disciplinary tradition, fall under population genetics. These dynamics of genetic change across time and space are particularly important to geneticists and are closely tied to selection events and past migrations. Such knowledge helps interpret molecular configurations and link them to specific phenotypic traits, thereby bridging the gap between genotype and phenotype. Oliviera et al (2018, 2–3) argue, for instance, that a genetically selected protection against a local pathogen in one population may, through migration and admixture, appear in a new region, where differences in ancestry between cases and controls reflect not sampling bias but a protective evolutionary mechanism.

15 See, e.g., Stoneking (2017) for a thorough, accurate and accessible treatment of molecular anthropology, the fundamentals of genetic variability, and evolutionary dynamics.

Beyond the strict medical genetics approach of using GWAS to identify genes that may play a role in specific pathologies, these population researchers are committed to understanding the role of different genetic configurations in immunological processes. To explore these links, and beginning with a connection between clinical phenotypes and epidemiological information based on visual traits, these practitioners produce sheets of genetic data that can be analyzed to identify loci in the genome related to this clinical information and phenotypes. Having identified these genomic locations, they must develop an understanding of the role of these loci as causal entities in producing differences in medical outcomes. With this problem in mind, the question researchers need to explore concerns dengue showing different forms and degrees of severity in populations with apparently different genealogical backgrounds. They attempt to clarify this connection by tracing the history of the genetic variants themselves and asking why such molecular differences exist, where they come from, and how following their trajectories might help explain the outcomes.

As noted above, GWAS tools typically identify genetic traits that may be relevant to certain diseases in population groups sharing similar genetic structures. But genetic structure similarity, which is crucial for GWAS to work correctly, is also tied to another feature of this method. GWAS configures a synchronic approach to genetics, in that it leaves time out of the analysis. It does not track genetic change over time—that is, it is not designed to incorporate evolutionary thinking into the interpretation of its results: mutation and selection processes are left at the door; what matters are the differences between the target populations (the cohort set of samples) under study and controls. This approach has limitations, as it disregards the theory of change that is the hallmark of population genetics. In fact, it is the trajectories of genetic transformations through time and the movement of populations through space—both of which allow researchers to relate geography, environment, and genetic change—that allow population geneticists to understand evolutionary processes, i.e., how certain genetic traits have been positively or negatively selected and how they have become present or absent in contemporary populations.[16] This knowledge can be of critical biomedical value and play a fundamental role in the search for clinical or pharmaceutical applications. If genetic and phenotypic data were related and made meaningful by considering the role of environmental pressures on past populations, differences could be read through evolutionary processes, and their molecular and metabolic consequences would be well positioned for study. Tracking genetic changes resulting from selection is made possible by ancestry information—in this case, by global ancestry. As DENFREE researchers proposed:

> Human populations are structured in three main groups, African, European and Asian. The independent selective pressures acting upon these groups can lead different genes to be selected in adaptation to the same pathogen, and those genes can interact in a common crucial pathway or in different pathways of additive importance to the disease process. (Sierra et al. 2017, 12)

16 On the role of evolutionary theory and thinking in the history of population genetics, see Reardon (2005).

While the genotype-phenotype relationship is likely the primary concern of geneticists focused on the medical dimension of genetic research, for population geneticists the problem is not merely whether certain genetic or molecular traits characterize a population or how frequently a profile appears. Another fundamental aspect—often absent in medical genetics—significantly shapes the work of population geneticists. Beyond genetic structure and profile frequency, these biologists are largely concerned with tracing processes of transformation across temporal frames. For population geneticists, evolution matters. They understand genetic traits and associated molecular configurations as products of evolutionary dynamics—that is, in Darwinian terms, as the outcome of transformation trajectories shaped by selective pressures and inheritance, and occurring over time at a particular rate of change. Genes—precisely because they are inherited and influenced by sexual-reproductive recombination—are seen as evolutionary products: the result of mutations and selection within populations characterized by a certain degree of variability. In sum, population scale and the temporality of mutations are the two fundamental aspects sustaining the evolutionary process, which is further guided by selection.[17]

I am arguing and attempting to demonstrate that it is this concern with evolution in Darwinian terms that renders the use of identity categories fundamental to the work of population geneticists.[18] For them, genes have a history—but that history can only be told in relation to the history of populations and their trajectories over generations, within which processes of biogenetic transformation, admixture, substitution, and adaptation can be observed. If variation and selection—the core premises of evolutionary theory—explain why certain molecular configurations are more common today in some populations and not others, and if genetic diversity arises through time-dependent processes, then the history of genes can only be pursued after the history of populations themselves.

How, then, can the history of DNA variation (or any molecule dependent on heredity) be explored, if not by identifying the populations that carry and transmit that variability through successive generations of sexual reproduction? For geneticists, there are clearly two productive ways to name such populations:

i. in relation to environment and geography, using categories of spatial location where specific molecular forms are found
ii. in relation to collective names for the populations themselves, often based on ethnolinguistic classifications

Geographic nomenclature typically refers to the places where populations live today rather than to hypothetical historical geographies, since the environmental origins that

17 In Darwinian biology selection is understood as "the only direction-giving factor in evolution" (Mayr and Provine 1980, 3n1).

18 Though tainted by the pernicious connotations introduced through early sociological appropriations, in contemporary biological sciences "evolution" simply means that living organisms change, that such change produces new species better adapted to shifting environments, and that this process is irreversible, meaning it does not revert to previous forms.

may have exerted selective pressures are rarely known with precision. Nonetheless, the geographic association with genetic configurations remains highly relevant, as heredity and evolutionary processes—such as drift, reproductive isolation, or selection on existing variation—usually occur in relation to ecogeographic conditions.[19]

It is not surprising, then, that the history of genes, as told by population geneticists, is partly expressed in a vocabulary that mirrors that used by historians to narrate population histories and distinguish between different peoples. These denominations are inevitably tied to names of geographical regions, cultural or linguistic classifications, and categories of political differentiation, all of which carry the symbolic weight of social identity.

Identifying Patterns and Tracing Connections

In relation to how geneticists name populations, two sets of scientific infrastructure are typically presented as fundamental to identifying genetic ancestry. The DENFREE genetics research program is no exception. The first set comprises chips with predefined SNP probes based on the HapMap,[20] used in genome-wide analyses to generate data on genetic variation. The second involves reference collections, which provide the comparative framework needed to estimate the proportion of different ancestries in analyzed samples.

I contend that to take advantage of these infrastructural and technological resources while engaging with evolutionary theory, the naming of present or past populations becomes a necessary step in producing knowledge from genetic data obtained in laboratory analysis. Time, along with geographical constraints (environmental factors, reproductive isolation), is central to the dynamics of genetic change in populations—commonly referred to as genetic drift. Thus, information that enables researchers to track the temporality of such processes is among the most valuable components of data. These naming practices—whether derived from fieldwork sampling or from primary and secondary historical sources—can provide the link connecting the present to historical events and contemporary populations to their historical predecessors. Published sources and interviews show that, in efforts to establish such connections and to relate contemporary populations to their genetic past and to history, ancestry has become one of the most useful and widely used concepts in recent decades.

In 2008, the American Society of Human Genetics (ASHG) issued a statement offering recommendations on ancestry testing. Following this, Fujimura and Rajagopalan

---

19 The correlation between genetic distance and geographic distance has long been considered one of the strongest influences on genetic variation, even in humans, who exhibit very low levels of intraspecific variation (Cavalli-Sforza et al. 1994, 121–125). Although some studies also show significant correlations with cultural and linguistic distances, geography—understood as spatial distance and natural barriers—has been regarded as the most relevant factor in explaining genetic distance (Tishkoff et al. 2009; Wang et al. 2012; Stoneking 2017, 130).

20 SNPs (pronounced "snips") are single nucleotide polymorphisms—variations at a single DNA position, either between individuals or between two chromosomes in one person (Stoneking 2017, 92). Chips (or SNP arrays) used in genomic analysis are "printed" with SNP probes that detect and determine the variants present in the analyzed samples.

argued that ancestry labels, though often necessary for calibrating population structure, are dispensable for epidemiological purposes (Fujimura and Rajagopalan 2011, 15–16).

However, it is reasonable to argue that these evolutionary explanations based on different background origins are indeed useful. I understand that ancestry labels matter to geneticists because they enable evolutionary interpretations of certain medical conditions. Geographic and historical information appears crucial for understanding evolutionary processes in relation to specific environmental conditions. But how, in practice, is this information important, and how does it relate to global ancestry categories? How is ancestry knowledge cognitively relevant?

The dengue research project offers a telling example. DENFREE geneticists observed that African ancestry conferred enhanced protection against dengue hemorrhagic fever compared to individuals of European background (Sierra et al. 2017, 12). In one of their published studies, DENFREE researchers discuss the possibility that another African-origin flavivirus—the yellow fever virus, which shows a 6.8 times higher mortality rate among Europeans—may have contributed to the selection of protective genetic variants, not only against itself but also against hepatitis C and dengue virus (DENV). Drawing on existing scientific evidence suggesting that this protective condition is linked to lipid metabolism, they use ancestry information to trace the link between DENV resistance and changes in lipid metabolism profiles:

> Signs of selection were already identified in West Africans for APOL1 and CD36 genes involved in lipid metabolism and probably driven by pathogen resistance. Our evidence adds two new genes to the differential lipid profile between Africans and Europeans, which play a role in infectious resistance. (Sierra et al. 2017, 12)

Without ancestry profile information on samples and populations, the cognitive articulation between lipid metabolism, infection outcomes, genetic profiles, and the specific genes involved would have been difficult to achieve. This information enables different sample sets analyzed in distinct research projects by separate teams to be related and compared, allowing solid statistical results to be collectively considered within broader epidemiological patterns. The classification of people has long been crucial for interpreting the epidemiological behavior of populations exposed to pathologies with unknown or unpredictable outcomes. One of the first steps in recognizing differential epidemiological behavior and identifying patterns of incidence, resistance, and susceptibility is developing methods for classifying and distinguishing groups within affected populations.

Without ancestry information, the findings on yellow fever and lipid metabolism profiles from other studies could not have been linked to this project. I argue that, without ancestry as a means to classify and relate genotypes, researchers would lack important clues about where to investigate differences in immune responses and their causes.

In the cited article, the DENFREE team explicitly stated that individuals with a certain lipid metabolic profile and a protective condition against some tropical viruses shared an evolutionary history with members of the Cuban cohort. They implicitly attributed this connection to information about the African ancestry of specific DNA segments and the loci identified within them (Sierra et al. 2017, 12–13). A link between two distinct sample groups from separate projects was established through ancestry information, allowing the DENFREE team to relate lipid metabolism studied in one project to two newly identified genes and to virus protection.

Without ancestry mapping, this relationship could not have been explained; no evolutionary rationale would have led to the hypothesis connecting DENV resistance to analogous protection conferred by lipid metabolic profiles. These links—between pathogen resistance, genes, and lipid metabolism—would have lacked their theoretical evolutionary basis. As the DENFREE team assert, referencing "intense selective pressure" on "genes involved in lipid metabolism" that was "probably driven by pathogen resistance" (Sierra et al. 2017, 12), it was the identification of a common evolutionary background across samples from different research projects that led to the hypothesis connecting lipid metabolism to dengue protection and linking this protection to the "genes" identified through genomic analysis of the DENFREE samples.

Genetic differences are not merely the result of random variation at specific genome locations. Variations do not all carry the same significance, as some are shaped by selective pressures. European, African, or Asian ancestry are not simply labels distinguishing molecular or genetic configurations; geneticists also relate these molecular traits to specific environmental and ecological habitats that may account for such differences, thereby creating a context in which selective pressures and evolutionary processes can be identified and interpreted.

Conclusion

While the concept of genetic ancestry, as used by population geneticists, brings together molecular and sociopolitical dimensions, it is precisely this configuration that makes it scientifically useful as a methodological tool. In the context of a dengue research project aimed at identifying and understanding genetic configurations that might explain differences in disease outcomes along resistance and susceptibility lines, genetic ancestry became central to understanding population-level patterns of clinical outcome. Despite the racialized connotations of epidemiological observations, genetic ancestry categories were not used to distinguish racial groups but to differentiate DNA segments and trace their possible origins. Though linked to the genomes of dengue patients, these ancestry labels served to distinguish chromosome tracts that could become targets of further research into the molecular causes of dengue clinical outcomes.

This genetic difference was interpreted through the differential hereditary transmission of identified loci, enabling researchers to establish ecogeographic origins and to understand why those DNA locations conferred protection against DENV. While race as a sociopolitical category has informed epidemiological observations, it was the transformation of such observations into genetic (molecular) ancestry that enabled geneticists to understand genotype-phenotype relationships in ways that were relevant to underlying immunological processes.

The geographic information associated with ancestry and population history became genetically meaningful because it enabled evolutionary interpretations, supported generalizations, and generated hypotheses for further inquiry. Genetic (molecular) ancestry nomenclature has followed paths shaped by historical, epidemiological, and geographic knowledge, while retaining links between labels and groups. Without these links, the relevance of such knowledge for interpreting genomic data through an evolutionary lens would be lost.

References

American Society for Human Genetics. 2008. "Ancestry Testing Statement." Accessed 24 June 2025. https://www.ashg.org/wp-content/uploads/2008/11/Statement-20081311-ASHGAncestryTesting.pdf.

Byeon, Yen Ji Julia, Rezarta Islamaj, Lana Yeganova et al. 2021. "Evolving Use of Ancestry, Ethnicity, and Race in Genetics Research—A Survey Spanning Seven Decades." *AJHG* 108 (12): 2215–2223. DOI: https://doi.org/10.1016/j.ajhg.2021.10.008.

Callon, Michel. 1984. "Some Elements of a Sociology of Translation: Domestication of the Scallops and the Fisherman of St. Brieuc Bay." In *Power, Action and Belief, Sociological Review Monograph*, edited by John Law, 196–230. London: Routledge & Kegan Paul.

Callon, Michel, and John Law. 1982. "On Interests and their Transformation: Enrolment and Counter-enrolment." *Social Studies of Science* 12 (4): 615–625. DOI: https://doi.org/10.1177/030631282012004006.

Cann, Rebecca L., Mark Stoneking and Allan C. Wilson. 1987. "Mitochondrial DNA and human evolution." *Nature* 325: 31—36, doi:10.1038/325031a0

Cavalli-Sforza, Luigi Luca, Paolo Menozzi, and Alberto Piazza. 1994. *The History and Geography of Human Genes*. Princeton: Princeton University Press.

Duster, Troy. 2006. "The Molecular Reinscription of Race: Unanticipated Issues in Biotechnology and Forensic Science." *Patterns of Prejudice* 40 (4–5): 427–441. DOI: https://doi.org/10.1080/00313220601020148.

El-Haj, Nadia Abu. 2007. "The Genetic Reinscription of Race." *Annual Review of Anthropology*, 36: 283–300. DOI: https://doi.org/10.1146/annurev.anthro.34.081804.120522.

Fujimura, Joan H., and Ramya Rajagopalan. 2011. "Different Differences: The Use of 'Genetic Ancestry' versus Race in Biomedical Human Genetic Research." *Social Studies of Science*, 41 (1): 5–30, DOI: https://doi.org/10.1177/0306312710379170.

Fujimura, Joan, H., Ramya Rajagopalan, Pilar N. Ossorio, and Kjell A. Doksum. 2010. "Race and Ancestry: Operationalizing Populations in Human Genetic Variation Studies." In What's the Use of Race? *Modern Governance and the Biology of Difference*, edited by Ian Whitmarsh and David S. Jones, 169–183. Cambridge, MA: MIT Press.

Fullwiley, Duana. 2007. "The Molecularization of Race: Institutionalizing Human Difference in Pharmacogenetics Practice." *Science as Culture* 16 (1): 1–30. DOI: https://doi.org/10.1080/09505430601180847.

Fullwiley, Duana. 2008. "The Biologistical Construction of Race: 'Admixture' Technology and the New Genetic Medicine." *Social Studies of Science*, 38 (5): 695–735, DOI: https://doi.org/10.1177/0306312708090796.

Fullwiley, Duana. 2014. "The 'Contemporary Synthesis': When Politically Inclusive Genomic Science Relies on Biological Notions of Race." *ISIS* 105 (4): 803–814. DOI: https://doi.org/10.1086/679427.

Fullwiley, Duana. 2024. *Tabula Raza: Mapping Race and Human Diversity in American Genome Science*. Oakland: University of California Press.

Gannett, Lisa. 2014. "Biogeographical Ancestry and Race." *Studies in History and Philosophy of Biological and Biomedical Sciences*. 47: 173–184. DOI: https://doi.org/10.1016/j.shpsc.2014.05.017.

Guzmán, Maria G. 2012. "Thirty Years after the Cuban Hemorrhagic Dengue Epidemic of 1981." *MEDICC Review* 14 (2): 46–51. DOI: https://doi.org/10.37757/MR2012V14.N2.11.

Guzmán, Maria G., G. Kourí, S. Vázquez, D. Rosario, J. Bravo, and Luis Valdés. 1999. "DHF Epidemics in Cuba, 1981 and 1997: Some Interesting Observations." *Dengue Bulletin* 23: 39–43, https://iris.who.int/handle/10665/148771

Guzmán, Maria G., Scott B. Halstead, Harvey Artsob et al. 2010. "Dengue: a continuing global threat." *Nature Reviews Microbiology* 8 (Suppl 12): S7—S16, doi:10.1038/nrmicro2460

Jaenisch, Thomas, IDAMS, Anavaj Sakuntabhai, DENFREE, Annelies Wilder-Smith, DengueTools. 2013. "Dengue Research Funded by the European Commission-Scientific Strategies of Three European Dengue Research Consortia." *PLOS Neglected Tropical Diseases* 7 (12): e2320. DOI: https://doi.org/10.1371/journal.pntd.0002320.

Koenig, Barbara, Sandra Soo-Jin Lee, and Sarah S. Richardson (eds.) 2008. *Revisiting Race in a Genomic Age*. New Brunswick and London: Rutgers University Press.

Lewis, Anna C. F., Santiago J. Molina, Paul S. Appelbaum et al. 2022. "Getting Genetic Ancestry Right for Science and Society." *Science* 376: 250–252. DOI: https://doi.org/10.1126/science.abm7530.

Lipphardt, Veronika. 2017. "The Body as a Substrate of Differentiation. Shifting the Focus from Race Science to Life Scientists' Research on Human Variation." *Varia Historia*, 33 (61): 109–133. DOI: https://doi.org/10.1590/0104-87752017000100006.

Lipphardt, Veronika, Gudrun A. Rappold, and Mihai Surdu. 2021. "Representing vulnerable populations in genetic studies: The case of the Roma." *Science in Context* 34: 69–100. DOI: https://doi.org/10.1017/S0269889722200002.

Mayr, Ernst, and William B. Provine. 1980. *The Evolutionary Synthesis: Perspectives on the Unification of Biology.* Cambridge, MA. and London: Harvard University Press.

M'charek, Amade. 2005. *The Human Genome Diversity Project: An Ethnography of Scientific Practice*. Cambridge: Cambridge University Press.

Nash, Catherine. 2015. *Genetic Geographies: The Trouble with Ancestry*. Minneapolis, London: University of Minnesota Press.

Nelson, Alondra. 2008. "Bio Science: Genetic Genealogy Testing and the Pursuit of African Ancestry." *Social Studies of Science* 38 (5): 759–783. DOI: https://doi.org/10.1177/0306312708091929.

Nelson, Alondra. 2016. *The Social Life of DNA: Race, Reparations, and Reconciliation after the Genome.* Boston: Beacon Press.

Oliveira, Marisa, Joana Ferreira, Verónica Fernandes, Anavaj Sakuntabhai, and Luísa Pereira. 2018. "Host ancestry and dengue fever: from mapping of candidate genes to prediction of worldwide genetic risk." *Future Virology* 13 (9): 647–655, DOI: https://doi.org/10.2217/fvl-2018-0073.

Panofsky, Aaron, and Catherine Bliss. 2017. "Ambiguity and Scientific Authority: Population Classification in Genomic Science." *American Sociological Review* 82 (1): 59–87. DOI: https://doi.org/10.1177/0003122416685812.

Pääbo, Svante. 2014. *Neanderthal Man: In Search of Lost Genomes*. New York: Basic Books.

Price, Alkes L., Nick J. Patterson, Robert M. Plenge, Michael E. Weinblatt, Nancy A. Shadick, and David Reich. 2006. "Principal Components Analysis Corrects for Stratification in Genome-Wide Association Studies." *Nature Genetics* 38 (8): 904–909. DOI: https://doi.org/10.1038/ng1847.

Rajagopalan, Ramya, and Joan H. Fujimura. 2012. "Making History via DNA, Making DNA from History: Deconstructing the Race-Disease Connection in Admixture Mapping. In *Genetics and the Unsettled Past: The Collision between DNA, Race and History*, edited by Keith Wailoo, Alondra Nelson and Catherine Lee, 143–163. New Brunswick, New Jersey and London: Rutgers University Press.

Rajagopalan, Ramya, and Joan H. Fujimura. 2018. "Variations on a Chip: Technologies of Difference in Human Genetics Research", *Journal of the History of Biology* 51: 841–873. DOI: https://doi.org/10.1007/s10739-018-9543-x.

Reich, David. 2018. *Who We Are and How We Got Here. Ancient DNA and the New Science of the Human Past*. Oxford: Oxford University Press.

Reardon, Jenny. 2005. *Race to the Finish: Identity and Governance in an Age of Genomics*. Princeton and Oxford: Princeton University Press.

Schramm, Katharina, David Skinner, and Richard Rottenburg. 2012. *Identity Politics and the New Genetics: Re/Creating Categories of Difference and Belonging*. New York, Oxford: Berghahn.

Sierra, Beatriz, Gustavo Kourí, and Maria G. Guzmán. 2007. "Race: a risk factor for dengue hemorrhagic fever." *Archives of Virology* 152: 533–542. DOI: https://doi.org/10.1007/s00705-006-0869-x.

Sierra, Beatriz, Petr Triska, Pedro Soares et al. 2017. "OSBPL10, RXRA and Lipid Metabolism Confer African-Ancestry Protection against Dengue Haemorrhagic Fever in Admixed Cubans." *PLOS Pathogens* 13 (2): e1006220. DOI: https://doi.org/10.1371/journal.ppat.106220.

Stoneking, Mark. 2017. *An Introduction to Molecular Anthropology*. Hoboken: John Wiley & Sons, Inc.

TallBear, Kim. 2013. "Genomic Articulations of Indigeneity." *Social Studies of Science* 43 (4): 509–533. DOI: https://doi.org/10.1177/0306312713483893.

Tishkoff, Sarah, Floyd A Reed, Françoise R Friedlaender et al. 2009. "The Genetic Structure and History of Africans and African Americans." *Science* 324 (5930):1035–1044. DOI: https://doi.org/10.1126/science.1172257.

Wade, Peter, Carlos López Beltran, Eduardo Restrepo and Ricardo Ventura Santos (eds.) 2014. *Mestizo Genomics: Race Mixture, Nation, and Science in Latin America*. Durham and London: Duke University Press.

Wailoo, Keith, Alondra Nelson and Catherine Lee. 2012. *Genetics and the Unsettled Past: The Collision of DNA, Race, and History*. New Brunswick, New Jersey and London: Rutgers University Press.

Wang, Chaolong, Sebastian Zöllner, and Noah A. Rosenberg. 2012. "A Quantitative Comparison of the Similarity between Genes and Geography in Worldwide Human Populations." *PLoS Genetics* 8 (8): e1002886. DOI: https://doi.org/10.1371/journal.pgen.1002886.

Winkler, Cheryl A., George W. Nelson and Michael W. Smith. 2010. "Admixture Mapping Comes of Age." *Annual Review of Genomics and Human Genetics* 11: 65–89. DOI: https://doi.org/10.1146/annurev-genom-082509-141523.

Global Power and the Circulation of Standards

Ambivalence about Race

Expert Opinions on Using Racial and Ethnic Categories in Clinical Research in Sweden

Anna Bredström and Shai Mulinari

Abstract *This chapter examines the localized ambivalence surrounding the use of race in contemporary biomedicine: how race is constructed as both valid/relevant and invalid/irrelevant in a specific national setting. It is based on interviews with twenty-six expert professionals in Sweden who are directly or indirectly involved in clinical pharmaceutical research, including clinical researchers, research nurses, pharmacists, regulators, statisticians, and professionals from the pharmaceutical industry and clinical trial companies. The analysis explores how the professionals narrate their experiences and understandings of racialized concepts and categories. Most interviewees express strong postracial sentiments by rejecting race as a scientifically valid and useful concept. However, our analysis shows that race is nonetheless present in their professional experiences and discourses, enacted in three ways: through the mundane, everyday use of racial categories in clinical trials; through the conflation of ethnicity and race; and through a population genetics discourse about the scientific usefulness of race. We conclude that, despite moral, political, and scientific arguments against race, many informants express ideas about racial differences, and we discuss whether this may be explained by a high level of trust in medical knowledge production and the institutions regulating this knowledge production, including drug regulatory agencies.*

Introduction

The website of the Swedish Medical Products Agency (MPA) publishes questions from the public together with replies from the drug regulator. Under the headline "About ethnic Background in Product labels of Pharmaceuticals,"[1] a person who calls themself Vanessa asks why some drug labels mention patient skin color. As an example, Vanessa points to drugs for high blood pressure whose product labels refer to uses in "black patients" (svarta patienter). While the question does not reveal Vanessa's opinion on the matter, the

---

1 https://fragor.lakemedelsverket.se/org/lakemedelsverket/d/om-etnisk-bakgrund-i-lakemedels-produktinformation, last accessed February 10, 2024.

MPA nevertheless replies: "We understand that choice of words such as 'black patients' may be perceived as inappropriate and denigrating," stressing that the agency "strives to describe patient groups in an equal manner, [and] where it is medically and scientifically acceptable" (our translation). After this caveat, the agency nevertheless goes on to explain that genetic differences between populations may influence uptake and breakdown of drugs in bodies and that this knowledge is important to pharmaceutical research. The agency also explains that the Swedish language product label is often a translation of the EU-wide English language label, in line with EU regulations on the matter, and that the use of concepts such as "black patients" may originate from foreign clinical studies that use other vocabulary or assign other meanings to concepts than what is customary in Sweden. As a final remark, the agency notes that vocabularies for describing ethnic background change over time and that some drug labels may include outdated concepts because the drugs have been on the market for many years.

The exchange between Vanessa and the Swedish MPA captures the key elements that make up the research problem addressed in this article, namely, the inherent ambivalence associated with the use of race as a concept and category in medical contexts in Sweden. On the one hand, the exchange illustrates how racialized concepts and categories such as "black" are used in Swedish medical contexts without definition or much further explanation. As the MPA explains, these concepts often originate from international clinical research categorizing patients by race and ethnicity (Mulinari et al. 2021), which then trickles down to Swedish—and, more broadly, European—clinical research (Mulinari and Bredström 2024b) and prescribing information (Mulinari and Bredström 2024a). On the other hand, the response by the MPA also reflects the official "post-racial" discourse (Goldberg 2009) in Sweden, which publicly chides notions of racial differences and translates race into the culturally more accepted concept of ethnicity. As in several other European countries, explicit talk of race in Sweden carries connotations of past racial biology and is therefore generally seen as negatively charged and pseudoscientific (Brännström 2018; Bartram et al. 2022). Thus, without knowing *why* Vanessa posed their question, the MPA feels the need to underline that the wording "black patients" may be experienced as "inappropriate and denigrating."

We aim here to delve into the ambivalence surrounding the concept of race in this specific context. Race is viewed as scientifically inappropriate and potentially demeaning, even as it is also perpetuated and normalized (Mulinari and Bredström 2024b). Our analysis is grounded in interviews with key stakeholders and experts engaged in pharmaceutical research and policy in Sweden. We engage their perspectives in dialogue with critical race studies that have explored the articulation of race in contemporary Sweden, and with social studies of science and medicine that have investigated the racialization of biomedicine in various national and disciplinary contexts. Through this investigation, our objective is to examine the localized circumstances under which race is constructed as *both* valid/relevant and invalid/irrelevant in contemporary biomedicine and to discuss its potential implications.

Methodology

Our research is centered on comprehending how race is conceived and constructed within contemporary biomedicine in Sweden. We adopt a poststructuralist understanding of race as socially constructed and perpetuated, including within medicine and biology, without being an a priori medical or biological "fact." However, the social constructions of race significantly influence individuals' lived experiences, including through healthcare. In Sweden, a growing body of literature has explored topics such as the presence of structural racism in healthcare institutions, affecting both healthcare professionals and patients (e.g., Hamed 2022; Wolgast et al. 2024). Critical race theorists highlight that contemporary constructions of race draw upon various notions of immutable differences, be they cultural, social, religious, or geographical, often substituting earlier ideas about biological distinctions between groups (Goldberg 2015). Nevertheless, some scholars have argued that the lingering role of medicine and biological research in shaping notions of race in the contemporary European context has received insufficient attention (Lentin 2020). This observation certainly holds true for the Swedish context, where many studies have concentrated on how static notions of culture have replaced ideas about inherent biological differences and hierarchies in racializing discourses (Schierup and Ålund 2011). Additionally, some critical race scholars advocate employing race (in Swedish: "ras") as an analytical tool and category for examining racism and prevailing discourses of whiteness and colorblindness in Sweden (e.g., Hübinette and Lundström 2014). However, there is limited research investigating the explicit and routine application of race as a biological concept in contemporary Sweden, as our study strives to do.

To explore the meaning-making around race in biomedicine in Sweden, we conducted qualitative semistructured interviews (Kvale and Brinkmann 2009) with twenty-six experts in the pharmaceutical scientific and regulatory fields representing different key actors. This included clinical researchers, research nurses, regulators, pharmacists, statisticians, and professionals from the pharmaceutical industry and clinical trial companies. The interview participants were strategically selected to represent different areas of expertise and experience. In addition, some snowball sampling led to additional participants, capturing the broader range of perspectives we were looking for. For example, while interviewing a key actor involved in pharmaceutical regulation at a national authority, we concluded that additional experience from within the same authority but covering other expertise areas (e.g., pharmacogenetics, product information) would add to the material we had already collected.

We specifically focused on clinical pharmaceutical research due to its relevance for discourses and practices around race and ethnicity in medicine (Kahn 2012; Mulinari et al. 2021). Clinical pharmaceutical research is often internationally collaborative, commercially driven, and subject to rigorous international and national regulatory standards. It is also a domain where experts in the field, as well as social scientists, patient activists, and regulators, have debated the use of race and ethnicity (Epstein 2007). Nevertheless, our previous research has shown that the application of race in Sweden extends beyond pharmaceuticals, encompassing various medical areas, such as dermatology and the measurement of lung and kidney function in physiology.

The interviews took place between July 2022 and December 2023, consisting of both individual and group interviews. Most interviews were conducted by both researchers. The project received ethical approval from the Swedish Ethical Review Authority (2021–02514), and we have deidentified names of the participants as well as workplaces, companies, and locations. Quotations have been translated from Swedish to English by the authors. We analyzed the transcribed interviews using principles of qualitative analysis (Silverman 2019). Specifically, we coded and thematized interview transcripts to capture the predominant ways in which participants narrated their experiences and understandings of racialized concepts and categories. Four main themes emerged. We will return to the themes after presenting the theoretical perspectives that guide our analysis.

Research Context

In addition to drawing inspiration from critical race theory, our research builds on science and technology studies, or STS, and medical sociological research that has scrutinized the "re-inscription of race" (Duster 2015) in contemporary biomedicine. Importantly, this renewed emphasis on racial differences in medicine has been driven to a large degree by the pursuit of more personalized, or at least more stratified, medicine in which a crucial step is imagined requiring studying human genetic variation within and between populations, with these populations categorized in terms of race, ethnicity, and nationality (Richardson and Stevens 2015). Critics have, however, questioned the use of race (along with ethnicity and nationality) as proxies for genetic variation, contending that human genetics research has contributed to the emergence of a new biological conception of race (Bliss 2012). This development has raised concerns for some scholars. Troy Duster (2015), for example, has described it as a "post-racial surprise," highlighting the irony of an increased interest in race in medicine coinciding with Western societies celebrating their own colorblindness and postracial ideology. But other scholars, such as Nikolas Rose (2007), have argued that the contemporary understanding of race in biomedicine should not be equated with earlier hierarchical notions of race. According to Rose, contemporary life sciences are less deterministic and more focused on the individual, enabling experts to better predict medical risks and optimize patient outcomes with biomedical technologies, including pharmaceuticals (see also Rabinow and Rose 2014). Moreover, some scholars have argued that race-based medicine can in fact enhance social justice, including health justice (Rotmi et al. 2013; Washington et al. 2022).

Promoting health justice is also the aim foregrounded by many who advocate for including specific racial and ethnic minorities in clinical trials, with policymakers and regulators in the United States leading the way (ibid. 2007). However, as shown by Steven Epstein (2007), in the United States this line of reasoning is readily upheld due to the "categorical alignment" between the social worlds of medicine, social movements, and bureaucratic administration, where racial categories, such as Black or White, are seen as similarly applicable across all realms. The categorical alignment is the result of a longer process of social movements seeking to improve the inclusion of ethnic and racial minorities (as well as women) in medical research in the United States (ibid.). However, a major challenge arises when these US racial categories fail to align with population categories and conceptions used elsewhere in the world. For example, in the 1990s, international efforts to harmonize pharmaceutical regulation and industry processes raised disagreements over racial categories, as Japan's representatives from its drug regulatory agency argued that the unique homogeneity of the Japanese population required that drugs intended for authorization in Japan be tested on a Japanese population. In particular, they questioned the relevance of the three main racial categories proposed by the US representatives from the FDA (Caucasian, Black, and Asian), asserting that this did not correspond to the Japanese understanding of relevant population differences between Asian populations (Kuo 2008).

In our research, we ask what happens to the use of racialized concepts and categories when they are applied in the Swedish context, which lacks the same "categorical alignment." Not only do the US racial categories fail to align with the more emic categories used in Sweden (e.g., migrants, Swedes), but the concept of race as such is, as mentioned above, more or less taboo in public discourse (Brännström 2018). Since the 1970s, there have been repeated attempts to eliminate the term "race" from Swedish law, with several official inquiries addressing this issue. The key message of these investigations is that since race lacks scientific validity and carries undesirable historical connotations, it should not be used within Swedish law and public policy (SOU 2015). Proposed alternatives, such as "assumptions about race," or the replacement of race with ethnicity, would emphasize the socially constructed and "false" nature of race (ibid). Nonetheless, erasing the concept of race has proven legally complex[2], although recent legislation, such as the Discrimination Act from 2008, exclusively refers to ethnicity, not race (SFS 2008: 567).

Our study also ties into STS research on standardization as a way of naturalizing and normalizing concepts and objects (Timmermans and Epstein 2010). Standardization is fundamental for establishing reliable evidence across scientific fields, with, for example, international diagnostic and classification systems being as pivotal in biomedical research as they are in clinical practice. Numerous studies have analyzed the standardization of racial and ethnic categories in biomedicine, particularly in the United States. A key actor in this context is the U.S. Food and Drug Administration (FDA), which "strongly recommends" using standard US racial and ethnic classifications in clinical trials for drugs intended for the US market (FDA 2016). It is also of interest for our arguments below that while FDA points out that "the categories in this classification are socio-political constructs and should not be interpreted as scientific or anthropological in nature" (9), potential group differences in response to drugs are nevertheless seen as both "attributable to intrinsic factors (e.g., genetics, metabolism, elimination)" and "extrinsic factors (e.g., diet, environmental exposure, sociocultural issues)." Importantly, the distinction between intrinsic and extrinsic factors refers to a conceptual framework that distinguishes between race and ethnicity, with race being an example of an "intrinsic" factor, and more specifically a *genetic* factor, while ethnicity covers both "intrinsic" and

2 Legislators argued that there was a risk that a shift in meaning would make it difficult to accurately translate international declarations. It could also have unintended consequences, in particular making it more difficult to define and combat racist behavior (Prop. 2017/18:59).

"extrinsic" factors (ICH, 1998; 10; see Mulinari et al. 2021). That is to say, the FDA articulates race as, ontologically speaking, both socially constructed and biological in nature, and, epistemologically speaking, both scientifically problematic and therapeutically necessary (Kahn 2012).

Another key actor is pharmaceutical companies, which strive to enroll racially and ethnically diverse populations in clinical trials using US categories to adhere to FDA standards and enhance their social credentials. For instance, in 2021 Pfizer published a summary detailing the diversity of participants in its US clinical trials since 2011, with the intention of using it as a benchmark for the company's future efforts in diversity, equity, and inclusion (Rottas et al. 2021). Nonetheless, while pharmaceutical research primarily operates on an international scale, and regulators and global companies are key players spanning national boundaries, several studies reveal that local cultures and institutions continue to shape outcomes (Merz and Williams 2018; Merz 2021; Petryna 2009; Thiers et al. 2008). This is consistent with STS research showing how the application of research and regulatory standards aimed at ensuring consistency across different domains, including across countries, is influenced by local social and cultural factors (Timmermans and Epstein 2010). In the context of race and ethnicity, the lack of shared definitions of these terms between countries has been noted in several studies, which also problematize the various ways in which researchers attempt to deal with this definitional and conceptual ambiguity (Smart et al. 2008; Williams 2018). In our own research, for example, we have shown how the English word "race" is often replaced by the Swedish word for "ethnicity" (etnicitet) in Swedish language regulator–approved product labels that summarize the evidence from clinical trials (Mulinari and Bredström 2024a); however, at the same time, racial concepts and categories are routinely used in clinical trials conducted in Sweden (Mulinari and Bredström 2024b). In the following, we continue this line of inquiry by analyzing the discourse of key actors involved directly (e.g., researchers) or indirectly (e.g., regulators) in clinical pharmaceutical research in Sweden in relation to the use of racial concepts and categories.

Findings

"There is no such thing as race"

In our interviews, a prevalent sentiment echoed especially by scientists was the resounding rejection of the scientific value of the concept of race. One scientist with many years of pharmacological research experience emphatically stated: "There is no such thing as race." Furthermore, when confronted with pharmaceutical regulatory guidelines that referenced racial differences, he expressed strong disapproval, dismissing such notions as outdated. He contended, "It makes no sense," and firmly asserted: "Race does not exist. There are no human races."

Another distinguished scientist, a specialist in heart diseases, agreed with this point, declaring: "There is only one race, the human race." He nevertheless admitted to collecting information about the race and ethnicity of research participants, albeit only when mandated by a company's clinical trial protocol (see Mulinari and Bredström 2024b), and very reluctantly. For him, using terms such as "black" was not only distasteful, but also deeply problematic from a methodological perspective.

> Z: The assessment [of race] is meaningless. It is often the patients themselves who assess [their race] or the staff. The patients assess according to the social context, not the genetic context. You are considered black if you live in a so-called black neighborhood where many black people live. [Regardless of] whether you are 100 percent black, both your mother and father are black, or 50 percent, or 10 percent.

He also returned several times to the idea that US categories are not relevant in a Swedish or European context:

> Z: "Are you from Hawaii?" Why would I be from Hawaii? Why don't you ask if I am from Gnosjö [municipality in Sweden] instead?
> Z: There is not a single African American in Sweden, by definition. And almost all of those [who are black] who are in Sweden are East Africans. And there is no [particular] genetic similarity between East Africans and West Africans.
> Z: Is a Spaniard in Spain white or Hispanic?

For this participant, racial and ethnic categories only made sense in their localized context. Thus, he suggested that US categories might be relevant in the United States, particularly for understanding social inequalities and their impact on health. However, beyond addressing social disparities in health, he considered these categories to be scientifically unsound concepts. He even saw a danger in their careless use, and he emphasized that "you are trying to construct a biological race which does not exist." Another participant, a senior manager at a private clinical research organization (which helps pharmaceutical companies with clinical trials) also voiced concerns about possible consequences. His main apprehensions were directed toward the future and the potential repercussions of generating race-based data:

> X: ... All these things [information on patients' race] that we rely on in ... pharmaceutical studies or [studies on] medical devices, or whatever we are looking at. There is an imminent risk that the data can be used in the wrong way. ...
> Anna: What do you mean by using it in the wrong way? Would you like to elaborate on that? What is the risk?
> X: I'm convinced that what happened in Germany can happen again. We should not talk politics here and now, but I'm terrified of where we are heading. Considering that everything is digitized today too. The traceability is there where you can ... Does it get into the hands of the wrong people at the wrong time ... You don't know what this kind of information could be used for.

The scientific unsoundness of race, apprehensions about medical racism, and worries about the mishandling of racial data were but some of the concerns and negative emotions voiced by the individuals we interviewed. Several participants found it challenging to discuss the topic of race altogether. For instance, one participant explained her reluctance to use the Swedish word "ras" for race when translating English-language prescribing information into Swedish, which is common practice in her line of work: "it's not used [in Sweden] and it feels very loaded, it feels like you would get complaints [if you use 'ras']." Another participant started discussing how she had "read something about hypertension and African Americans" but hesitated and commented, "it's hard to know what word to use." She feared that she might unintentionally express herself inappropriately when discussing racial differences. Others took a normative stance, emphasizing that in Sweden, phenotypic features like skin color should not matter. One participant stated, "As we see it in Sweden ... It doesn't matter if you are red, yellow, or blue." Similarly, another participant asserted:

> K: I'm very much like ... "No, I don't see the difference between people." If I would start looking at what they are just because [of skin color] ... No, that's not me!

While it is important to note that not all participants held identical views, the discomfort experienced by all participants in one way or another underscores the salience of a "postracial" frame from within which they spoke, i.e. the widely held belief that race is deeply problematic scientifically and morally, and intimately tied to historical atrocities. However, notions of race still surfaced in the participants' narratives in various and sometimes contradictory ways. Below, we analyze three distinct modes of articulation of race.

A Standard Practice amid Silence

The first mode requires some sorting between what at first sight appear to be several embedded paradoxes; for example, how the use of racial categories is part of everyday routine in pharmaceutical clinical research yet unheard of by some professionals. Additionally, one of the initial surprises revealed in our interviews was that many of the participants had not given these issues much consideration before our discussions, despite their strong negative opinions about race and the value of racial categories, and even though some were aware of their routine use in clinical research. We might have expected that their impassioned responses during the interviews would have translated into robust debates and discussions with colleagues in their daily professional lives, but the narratives pointed in a different direction—towards silence and irrelevance of the subject matter. One participant with over two decades of experience of policy making in the pharmaceutical industry in Sweden, for instance, admitted that despite being aware that the recording of race and ethnicity was part of the protocol of many clinical trials conducted in Sweden, he had not devoted much thought to it.

This sentiment was echoed by several other professionals, such as staff involved in translating prescribing information into Swedish and individuals in pharmaceutical marketing roles. During a group interview, two participants responded with, "I had never thought about it to be honest," and, "Me neither," while another participant remarked that our discussion had "put ideas in their heads." For this last participant, the mere presence of racial categories in regulator-approved Swedish prescribing information was entirely novel information, and she was taken aback by the examples we provided (see Mulinari and Bredström 2024a).

This lack of awareness was not true for all participants, though, as several of them were indeed directly involved in categorizing patients by race and ethnicity in clinical trials. In these cases, ignorance was replaced by routinized but silent practice. All the participants with first-hand experience of applying racial and ethnic categories in clinical trials repeatedly pointed to the United States as the driving force behind the imposition of these categories in clinical research. One participant bluntly stated: "It's the United States that runs the show." These participants further described the role of the clinical trial protocol in which pharmaceutical companies instructed them to record the ethnicity of the patient as "Hispanic or non-Hispanic" and the race as "American Indian or Alaskan Native," "Asian," "Black or African American," "Native Hawaiian and other Pacific Islander," or "White." None of the clinical trial personnel we interviewed had experience with racial or ethnic categories tailored to a European or Swedish setting (see Bartram et al. 2023; see also Smith et al. 2022). Either the US categories were used, or race and ethnicity were not addressed at all.

A research nurse with extensive experience in completing the demographic information in what are called "case report forms" with clinical trial participants confirmed the challenges of this categorization. She described a scenario where a patient who was adopted from another country faces the dilemma of determining their appropriate category. In such cases, she actively collaborates with the patient and, if needed, resorts to online research to resolve the classification:

> Y: ... Sometimes you end up in a discussion and [they] just say, "Well, I'm from that country and I'm adopted and from there, what will I be?" And then it can be like this; then you're standing there and have to google it in the end. ... It can be a bit like this; you have to be a bit of a detective sometimes...
> Anna: And what do you do if someone says that "I have a black mother and a white father," for example?
> Y: Well, then I usually ask like this: "What do you identify as?" Then they can choose [for themselves].

However, in contrast to the United States, where diversity, equity, and inclusion issues are prominent concerns to stakeholders and experts, there appeared to be a conspicuous absence of discourse on these issues in the Swedish context. Notably, there are no Swedish guidelines for promoting racial or ethnic inclusivity in clinical trials, whether in private or public organizations, and we found no evidence of active recruitment based on race or ethnicity.

To sum up, the first mode in which notions of race were articulated in the interviews was as a routinized, everyday practice that neither provoked significant recognition nor concern. We interpret this mode as the outcome of a standardization process in clinical research in which the US categories are simply transposed onto other local contexts where explicit race talk is uncommon or even taboo due to postracial norms, leaving professionals to navigate silently through the resulting difficulties (Mulinari and Bredström 2024b).

Conflating Ethnicity and Race

The second mode of articulation was more familiar to everyone involved, including us as researchers, and it concerned the way race was replaced by ethnicity, sometimes alongside other markers of sociocultural difference (e.g., culture, nationality, religion, migration background or status), in resonance with a postracial discourse (Jonsson 2003; Lentin and Titley 2011; Lentin 2020; Bredström and Mulinari 2022). "We don't talk about race, but we do talk about ethnicity" was a recurring remark from the participants. In many cases, "ethnicity," in their vocabulary, referred simply to notions of cultural differences. In relation to pharmaceutical research, the participants mentioned, for instance, studies of medication compliance and cultural differences in attitudes towards medical treatments, or differences in diet and social habits among different ethnic groups that may affect the use and uptake of drugs.

While participants seemed more comfortable talking about cultural differences, it is worth noting again the absence of these more emic notions of "differentiating markers" in clinical studies and policy in Sweden. A research nurse involved in clinical trials confirmed, for instance, that in terms of social categories she had never come across a study that asked about country of birth or nationality, categories that are used in Swedish official statistics and epidemiological research.

However, what is particularly interesting to our research on race is that the interviewees sometimes simply translated ethnic categories into racial ones (see also Mulinari and Bredström 2024b), as in this case:

> D: We never talk about race in Sweden ... we've never used that word, but we have perhaps said ... talked about ethnicity, that you have a ... that you are Caucasian, that's something that occurs quite often and that becomes relevant in this environment because we are Swedes and so on. (Policy maker, Pharmaceutical Company)

In his way of reasoning, therefore, the ethnonational term "Swede" could be subsumed under the racial term "Caucasian," which in turn is often used as a synonym for "white." Sometimes this implied shifting from ethnicity as a purely sociocultural phenomena to an implicit or even explicit understanding of ethnicity as something biological. Thus, in addition to the first mode, where race is enacted through the mundane and casual use of standardized racial categories in clinical trials, race is also articulated in relation to a postracial framework where ethnicity can become a proxy for race either by direct translation back and forth between ethnicity and race, or by a "biologization" of ethnicity.

Race through Genetics

Another participant, however, made a point of distinguishing between ethnicity as sociocultural and race as biological by referring to a key international pharmaceutical regulatory document (the "ICH-E5") that defines ethnicity and race as distinct concepts (ICH, 1998). This participant is involved in pharmaceutical regulation at a national authority. When explaining the relevance of studying potential racial and ethnic differences, the participant referred to the distinction—also made by the FDA (see above)—between intrinsic (e.g., genetics, metabolism) and extrinsic (e.g., environment, culture) "ethnic factors":

> B: Yes, there is, I can show this table [referring to a table in the regulatory document ICH-E5 adopted by both the European Medicines Agency and FDA on the role of intrinsic and extrinsic "ethnic factors"]. So, there's what's intrinsic and extrinsic. And what I work with ... it's all intrinsic. So, it depends much more on the genes than on external factors. Yes, external factors can play a role. And that's where I make the difference between race and ethnicity [since the ICH-E5 table defines race as only intrinsic]. Because you can be, say, adopted from another country and then be of a different race. Whereas your ethnicity reflects where you live.
> Anna: So [race is] simply biological differences?
> B: Yes, biological.

As seen in the quote, when conceptualizing racial differences, the participant refers to "genes" and biological differences. Similar references to important genetic differences between populations popped up from time to time in the interviews, often presented as a scientific "fact." The main reference was to genetic differences in general, but there was also talk of differences in specific drug metabolizing enzymes, and the possibility of differential vitamin D production between groups due to differences in skin pigmentation, which in turn were genetically explained. Most participants mentioned that some people are slower or faster at metabolizing some drugs, which can have major health effects. While some participants emphasized that there are slow and fast metabolizers in all populations, many also pointed out that there is a greater frequency or likelihood of one or the other in different populations.

Importantly, as has been explored by many social science scholars, in such geneticized explanations, race often serves as a proxy for genetic differences (Bliss 2012; Duster 2015), and so it was in our case. Thus, there are several examples where ideas about genes were translated into the more racially coded notion of "blood":

> Z: Many of us have a lot of genetic influences [genetic heritage] that we have no idea about. I can have Asian or Southern European blood in me even if I don't know about it.

The third mode of articulation of race in the interviews was thus through ideas about population genetic differences. One of the scientists who vehemently opposed the idea that there is such a thing as race argued that there was still value in dividing the world into populations based on genetic similarity that were unmistakably similar to some traditional racial categories. Thus, he claimed that while Africans were too heterogeneous, "Asian" people were genetically homogeneous enough to be categorized as a group for the purpose of pharmaceutical research.

On the other hand, several participants pointed out that the US racial and ethnic classification would not accurately capture what they then saw as "real" genetic and enzymatic differences between populations. However, no one had a better suggestion on how to reliably label these differences:

> Z: If we agree on a global classification that is very rough. Then I could imagine that we would have ten classes or continents in general. It could be some kind of description and there might be some kind of genetic link. If my parents and I come from Europe or Africa or Asia ... you could perhaps divide into three areas. So, it might be possible to find some kind of benefit from that.

Others were less careful and more or less conflated genetic variation with particular ethnic or racial groups, nations, or geographic regions. Some participants with extensive international experience referred, for instance, to race and ethnicity as proxy for what they saw as regional differences globally, differences that presumably would affect metabolism and thus should inform dosing instructions of some drugs. Another participant, who manages clinical trials at a pharmaceutical company, talked about how studies conducted in Africa make it easier to fulfil the FDA requirement of including a certain number of patients of a particular racial group. Black people in Africa, in this case, would presumably be accepted as genetically equivalent to African Americans, a position that many geneticists would ostensibly question. The point, however, is not to question the scientific validity of this claim, but to use it to exemplify how, in certain circumstances, genetic differences are articulated in ways that reinforce rather than challenge the existence of race.

Discussion

"If the aim is to have equal health, equal care and so on, then maybe you have to acknowledge or accept that there are different genetic predispositions," said one of the participants towards the end of our interview. Like most participants, she had talked about how race is an unscientific concept and how difficult it is to use given its historical baggage. However, after talking at length about discrimination in Sweden's health system, and the problem of unrepresentative clinical research data, she concluded by saying that while it is difficult to move beyond the negative connotations attached to race, knowledge about biological differences between groups must be pursued. *It all depends on what the knowledge of biological differences is used for*, she argues.

This participant is one of the few who comes close to the "inclusion-and-difference" paradigm that Epstein (2007) eloquently described as the existing biopolitical framework in the United States for understanding and dealing with social inequality through biomedicine. As Epstein shows, this paradigm suggests that being inclusive in medicine requires attention to immutable biological differences, including race, and he describes the processes through which clinical trial participants are recruited based on race and ethnicity, and how treatment recommendations are developed for specific racialized groups.

In the Swedish case, by contrast, strong discourses on racial or ethnic inclusivity and difference are absent, and so is the practice of recruitment based on race and ethnicity. Furthermore, the dissimilar categories used in broader Swedish society (Swedish, immigrant, Muslim, etc.) vs. pharmaceutical research (white, black, Asian, etc.) means that the categorical alignment that forms the basis of the paradigm in the United States is missing. But this does not mean that race is absent in clinical contexts in Sweden. On the contrary, as we have seen in this paper, race is articulated and made relevant in various ways.

Our findings confirm that race continues to exist as an "absent presence" in Europe (M'Charek et al. 2014), and we also adhere to the critical race argument that race can be enacted through notions of ethnicity and culture in present-day Europe. In our research, however, we found evidence of ambivalent ideas about biological racial difference that we believe are not fully accounted for in the critical race analysis of postracial discourse. For instance, in the first mode in which race is present in the interviews, race is neither circumvented nor replaced by more culturally appropriate concepts such as ethnicity. Instead, it is applied in a rather blunt way, simply as race, in contexts such as clinical trials. We see this as an outcome of successful standardization in clinical research that silences existing tensions and leaves little room for alternatives. In a second mode, race is articulated via ethnicity to refer to supposedly immutable biological differences similar or identical to race. In a third mode, race is also "replaced," albeit not with ethnicity, but with genetics. In this mode, population genetic differences along perceived racial and ethnic lines are referred to as scientifically valid and relevant facts, which is paradoxical given the informants' strong rejection of the scientific value and usefulness of race.

By focusing on the ambivalence of race, where biological differences are simultaneously rejected and recognized, and where biological differences are put forward in a discursive national setting that otherwise focuses on culture, our study contributes to research highlighting how ambiguities are embedded and handled in scientific practices, including practices related to racial classifications (Panofsky and Bliss 2017; Will 2009; Vyas et al. 2020). Conceptualizing ambivalence and inconsistencies is a common thread in STS, and one important analytical point to be made is that the participants seemed to resolve the inconsistencies and contradictions of using concepts and categories that do not seem to fit—and indeed the paradox of using racial categories while resisting the notion that race exists—by attributing the use of racial categories to regulatory and corporate mandates beyond their control. Relatedly, several of the participants expressed great trust in the *institutions* of biomedical research in general and pharmaceutical regulation in particular. For example, some participants expressed confidence that if the FDA requires racial classification, it must somehow be based on solid scientific evidence, although they themselves could not point to that evidence. As one participant put it: "It's based on facts. And then the company just has to comply and include the information [on racial differences]." She also explained that if the study shows that a certain drug is metabolized differently by a certain group, then this information needs to be communicated to the healthcare provider and the patient. "It's about patient safety," she concluded. The fact that such trust in medical knowledge production and the institutions regulating this knowledge production triumphs over scientific, political, and moral objections urges us social scientists to continue investigating the biopolitics of medical research and to consider this research's pivotal role in the contemporary politics of race.

Acknowledgments

The research is funded by the Swedish Research Council (2019–03310; 2023–06166). We would like to thank the reviewers, our colleagues at the Institute for research on migration, ethnicity and society (REMESO) at Linköping University, and the participants in the workshop "Mere labels, racial concepts, or scientific methods? The politics, epistemologies and ethics of naming populations in the life sciences and beyond," Freiburg 30–31 March 2023, for valuable comments and suggestions.

References

Bartram, Isabelle, Laura Schnieder, Nils Ellebrecht, Florian Ruland, Tino Plümecke, and Andrea zur Nieden. 2023. "Categorizing People in the German Life Science: A Systematic Literature Review of Classifications of Human Diversity." *Discover Social Science and Health* 3 (4). DOI: https://doi.org/10.1007/s44155-023-00033-5.

Bliss, Catherine. 2012. *Race Decoded: The Genomic Fight for Social Justice.* Redwood City: Stanford University Press.

Bredström, Anna, and Shai Mulinari. 2022. "Conceptual Unclarity about COVID-19 Ethnic Disparities in Sweden: Implications for Public Health Policy." *Health* 27 (2): 186–200. DOI: https://doi.org/10.1177/13634593221074866.

Brännström, Leila. 2017. "The Terms of Ethnoracial Equality: Swedish Courts' Reading of Ethnic Affiliation, Race and Culture." *Social & Legal Studies* 27 (5): 616–635. DOI: https://doi.org/10.1177/0964663917722827.

Duster, Troy. 2015. "A Post-genomic Surprise: The Molecular Reinscription of Race in Science, Law and Medicine." *The British Journal of Sociology* 66 (1): 1–27. DOI: https://doi.org/10.1111/1468-4446.12118.

Epstein, Steven. 2007. *Inclusion: The Politics of Difference in Medical Research.* Chicago: The University of Chicago Press.

Food and Drug Administration. 2016. "Collection of Race and Ethnicity Data in Clinical Trials: Guidance for Industry and Food and Drug Administration Staff." Accessed June 20, 2024. https://www.fda.gov/media/75453/download.

Goldberg, David Theo. 2009. *The Threat of Race: Reflections on Racial Neoliberalism.* Chichester: John Wiley & Sons.

Hamed, Sarah. 2022. *Healthcare Staff's Racialised Talk: Examining Accounts of Racialisation in Healthcare.* Uppsala: Uppsala Universitet.

International Conference on Harmonisation of Technical Requirements for Registration of Pharmaceuticals for Human Use. 1998. "ICH Harmonised Tripartite Guideline: Ethnic Factors in the Acceptability of Foreign Clinical Data E5 (R1)." Accessed June 17, 2024. https://database.ich.org/sites/default/files/E5_R1__Guideline.pdf.

Jonsson, Stefan. 2004. "Rasism och nyrasism i Sverige 1993–2003." In *Rasismer i Europa: kontinuitet och förändring: Rapport från forskarseminariet* 5 november 2003, edited by Katarina Mattsson and Ingemar Lindberg, 45–77. Stockholm: Agora.

Kahn, Jonathan. 2012. *Race in a Bottle: The Story of BiDil and Racialized Medicine in a Post-genomic Age.* New York City: Columbia University Press.

Kvale, Steinar, and Svend Brinkmann. 2009. *InterViews: Learning the Craft of Qualitative Research Interviewing.* Los Angeles: Sage.

Kuo, Wen-Hua. 2008. "Understanding Race at the Frontier of Pharmaceutical Regulation: An Analysis of the Racial Difference Debate at the ICH." *The Journal of Law, Medicine & Ethics* 36 (3): 498–505. DOI: https://doi.org/10.1111/j.1748-720X.2008.297.x.

Lentin, Alana, and Gavan Titley. 2011. *The Crises of Multiculturalism: Racism in a Neoliberal Age*. London: Zed Books.

Lentin, Alana. 2020. Why Race Still Matters. Cambridge: Polity Press.

M'charek, Amade, Katharina Schramm, and David Skinner. 2014. "Technologies of Belonging: The Absent Presence of Race in Europe." *Science, Technology, & Human Values* 39 (4): 459–467. DOI: https://doi.org/10.1177/0162243914531149.

Merz, Sibille. 2021. "Global Trials, Local Bodies: Negotiating Difference and Sameness in Indian For-Profit Clinical Trials." *Science, Technology, & Human Values* 46 (4): 882–905. DOI: https://doi.org/10.1177/0162243920963813.

Merz, Sibille, and Ros Williams. 2018. "'We All Have a Responsibility to Each Other': Valuing Racialised Bodies in the Neoliberal Bioeconomy." *New Political Economy* 23 (5): 560–573. DOI: https://doi.org/10.1080/13563467.2017.1417368.

Mulinari, Shai, and Anna Bredström. 2024a. "'Black Race,' 'Schwarze Hautfarbe,' 'Origine africaine,' or 'Etnia nera'? The Absent Presence of Race in European Pharmaceutical Regulation." *BioSocieties* 19 (5): 19–36. DOI: https://doi.org/10.1057/s41292-022-00291-7.

Mulinari, Shai, and Anna Bredström. 2024b. "Race in Clinical Trials in Sweden: How Regulatory and Medical Standards in Clinical Research Trump the Post-racial Discourse." *Sociology of Health & Illness* 46 (2): 315–332. DOI: https://doi.org/10.1111/1467-9566.13709.

Mulinari, Shai, Andreas Vilhelmsson, Piotr Ozieranski, and Anna Bredström. 2021. "Is There Evidence for the Racialization of Pharmaceutical Regulation? Systematic Comparison of New Drugs Approved over Five Years in the USA and the EU." *Social Science & Medicine* 280 (1). DOI: https://doi.org/10.1016/j.socscimed.2021.114049.

Panofsky, Aaron, and Catherine Bliss. 2017. "Ambiguity and Scientific Authority: Population Classification in Genomic Science." *American Sociological Review* 82 (1): 59–87. DOI: https://doi.org/10.1177/0003122416685812.

Petryna, Adriana. 2009. *When Experiments Travel: Clinical Trials and the Global Search for Human Subjects.* Princeton: Princeton University Press.

Rabinow, Paul, and Nikolas Rose. 2006. "Biopower Today." *BioSocieties* 1 (2): 195–217. DOI: https://doi.org/10.1017/S1745855206040014.

Regeringskansliet. 2015. Ett utvidgat straffrättsligt skydd för transpersoner m.m. (SOU 2015:103). Accessed June 17, 2024. https://www.regeringen.se/rattsliga-dokument/statens-offentliga-utredningar/2015/12/sou-2015103.

Regeringskansliet. 2017. Ett utvidgat straffrättsligt skydd för transpersoner (Prop. 2017/18:59). Accessed June 17, 2024. https://www.regeringen.se/rattsliga-dokument/proposition/2017/12/prop.-20171859.

Richardson, Sarah S., and Hallam Stevens, eds. 2015. *Postgenomics: Perspectives on Biology after the Genome.* Durham: Duke University Press.

Rotimi, Charles, Daniel Shriner, and Adebowale Adeyemo. 2013. "Genome Science and Health Disparities: A Growing Success Story?" *Genome Medicine* 5 (61). DOI: https://doi.org/10.1186/gm465.

Rottas, Melinda, Peter Thadeio, Rachel Simons, et al. 2021. "Demographic Diversity of Participants in Pfizer Sponsored Clinical Trials in the United States." *Contemporary Clinical Trials* 106. DOI: https://doi.org/10.1016/j.cct.2021.106421.

Silverman, David. 2019. *Interpreting Qualitative Data*. London: Sage.

Smart, Andrew, Richard Tutton, Paul Martin, et al. 2008. "The Standardization of Race and Ethnicity in Biomedical Science Editorials and UK Biobanks." *Social Studies of Science* 38 (3): 407–423. DOI: https://doi.org/10.1177/0306312707083759.

Smart, Andrew, and Kate Weiner. 2018. "Racialized Prescribing: Enacting Race/Ethnicity in Clinical Practice Guidelines and in Accounts of Clinical Practice." *Sociology of Health & Illness* 40 (5): 843–858. DOI: https://doi.org/10.1111/1467-9566.12727.

Smith, Zachary, Emily Botto, and Kenneth Getz. 2022. "Quantifying Diversity and Representation in Pivotal Trials Leading to Marketing Authorization in Europe." *Therapeutic Innovation & Regulatory Science* 55 (5): 795–804. DOI: https://doi.org/10.1007/s43441-022-00421-0.

Sveriges Riksdag. 2008. Diskrimineringslag (2008:567). Accessed June 17 2024. https://www.riksdagen.se/sv/dokument-och-lagar/dokument/svensk-forfattningssamling/diskrimineringslag-2008567_sfs-2008-567.

Timmermans, Stefan, and Steven Epstein. 2010. "A World of Standards but Not a Standard World: Toward a Sociology of Standards and Standardization." *Annual Review of Sociology* 36 (1): 69–89. DOI: https://doi.org/10.1146/annurev.soc.012809.102629.

Thiers, Fabio A., Anthony J. Sinskey, and Ernst R. Berndt. 2008. "Trends in the Globalization of Clinical Trials. *Nature Reviews Drug Discovery* 7 (1): 13–14.

Vyas, Darshali A., Leo G. Eisenstein, and David S. Jones. 2020. "Hidden in Plain Sight: Reconsidering the Use of Race Correction in Clinical Algorithms." *New England Journal of Medicine* 383 (9): 874–882. DOI: https://doi.org/10.1056/NEJMms2004740

Washington III, Charles, Matthes Dapas, Arjun Biddanda, et. al. 2002. "African-Specific Alleles Modify Risk for Asthma at the 17q12-q21 Locus in African Americans." *Genome Medicine* 14 (112). DOI: https://doi.org/10.1186/s13073-022-01114-x.

Will, Catherine. 2009. "Identifying Effectiveness in 'The Old Old': Principles and Values in the Age of Clinical Trials." *Science, Technology, & Human Values* 34 (5): 607–628. DOI: https://doi.org/10.1177/0162243909340261.

Williams, Rosalind. 2018. "Enactments of Race in the UK's Blood Stem Cell Inventory." *Science as Culture* 27 (1): 24–43. DOI: https://doi.org/10.1080/09505431.2017.1322054

Wolgast, Martin, Sima Nurali Wolgast, and Henrik Levinsson. 2024. "Effects of Patient Gender, Ethnicity, and Socioeconomic Status on Psychiatric Assessments: A Vignette-Based Experimental Study." *Scandinavian Journal of Psychology*. DOI: https://doi.org/10.1111/sjop.13004.

A Matter of Getting Used to It?

Doing Medical Science with Race in a Race-Cautious Country

Nils Ellebrecht

Abstract *This study investigates the contentious role of racial classifications in medical research in Germany. In contemporary Germany, discussions about race are approached with caution, and the German word 'Rasse' is considered taboo when referring to human groups. At the same time, racial classifications continue to be used in the field of medical research, primarily due to their undiminished popularity in the United States, which gives them global circulation via English as a lingua franca of science. This contributes to their ongoing disciplinary compatibility. Consequently, race remains a tacit yet enduring scientific concept in medical research institutions in Germany. Where controversies do occur in the field, they mostly center on terminological issues and how to avoid them; rarely do they go further. This study ideal-typically distinguishes between three medical actors and how they deal with these difficulties. Confronted with racial classifications in research, medical-science novices (1) are compelled to reconcile their learned social conventions with the scientific standards in their field of research, leading to terminological vagueness and conceptual uncertainty. Established researchers (2), by contrast, differ in understanding race either as a global biological observation schema or as a local schema of group differentiation whose use and meaning is nationally bound, while crossover researchers (3) point to the social risks of science working with racial classifications.*

Introduction

Syringomyelia, sarcoidosis, shigellosis: various fields of medicine use specific criteria, technical terms, and classification systems to distinguish among ailments they consider relevant. Because of this high degree of specialization, many diseases and patient groups are familiar to only a handful of experts and remain largely unfamiliar to physicians in other fields. Such specialized forms of sorting contrast starkly with commonly made *human distinctions* such as gender, age, or descent. Not only do all medical disciplines work with the latter type of categories; they also circulate outside and independently of medicine and science. Ubiquitous everyday categories for distinguishing between different people are characterized by their use in a wide variety of situations and social worlds.

This is precisely what makes them categories at all, distinguishing them from both ad hoc and seemingly arbitrary distinctions, on the one hand, and from disambiguated, precisely defined classifications that are predominantly of interest to experts, on the other (Hirschauer 2023). This article will engage with a differentiation schema—that of Rasse/race—that has long been a tangled mixture of both: that is to say, of an everyday category and a scientific classification which, over a long period, sought to find, and claimed that, a precise and systematic scientific distinction could be made between races. The matter, however, is even more complicated in Germany, as this schema in fact includes a third thing: a taboo. As such, it is avoided by both scientists and society at large. This does not mean that it has entirely disappeared; it hiding under the surface, as something known, but prohibited. This paper explores how medical scientists in Germany navigate the triple meaning of Rasse/race—as a scientific classification, a common descriptor of personal characteristics, and a taboo.

Given the social significance of everyday categories for distinguishing between different people, one might think—or wish—that the medical sciences would first clarify the implications of their own use of these categories. However, unlike sociology (Bourdieu 2002, 484–87, 2018; Wacquant 2022), these disciplines show little interest in reflecting on such categories as part of social orders and considering how they themselves co-produce and reify such classifying schemas by adopting them as people or population descriptors. The medical disciplines show little sensitivity in understanding such categories as expressions of social power and the relations comprising it, in noticing associated stereotypes, or in recognizing the blind spots that these observational schemas entail. The object of medical observation is the *sick individual* (Luhmann 1990, Gibson/Boiko 2012), not so much the *sick population* (Rose 2001), and even less so an innately antagonistic society (Marchart 2018), with its semantic struggles and social conflicts over distribution, recognition, and identity. Questions of power only come into view in the medical sciences when they contribute to identifying the risk factors for a particular disease in people who fall into a certain category of persons. The social effectiveness of the category itself, however, is left out of the equation.[1]

And yet: if there is one human classification whose use in the life sciences is controversial due to its negative social and historical significance, and that has even been largely rejected in anthropology, human genetics, and other life sciences, then it is that of Rasse/race. Genetic transitions between races are described as so fluid, gradual, and seamless[2] that the vast majority of human geneticists today no longer understand races as distinct groups, as useful units of analysis. As one research group, writing in *Science*, summarized: "[R]acial classifications do not make sense in terms of genetics" (Yudell et

---

1 The different conceptualizations of race or the different ways of dealing with racial classifications in medicine and epidemiology cannot be discussed in detail here. For a comparison of the two bioscientific disciplines, the works of Andrea zur Nieden et al. (in this volume) and zur Nieden (2025) provide a good reference. These are devoted to epidemiological studies with German participation, which were investigated as part of the same meta-study that also formed the basis of this article.

2 Frank B. Livingstone (1962) succinctly summarized that there are no genetically distinct human groups, but only genetic trait differences in the formula: "There are no races, there are only clines."

al. 2016, 565).[3] Classifying human collectives in terms of race is considered historically outdated, as it lacks any genetic correlate.[4] However, although scientific evidence today speaks clearly on this matter and the theory of the existence of races has been refuted, social knowledge, popular belief, and even their scientific application does not automatically dissolve as a result.

Authorities and researchers have reacted differently to the lack of genetic evidence. While in most parts of the world racial classifications have been abandoned altogether, in others race has come to be understood as a category of social affiliation that can be personally meaningful and ascribed to oneself and others. Such an understanding of race as a social force rather than a biological fact means that the responsibility for the correct racial classification of a person is no longer left to the natural scientist, but is placed in the hands of those who are the subjects of research. Since race has become a subjective criterion, today only the category claimed by individuals is considered valid. Race is now—at least in the United States—a personal category self-reported by individuals. As such, patients in medical studies are now asked to state which race they consider themselves belonging to. This procedure takes into account the conviction that there are no objectively valid racial characteristics that can be observed from the outside and grasps race as a genuine feature of identity, as a social identity.

At first glance, this subjective self-ascription shields the concept of race against an essentialist, even naturalizing conceptualization. At the same time, however, it becomes a precise, distinct classification (If, for a moment, one disregards the possibility, claimed by a growing number of people, of classifying themselves as mixed race or as members of two or more races). Moreover, as this article will also show, this change in classificatory authority does not prevent the further course of research from pretending that these categories are biologically discrete units.[5] And finally, the patient is not always consulted during the classification process; often it is a medical actor who silently determines a patient's race by means of a visual diagnosis. It may therefore be noted that the procedure of self-reporting has not only shifted, but even exacerbated, the trouble with race. While the classification has now officially been completely subjectivized the dataset, once collected, offers little protection to individuals against being objectified and instrumentalized as a point of data taken to represent a fact in the natural world.

Racial thought remains efficacious, albeit in different variants: as a suppressed, yet latent belief; as tacit conviction; as an identity that is ascribed or/and adopted; as discursive strategy; as a self-reported category documented in official or scientific reports;

3 In the same breath, however, human genetics has switched to ancestry analyses. However, the extent to which these replicate racial classification or use and confirm ethnic typologies is disputed. On the one hand, genetic ancestry research works with admixture analyses, i.e., it no longer assumes that pure types are to be found. On the other hand, it can be assumed that the historical existence of the latter is implied when talking about admixture. Cf. the work by Ricardo Gomes Moreira in this volume.

4 Recently emphasized by four German zoologists and life scientists in the "Jena Declaration" (Fischer et al. 2019).

5 Some studies even control for whether the self-reported race was stated correctly by comparing it to genetic information from the respondent. This was the case in one of the studies (Schenk et al. 2020) from our sample (see below).

and, last but not least, as a false belief or myth that must be fought against. Sociology, therefore, has not completely rejected the concept of race. Many sociologists argue that races still exist as *social facts* (Durkheim 1982, 50–59) or social constructs. In Max Weber's definition (1972, 237), races are a special case of ethnic groups. As ethnic groups, they exist insofar as they are believed to be real; they are communities of descent in which people subjectively believe. This belief gains social reality not only when other people are categorized and described as members of foreign races, but also when individuals use this external identification to describe themselves. The socially effective reality of imagined races begins where the belief in them has real consequences in action (Thomas and Thomas 1928), i.e., when Ego considers their own race or that of Alter in their actions. This approach of social constructionism, as taken by sociologists, is therefore primarily concerned with understanding race as a "social invention," an imaginary taken by individuals and communities as a basis for action.

This sociological concept of race can be extended in many directions. Loïc Wacquant, for instance, emphasizes the *illusory nature of race* by distinguishing it from other established social classifications. While other social classifications, he argues, in addition to their symbolic reality, have "a self-standing material foundation independent of cognition: class (the mode of production), gender (the mode of reproduction), age (the unfolding of biological life), citizenship and nationhood (affiliation with a state)" (2022, 74), race remains a mere cognitive construct, in short: an ideology. Yet, racial constructions categorically deny their own fictionality by consistently asserting themselves as eternal, natural entities. This rejection of any historicity of their own makes races special ethnic groups. Furthermore, Robert Brubaker warns against an improvident *groupism*—"the tendency to treat ethnic groups, nations, and races as substantial entities to which interests and agency can be attributed" (2002, 164). All too often, sociologists are still subject to this tendency. Races, however, remain merely cognitive categories; even if one often thinks and speaks of races as racial groups, there are no such real groups "behind" these categories. Cognizant of this fact, Adam Hochman even goes one logical step further than Weber, Wacquant, and Brubaker by casting doubt on the assumption of the social existence of ethnic groups in general. At most, he argues, it is "*racialized groups*—groups misunderstood to be races—[that] are real" (Hochman 2021, 459).

"Today and in the future, not using the term race should be part of scientific decency."

Although sociology and human genetics share the same conviction—that there is no biological essence of race—these two disciplines take contradictory positions regarding the significance of races. Unlike human genetics, which can easily demand of its own discipline that "[t]oday and in the future, not using the term race should be part of scientific decency" (Fischer, 2019), sociology cannot treat races as scientifically insignificant—precisely because races are not socially insignificant. As long as racial classifications are an important and disputed part of social reality, sociology cannot comply with such a demand.

Both perspectives and approaches are scientifically justifiable; scientific misconception and societal efficacy are not mutually exclusive.[6] However, the lack of a genetic correlate does not mean that no biological differences can be statistically measured between the races. There are two main reasons for this. The first is methodological. Many medical studies commit a scaling error when they use racial classifications to statistically calculate genetic or phenotypic differences. A metric variable (human variance) is scaled nominally (race typology). If the scaling error is not recognized, significant differences can still be statistically claimed. Our study dealt with several research papers that demonstrate significant genetic differences between the races. These consider, for example, the shape of the skull, the eye, the skin, or the liver. All these studies forget that the sharp boundaries between the races are arbitrary. The second reason is not a methodological error but is due to the interaction of social inequality and health disparities. The social significance of everyday categories of personal identity is also expressed physically and therefore can be measured medically. The unequal distribution of resources, lifestyles, experiences of discrimination, and many other factors or conditions differ between people assigned to different everyday categories of personal identity and extend into the biology of the body; ultimately, this is also expressed in terms of health. This is undisputed; the only question is how much weight one sociodemographic category (e.g., socioeconomic status) has over another (e.g., ethnicity).[7]

How racial classifications are used does not only depend on the scientific conceptualization of race, but also on the very different ways in which these classifications are structured around the world. It is not just that different races are recognized in different countries and, correspondingly, that official classifications differ: Brazil, for example, identifies a race called *pardo,* loosely meaning brown or mixed race, the United States, on the other hand, does not have such a category, but allows people to be assigned to more than one race in surveys. Above all, however, in many parts of the world there are no racial classifications in use at all; in fact, the word race and racial classifications are socially ostracized in some countries (Morning and Maneri 2022). Many European nations, for instance, are regarded as or regard themselves as postracial, color-blind, or race-mute societies. Against the background of National Socialist atrocities during Germany's Third Reich, the idea of races existing is officially considered to be scientifically untenable, morally reprehensible, and politically dangerous. With this historical lesson in mind, race, or rather the respective translation, such as the German term Rasse, has become an "ugly word," a taboo, a term one doesn't say out loud, to explain the social world around oneself. International studies, should they resort to race to classify their

6 Strictly speaking, however, both positions overlook or deny the other. Sociologists must acknowledge that race is not only a political project, but has a biological dimension (M'charek 2023). A strict ontological self-limitation to "the social" obscures the view of the plural and overlapping modes of existence of race (e.g., Latour 2013, consider also Knorr-Cetina 1989). Geneticists, as well, take the easy way out, if they imagine that abandoning the genetic relevance of race also disposes of its social significance. Rather, they should ask themselves how much race is still contained in their concepts—for example (biogeographical) ancestry.

7 The significance of race as a social health factor is sometimes doubted, and class affiliation or socio-economic status is considered to be much more relevant. For the discussion in Germany, see Schenk 2016, 70–71. For more information on this topic, see the study by Krieger et al. 2005.

patients and subjects, are therefore faced with the fundamental challenge of having to deal with different taxonomies and acceptance levels depending on the social context.

This brief discussion has already made clear how fraught it might be to use of racial classifications in the life sciences. As Germany is considered a paradigmatic example of a *race-cautious* society[8], researchers working in Germany should be particularly affected by this issue. The German word for race, *Rasse,* is no longer viewed as a concept that can be legitimately applied to the categorization of people, whether socially or scientifically. The affirmative use of Rasse to describe groups of people has largely disappeared from everyday language, public usage[9], and life-science discourse (Lipphardt 2009; zur Nieden 2014; Mulinari and Bredström 2024a; DeZIM 2023, 193). Unlike in the United States, there are no regulations in Germany stipulating a distinction between races or ethnic groups in scientific studies. Nevertheless, our research group was able to identify racial terminology and racial assignments of patients or subjects in a series of medical science studies conducted in collaboration with German research institutions (Bartram et al. 2023).

This paper examines these medical studies from the perspective of the sociology of science. We begin with three questions: *Why do medical scientists in Germany conduct research with race? How do they deal with racial classifications? And what conflicts do they encounter in the research process?* We are particularly interested in the reasons researchers have, contrary to Germany's postracial self-image, for continuing to use racial classifications, and how they manage the balancing act between societal demands and the explicit use of racial categories in their research. How do they reconcile the medical-scientific meaning of race with the race-skeptical social context? Will the two usages remain, as Mulinari and Bredström (2024b, 327) with Annemarie Mol (2007) assume, "two divergent enactments of race" circulating and being cultivated separately in their own social spaces? As a scientific insistence on race existing next to a lived taboo? Or do medical scientists who are required to process both ways of meaning themselves harbor reservations? Have they observed tensions and conflicts? Is it even possible to recognize the contours of an arena in which members of this and other social worlds are struggling over the correct meaning and use of race?

8 In their recent study, Morning and Maneri (2022, 39) distinguish between race-conscious— "or 're-alist' to its proponents"—and race-skeptical— "or perhaps 'cautious' to its adherents"—positions. I extend this distinction to differentiate between societies that officially use racial classifications and those which are "race-mute" (Jugert et al. 2022) and regard themselves as color-blind or postracial. Since, as we will see, a strong linguistic taboo has been established around the term *Rasse* (Chin 2017), Germany is an extreme case of a race-cautious society—at least on the level of discourse.

9 When the word *Rasse* is used, it is mainly found in historiographic studies or in connection with the question of whether the word *Rasse* should be removed from legal texts. The latter also shows that there are calls to remove the word from public use. Beyond this, it is completely unproblematic to talk about "Pferderassen" (horse breeds) and "Hundrassen" (dog breeds) and, in this context, about "reine Rassen" (pure breeds) and gemischte Rassen (mixed-breeds). The linguistic distinction between breeds and races, as it is made in English, is rather unfamiliar in the German context (Heintz 2024).

Method

At the heart of this paper are 17 interviews with scientists who were affiliated with German research institutions at the time of publication of a medical article that used a racial classification. These articles, published between 2019 and 2021, were identified within a metastudy that screened 546 research papers from all life science disciplines for the human classifications they apply (Bartram et al. 2023). For the present study, we selected only these 53 articles that used the terms "race" or "racial" in their study design and whose first or last author is affiliated in Germany to a medical department (this was our criterion for a medical paper).

The 53 studies included correspond to a complete survey of the three selected publication years for the aforementioned criteria.[10] We can assume that, for the period chosen, we have found almost all published medical studies in which researchers at German institutions worked using concepts of race. All of these studies were written in English and, thus, used the English term "race." No article applied the German term *Rasse.* Only four of these studies were conducted and coauthored by researchers who were all exclusively affiliated with German research institutions; all other articles were written by international teams.

Table 1: Illustrative examples of classifications and categories used in the studies examined

Publication	**Data source**	**System of racial classification**	**Race descriptors used in the study**
Ankerst et al. 2018	primary data	N/A	Black or African American, Other, Unknown
Hardtke et al. 2018	primary data	Brazil census	White, Black, Brown
Hofmann et al. 2018	Osteoarthritis Initiative (OAI)	United States Office of Management and Budget (OMB) (assumed)	Caucasian, Other Non-white, African American
Kromrey et al. 2018	Study of Health in Pomerania (SHIP) + primary data	N/A	Caucasian
Suda et al. 2021	primary data	N/A	Caucasian, Asian
Trong et al. 2018	Human Connectome Project	OMB (assumed)	Asian, African American, Caucasian
Uhlig et al. 2020	US National Cancer Database (NCDB)	OMB	White, African American, Others

10 A list of these 53 articles is available on request.

Statistical analyses of this corpus provided a first impression of the scope of racial classification (tab. 1) and the factors contributing to its usage. However, they do not tell us much about the social context, the discussions within the research teams, and the motives that guided their decisions. For this, we subjected the medical articles to qualitative analyses of and conducted interviews and conversations with 17 authors (see figure 1 for our recruiting strategy). For the following analyses, it is of utmost importance to bear in mind that all the results we derived from the empirical data concern articles and researchers who have worked with racial classifications in recent studies. Thus, we cannot say anything about, for instance, scientists who decided against or never thought about working with race at all.

Figure 1: Participant recruitment strategy

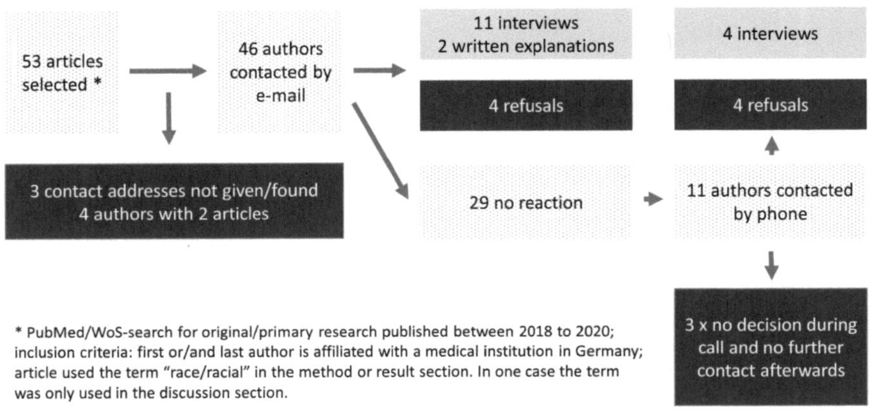

The analysis was carried out using grounded theory and took a highly inductive approach to the formation of codes and concepts. Relations among codes worked out were initially traced using Juliet Corbin's and Anselm Strauss's coding paradigm (1990, 96). The core phenomenon in our study was "approaching the delicate category *Rasse*/race in medical research," to which other codes were placed in relation as causal conditions (e.g., motives, aims); contextual conditions (e.g., social context: race-cautious society); intervening conditions (e.g., field role, social world position); strategies (e.g., word substitution, justification, voice); and consequences (e.g., affirmation, accustoming, conflicts, concerns). In developing our theoretical framework, we constructed four ideal types (Kelle/Kluge 2010). Each ideal type is tightly linked to the role of the interviewee in the scientific field. We therefore distinguish the (1) the *medical novice* from (2) the *established medical scientist* and (3) the interdisciplinary *crossover scientist*. A fourth type is a special case that is largely independent of its role in the scientific field. Due to its significance for our study, it is presented as a further type: (4) the *racialized researcher*. Each ideal type is distinguished by a specific configuration of the five coded dimensions. These are: (I) scientific role, motives and research agency; (II) approach to the Rasse/race distinction; (III) self-positioning towards/in the racial classification (figure 3); (IV) conceptualization of race; (V) science–society relationship.

For further systematization and theory building, we mapped these four types and other actors mentioned by the interviewees within the social worlds and social contexts involved (figure 2). This method is based on Adele Clarke's arena mapping (Clarke, Friese, and Washburn 2018). Arenas are shared discourse spaces in which "various issues are debated, negotiated, fought out, forced and manipulated" (Strauss 1978, 124). Social arenas are not only the locus of small, isolated quarrels. The controversies found within them reflect divergent interests, political positions, and various perspectives with which specific groups, organizations, or entire social worlds who are mobilized on behalf of specific causes generally identify (Clarke 1990). In this respect, arenas are never quiet places, but always places of voice (Star and Strauss 1999). The concept of the noisy arena thus stands in a certain contradiction to the assumed race-muteness of German society.[11] At the end of the article, we address the question of the extent to which the field we examined is actually part of an arena.

Finally, it is important to point out a methodological-theoretical reflection that arose in the course of arena mapping. It concerns the uncertain status of social contexts. For the sake of simplicity, social contexts could be understood as separate social worlds. Similarities between worlds and contexts certainly exist; however, contexts are far more difficult to distinguish from other contexts, as well as from social worlds: Where does a national or sociocultural context begin and where does it end? Where does the postracial self-conception that many European countries are said to have border on other, on different self-conceptions? For Strauss (1978, 122), every social world is characterized by a "primary activity" that is performed in specific locations, often set up precisely for this activity, and with the help of dedicated technologies. The research work of medical scientists, for example, takes place predominantly in laboratories and university clinics, where they have access to specialized instruments they need for their specific fields of medicine. In close reference to this notion of social worlds, but with a stronger emphasis on the formation of latent "naturalizations," Bowker and Star (1999, 294) speak of social worlds as "communities of practice." Both Strauss's, and Bowker's and Star's, focus on a primary activity precludes the conceptualization of national, Western European, or other cultural contexts as social worlds. Nor do we find it helpful to think of contexts as social worlds. Rather, we argue that social worlds do not, as social world/arena maps graphically suggest, exist in a vacuum, but are embedded in (multiple) social contexts that vary according to time and place. Due to their lack of expertise and technical specialization, social contexts are less than social worlds. But at the same time, they are more than them, because they are more heterogeneous and boundless than the latter. Though the boundaries of social worlds are not carved in stone and are drawn and fostered by their members (Gieryn 1983), they prove to be much sharper than those of a cultural area.

---

11 Similarly, the public debate over the value and meaning of the term *Rasse* in German law, which has been going on for several years, refers to both the arena and the taboo. Some are calling for the word Rasse to be removed from legal texts. It is used there to formulate a general protection against racial discrimination. The main argument here is that the very mention of the term grants it a reality that it does not possess. The public controversy therefore also reveals a strategic expansion of the conceptual taboo.

Figure 2: Social world map from the perspective of scientists having worked with race in their medical research

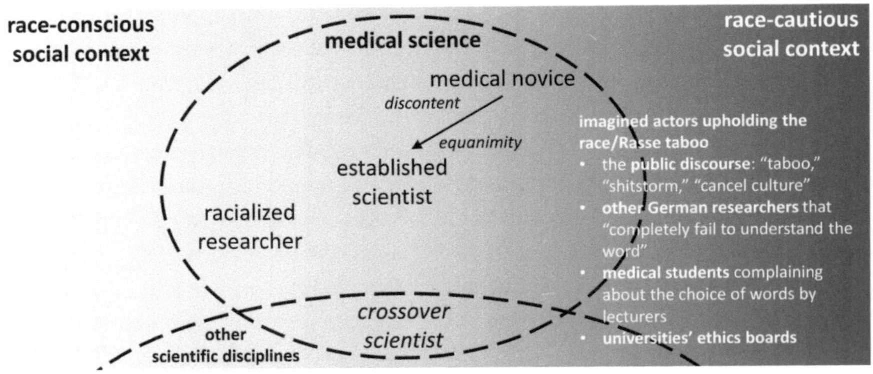

How Novices, Established Researchers, and Crossover Scientists Approach Racial Classifications

The results presented below are divided into three sections, each of which focuses on one of the identified types and the way they typically deal with racial classifications. In the first two sections, the (3.1) *discontent of medical novices* with race is contrasted with the (3.2) *equanimity of established medical scientists*. Of the 17 authors we interviewed, 6 have been working in research for more than 10 years. They are senior physicians or professors of medicine aged between 42 and 65 who, as *physician-scientists*, occupy the intersection of medicine and science. They represent the medical researcher establishment, not only because of their extensive research experience and their leading position in the field, but also because racial classifications represent a heuristic for them with which they work largely naturally and matter-of-factly.

According to Howard Becker (1966), the establishment can also be described as *entrepreneurs*, as they are particularly committed to their world, are actively engaged in it, and group themselves around its center. In this position, they in turn mobilize those around them (Clarke 2012, 86), including doctoral students and interns/resident doctors. The latter constitute a second group of interviewees who represent the medical novitiate. As novices, the interviewees are dependent on the established researchers, as the latter guide them, support them throughout their careers, promote them, evaluate them, hire them, etc. Four of the interviewees largely correspond to the type of medical novice. On the one hand, they are characterized by their early to mid-career phase at the time of publication (doctorate completed, postdoctoral (*Habilitation*) degree not yet taken or not considered). For some of them, the time spent in an academic institution remains a brief stopover on the way to becoming practicing physicians. This career phase corresponds to the comparatively young age of the novices at the time of their interviews (≤ 37 years). Central to our study is the novice's characteristic discontent with race. Compared to the established scientist, the medical novice has not become habituated to the customs of medical science, at least as far as the matter-of-fact handling of race is concerned. Unlike in the case of the established scientists, racial classifications are not proven or familiar instruments for the novice. So far, they have mainly known race and Rasse as discredited and dangerous terms.

While ten interviewees could be considered to largely represent one of these two types, two of the medical scientists interviewed cannot be clearly categorized. One is discussed in subsection (3.1.2) *Loyalty and Voice*, the other in (3.2.2) *Race in the Lab: Research and Identity*. These sections show how different the relationship between scientist and racial classification can be in terms of self-positioning, motive, and aim of research. The final section of the main part (3.3) *Crossover Scientists: Science and Society describes a third type of scientist.* Crossover researchers work in scientific disciplines outside medical science and have no basic medical training. However, they conduct research as part of medical-science teams and sometimes contribute views on race and racial classifications from their own fields. The way they approach race emphasizes the social responsibility of science.

The Novices' Discontent with Race

The interviews clearly show how differently racial classifications are dealt with, depending on whether the interviewees are medical novices or established medical researchers. For many young interviewees, the work that brought them to our attention was conducted as part of their doctoral studies or shortly afterwards. During this phase of academic socialization, they often conducted research abroad, belonged to an international project group, or had access to US medical databases. It was in this context that they came into contact with racial classification for the first time. In the interviews, they look back on this situation and describe how the initial contact caused them discomfort and irritation. Dr. Robin (a 35-year-old junior physician[12], para. 86) said, "the term *race*[13] also initially struck me as negative. And I had to work on understanding it first." In some cases, the discomfort persists to this day: "As I said, I always feel uncomfortable using the word *race* because it is so closely related to the German *Rasse*[12] in terms of terminology. I find that extremely difficult, especially for me as a German" (Dr. Müller, physician, 35, para. 22). At the end of the acquisition phase, race did not become a natural concept of medical science for every novice. Some interviewees stated that they still had difficulties with it at the time of the interview.

The initial irritation derives from the proximity of the word "race" to the German *Rasse*. For the novice, both words are "closely related"; one term triggers associations with the other. In particular, race is associated "with our National Socialist past" (Dr. Müller physician, 35, para. 82). Moreover, race is not only a "classification, but also a hierarchy" (ibid, para. 36) of groups of people. It asserts a natural order that is given among people. It is this "incredibly negative connotation" (ibid., pos. 36) of *Rasse* that is invoked by the term race and is perceived as unpleasant. Morning and Maneri (2022, 85–101) found almost identical associations in their study of Italian students; their American interviewees did not have these.

---

12 To increase anonymity, the ages of interviewees are all rounded to the next multiple of five.

13 Note: Whenever the word race or *Rasse* appears in the interviews, it is left untranslated.

Many interviewees find it audibly difficult to pronounce Rasse, especially when it is expressed for the first time in the interview, which was always left for the interviewee to do. In addition to short breaks, aposiopesis, and filler sounds, apologetic interjections also herald the term and at the same time defuse its use by performatively bracketing the breach of language taboo:

> "um, uh (.) yeah, you aren't allowed to, you—I believe—don't say *Rasse* anymore, right?" (Dr. Juli, 40, para. 10)
> "um, how, how d-. What do you call it now? There's no longer a term for *Rasse*, something, like, I don't even know the PC [= politically correct] term." (Prof. Kreh, 55, para. 43)"in the literature, that, yes, race, you don't really want to say *Rasse* as a German" (Dr. Müller, 35, para. 10)

The concept of cognitive dissonance can be used to explain the exact grounds of this discontent. Cognitive dissonance describes an inner discomfort that is caused by two conflicting elements of knowledge in an individual (Harmon-Jones and Mills 2019). Therefore, it is about more than just a bad feeling associated with a certain term. The negative connotation of Rasse that the use of "race" evokes only explains half of the novices' discontent. The other half is due to the liberal use of race by other researchers or in medical databases. As "race" is treated as a self-evident and, at first glance, unproblematic scientific classification in international research groups and databases, it stands alongside other terms such as "age," "class," or "sex" on an equal footing. In the interviews, however, its use in non-German contexts does not provoke disgust or rejection; it is not interpreted as a breach of a taboo. Rather, the novices are troubled by the contradiction they experience between their own feelings and their absence in others: "I just realized that the use of the word is totally normal, including to stratify the analyses. I didn't have the feeling that it was a controversial topic there [in the United States], at least in the study group. That would be the case here for sure. Here in Germany, it would already start with just collecting data on *Rasse*." (Dr. Müller, physician, 35, para. 50)

For cognitive dissonance to occur, a further step is eventually required, with which the mere observation of different usage patterns acquires meaning for one's own behavior. Only when internalized beliefs come into conflict with new, alternative action guidelines does this ultimately arouse cognitive dissonance (Harmon-Jones and Mills 2019, 13). Now one not only *knows* that one's own values and feelings are in conflict with a different, foreign standard of behavior, but one is also forced to *choose* between these two—and what's more, to use terms and concepts that one's own behavioral repertoire did not previously include or permit. In this situation, a tension arises between two incompatible behavioral expectations. This is where the theory of cognitive dissonance can be applied, asking how the resulting psychological tension can be reduced. How can a consistent self-image be kept in view of the inconsistency between (old) attitude and (new) behavior?

Workaround Strategies

So, how do novices deal with the demands placed on them? The answer can primarily be found in the publications of the researchers we interviewed. Scientific publications contain and document a large number of decisions. These can no longer be reversed by the authors, but they can be explained, defended, or critically appraised—for instance, in the interviews we held with the authors. One younger author we approached declined to be interviewed, referring to her own lack of expertise on this issue.

A common solution to deal with the conflicting requirements is to switch from the term "race" to "ethnicity." When asked about this modification, which was not consistently implemented in the publication, one interviewee described the internal conflict behind it and the way it was dealt with in the publication, which in retrospect she disqualified as inconsistent:

> Well, I changed to ethnicity, yes. I think at some point..., I thought that this thing about race was a bit too..., but this is inconsistent, of course, so I have to..., I wouldn't do that anymore; so, if the database says to call it Black and White, then of course you should just call it Black and White, because then you're basically the least vulnerable, because it's no longer your own formulation, right? (Dr. Juli, physician, 40, para. 48)

The cognitive dissonance is expressed in the search for expression, and even more clearly in the rapidly changing weighing up of arguments (yes, but, I have to, I would now). The novice is struggling with her terminological deviation and sees it as a scientifically inconsistent view of the usage in the database. Even if she does not express it directly, the motives of her action are nonetheless clear: "this thing about race" ("*das mit dem Race*") was "a bit too..." In doing so, the interviewee distances herself from the term by means of singularization ("this") and objectification ("thing"). Secondly, by using the adverbial "a bit too..." she shows that "race" in some way falls short of or oversteps the bounds of appropriateness or acceptability, with a more precise adjectival determination in this case left hanging. Either way, there is something excessive about "race." In her eyes, "ethnicity" would be the measured, socially palatable alternative.

She pays for the decision she made to refrain from using the term race, which was correct in terms of her own value system, by making herself scientifically vulnerable. At the end of a series of aborted sentences, she, a novice, arrives at the view that retaining the given term race is the less risky communication strategy, because that way the choice of word cannot be attributed to her. In the subsequent paragraphs, however, the interviewee defends her use of race independently of this. Throughout the interview, she weighs different terms against each other, first stressing "the difficulty of talking about *race* in German" (Dr. Juli, para. 34), but then commenting rather dismissively that "ethnicity" is just "a pretty word." Finally, she returns to the use of the term "race" and its German translation as *Rasse*: "Well, it doesn't help much to avoid saying things. Um, yeah, maybe it's more um..., yeah more... I ... I mean I have a real problem with it. It's really hard." (Dr. Juli, 40, para. 38).

Later we will see that established medical scientists separate the scientific use of racial classifications more sharply from its moral-political dimension. In the case of Dr. Juli, these two meanings have not (yet) been divorced from each other, and there is no clear dividing line. This may have led to the publication occasionally switching from the touchy term race to the less controversial term ethnicity, or at least this is what the explanations in the quoted excerpt suggest. In retrospect, she criticizes her own usage as inconsistent from a scientific point of view. But even if the switching is not without contradiction, it can be understood as an early attempt to deal with the inner conflict. At first glance, this maneuvering may not appear to be a viable long-term solution. But similar to ignoring, suppressing, and then forgetting the experienced inner conflict, it offers a workaround for cognitive dissonance (Festinger 1957).

When race has become completely replaced by other terms, the terminological workaround can certainly be said to have become a strategy of purification. Frank Reeves (1983) speaks of "sanitary coding." It can be assumed that some publications simply replaced the term "race" with the alternative designation "ethnic group." In these cases, the reader might not notice the change has been made (and consequently these publications were successfully hidden from our search strategy and could not be examined for this study.) Other interviewees reported that they translated race as ethnicity when required to for German abstracts, or completely dispensed with generic terms such as ethnicity, race, or *Rasse* and only mentioned the categories of classification. The latter represents a third strategy of terminological displacement, which still speaks of Whites, Asians, Caucasians, or other typical categories for race, but no longer identifies the classificatory system.

Loyalty and Voice

None of the young scholars officially expressed their uneasiness in their research team or discussed it with their professors. However, the topic was not raised by other research group members either. The novices often saw themselves as learners, they joined established research groups as newcomers or saw their doctoral topics as "commissioned work" given to them by the principal investigator (physician, 35, para. 48). In Hirschman's words (1970), they remained *loyal*. In no case was it the novice themselves who decided to use the racial classification in the study. Consequently, their research agency may be characterized as weak: "So the topic was given by the boss: 'We want to do a paper on race. And then we did it. So, there wasn't actually much discussion (laughs)" (ibid., para. 20).

Between the ideal-typical medical novice and the established researcher, there are individual actors in our sample who cannot be clearly assigned to one side or the other. They demonstrate various transitional phenomena between the two statuses. This includes the only case in which a German physician, Dr. Muske, explicitly questioned the use of race within the research group. It demonstrates that the likelihood of an individual protesting is closely linked to their academic position. As the study coordinator, Dr. Muske is not only in a position to do so, but she is explicitly expected to choose the voice option (Hirschman 1970) and critically contribute her moral and epistemological irritation:

> I had several conversations on the phone in which I said I didn't understand what is behind these [racial] classifications or distinctions in [country]. And then [the international research colleague] said it would be quite normal in [her country], it would be like that. [...] That would be generally valid. And then we said, well, it's her population, she is familiar with it, and she introduced these descriptions. And then we sort of bowed to that and said, if she assures us that this is the case, then we'll do it. In Germany, it would always be a bit more problematic to use the word *race*. (Dr. Muske, 50, para. 14)

Unlike the novice, who remains loyal, the scientist is able to resolve her cognitive dissonance by delegating her professional authority. She fulfilled her responsibility by expressing her misgivings and ensuring that the classification was accepted in the country where the data was collected. In the end, even the explicitly voiced discontent does not lead to an intervention in the research design.

Race as a Diffuse Concept

In the course of their work, many of the young researchers have learned that race is a variable that can render statistically significant differences in the outcome of a study visible (whether this relates to incidence, severity, the course of a disease, a treatment decision, etc.). However, this does not yet clarify what these differences are due to and why differentiation by race shows differences. When we asked these researchers about this, the meaning of race remained open and diffuse. On the one hand, as one novice emphasizes, race is bound to correlate "through a more or less systematic disadvantage in the US with low income, with poorer insurance and thus presumably also with a poorer [...] outcome." Race is thus used as a proxy to capture exposure to "systemic racism" (Lett et al. 2022), which is expressed in the form of adverse results on several levels. On the other hand, according to the same interviewee, "race simply also has direct genetic effects" (Dr. Müller, physician, 35, para. 24 and 48) and is therefore also used as a proxy for genetic differences (Root 2003). The different medical outcomes of different racial groups can be read as the result of cultural habits or social structures, but there is also the possibility that they reflect genetic differences. After outlining a similar list, in which she cites first the "genetic component" and second the "poorer insurance status, lower income, lower education," a young physician concludes: "So, it's a bit diffusely multifactorial as to why this variable [= race] is important. But I think these two reasons are among the most important (Ms. Robin physician, 35, para. 30)." In these studies, race is being transformed from an *explanans* into an *explanandum*. It is underdetermined and contains everything all at once. In this sense it comes very close to the broad concept of "population" (Reardon 2009, 34–36). Any attempt to answer the question of the nature of the differences that racial classification has made apparent becomes, in the same breath, a statement about the nature of race.

Beyond the significance of the classification for medical science, novices also point to the public significance of race in other social contexts—particularly, but not only, in the United States—in order to explain why studies from or about these countries should examine race effects. It is argued that these countries are "melting pots", i.e., societies with highly diverse populations. Germany, by contrast, is still seen as much more homogeneous. Moreover, some interviewees themselves distinguish between postracial and race-conscious societies: while "in Germany we simply don't have this history of dividing people up by race in this way; in America it is very, very much on people's minds" (Dr. Kurz, 35, para. 28). In principle, the concept is handled "more freely, more liberally in the US than here" (Dr. Berry, 40, para. 48). During research stays in the United States, one learns "that race can be used in a completely value-free way in some settings" (Dr. Robin, 33, para. 86). At these points in the interviews, the social context identified often blurs with the significance of the classification for the medical sciences. The distinction between meaning and use of the classification in society and in medical science is rarely rigorously maintained.

Several novices observed that medical science journals sometimes request or require their authors to record the race and/or ethnicity of their study subjects, with the aim of ensuring that studies are representative of the general population. This is, in other words, an *egalitarian* measure that seeks to ensure equal consideration and treatment by including a range of social groups, thus being able to take into account their (health-related) *differences* (Epstein 2010). In addition to the scientific interest in racial discrepancies, there is also a sense in which considering racial perspectives in a piece of work can raise its profile: "Additionally, it must also be said that it is important to the journals. So, of course we include it because we think it makes scientific sense, but, to put it bluntly, including it also sells well. I don't want to say that it's en vogue, but it's simply something that is incredibly important in the United States in particular" (Dr. Müller, 35, para. 24).

3.2 Scientific Equanimity or A Matter of Getting Used to It? Racial Classifications as a Region-Specific or Global Observation Schema

A difference in perspective is evident between medical novices and medical researchers with experience of race, first and foremost with regard to the position that German ethics committees are assumed to take in this matter. Novices tend to assume that ethics committees adopt and institutionally represent German skepticism towards race in their statements. One young physician interviewed at the beginning of our study expressed the opinion that a German "ethics committee would never let us collect a database that contains such blatant *Rassenklassifikationen* [racial classifications]" of the kind that are standard for US databases (Dr. Schmitt, physician, 35, para. 28). We ran with this suggestion and asked all subsequent interviewees for their views. Opinions differed greatly: while novices were willing to take the idea seriously and to weigh up the pros and cons, race-experienced researchers stated they found the very idea that ethics committees could reject a research project merely because it proposed to gather data on race "odd" (Prof. Karstens, 65, para. 62) and unlikely. The very possibility of such a decision, as raised by the question, was so irritating in some cases that concern was expressed that "the discussion in Germany is in danger of taking a major turn for the worse" (ibid., para. 60). Having used racial classifications in their studies for decades, established researchers emphasized that the collection of such data had never been criticized in the past. They would not expect this as long as the study and survey were reasonably justified and the German term *Rasse* was not used.

These contrasting views on the opinions of ethics committees regarding race make clear that in the transition from novice to race-experienced medical scientist a different, more tolerant view of racial classification emerges. Many established medical researchers who have been working with racial classifications for a long time routinely distinguish between race and Rasse by placing the former in the service of science and emphasizing the political dimension of the latter. One professor (Prof. Hausinger, 60, para. 42), for instance, emphasizes when asked about the meaning of the term *Rasse*:

> The question is, do you mean the term or do you mean the subject? As far as the term goes, we don't ask about *Rasse* in German, but about ethnic background or ethnicity. This is partly for political reasons, if you like. But we have to grasp the subject as such. In medicine, regardless of the subject area, we have to know for the reasons already mentioned whether someone has a sub-Saharan African genetic background, ethnic background, or a Native American background or, if you prefer, Indigenous and so on. That is a basic requirement for certain frequencies of diseases, side effects of diseases, comorbidity and so on. That is so essential that there is no discussion about it at all in medical research. These are recorded and named.

The clear distinction between term and topic takes into account the demands of both sides, namely, the German public and medical science. As a representative and spokesperson of his social world ("we"), the established researcher explains and represents the rules and principles that apply within it. For him, *Rasse* refers objectively to differences associated with descent or lineage whose relevance for and in medicine is undisputed; it would, he believes, be downright negligent not to pursue this topic scientifically. At the same time, for him *Rasse* cannot be used due to its politicization. Nevertheless, he argues, this is the only influence that society is allowed to exert on medical science here—only the choice of words is affected, not the scientific approach. The concealment strategies he mentions here are, then, similar to those described above. In German texts, he claims, one does not speak of *Rassen*, but of *Ethnien*; in English articles, however, the term "race" can be and is still used to designate the classification.

This internalized distinction shifts the locus of the dissonance from the conscience to the social realm. A physician who spent several years conducting research abroad describes how this reversed the difficulty she faced. While the use of racial classification was alien to her at first, she notes, and getting used to it presented a personal challenge, the same was now also true of the race-skepticism prevalent in Germany that she encountered after returning:

> I've realized that my time in the US has made me much more accustomed to recording race. It has become a matter of course. But in German, it's often difficult to find the right terminology. I once gave a lecture to students. It was about [medical topic]. And I said that *Schwarze* [Blacks] are more affected and that this is certainly also due to the fact that they come more frequently from socially deprived groups and that, for example, allergens and the like play a greater role in the household, but that there are probably also genetic factors. I explained this accordingly. And then I received an email from students [...]. They felt it was racist that I mentioned a group that is more affected. So, there seems to be a completely different sensitivity here. And you have to be very careful what terms you use. There is no conventional concept. (Dr. Bosch, physician, 45, para. 49)

Dr. Bosch had moved beyond her initial discomfort. Conversely, however, her new self-conception, her new way of working, now clashes with the speech rules that apply in Germany, which her students are sensitive about transgressing. Similarly, another senior physician describes how she was "once really taken to task by a student during a lecture because I had apparently and unconsciously not expressed myself correctly. That was extremely upsetting" (Dr. Schmitt, physician, 60, para. 47).

Both doctors are examples of established medical researchers who have been working with racial classifications for a long time—be it, as in the first case, because they conducted research in the United States for an extended period, or, as in the second case, because it has long been common practice in their medical field to classify phenotypic characteristics on the basis of race. For them, race and racial categories are no longer burdened by the semantic weight of the word *race*. In the interviews, medical researchers cognizant of race as a factor in their work have a medical-scientific concept of race and distinguish it from *Rasse*, which they consider to be a politically inacceptable and publicly unusable term. Younger physicians also quickly learn to distinguish between these two contexts of meaning, even though they still have vivid memories of their own discontent with race. Hence, they distinguish race from *Rasse*. At the same time, they are beginning to realize that the use of *race* is not welcomed in Germany, as the scientific value of the classification has not been disentangled from the negative political associations of *Rasse* here: "There are people who completely fail to understand this term [= race], because they are not involved in the scientific discussion that takes place in the United States or generally outside of Europe. They find that [= the term race] hard to swallow at a congress" (Dr. Robin, 35, pos. 48). Medical scientists do not necessarily learn to deal with racial classifications with equanimity during their own careers; this is acquired through research with racial classifications.

Race: A Global Human Classificatory Schema or a Regional Peculiarity?

Although researchers who have gained experience in working with the concept of race make a very sharp distinction between the political sphere and scientific applications, this commonality does not lead to a uniform use of racial classifications. Two forms of understanding and use can be observed. For one group of researchers, race remains a regionally defined concept that cannot be transferred to different social contexts; others, however, understand it as a globally applicable template, a global classification schema, though the American version has become the universal template[14]. For these researchers, racial classification is a schema suitable for observing the entire world. In the interview excerpt quoted above (Prof. Hausinger, 60, para. 42), we can recognize the ideal type of this view. As a rule, it goes hand in hand with the assumption that races will exhibit genetic differences.

Researchers justify the use of genetically conceived race typologies with reference to their long-standing use within their discipline. One researcher, for example, sent us a foundational text in her field, showing that differentiation by race has long been common practice in her discipline, and, in her opinion, for good reason: the text argues that

14 Wacquant (1997, 224) argues that the US version of race nowadays "has been universalized as the template through which analyses of 'race' in all countries and epochs are to be conducted." Above all, he is concerned that the US version "is suffused with U.S. folk conceptions of 'race'" (ibid., see also: Morning/Maneri 2021, 46). That is not so much my point here. My observation is that the formal classification schema is often universalized and used as a template.

there are statistically significant anatomical differences between Asians, Europeans, and Africans that are important for surgery and treatment.

Neither the textbook nor the researcher contested the objectivity of the racial categories they used. Our efforts to break down this typological reasoning in a joint discussion, by pointing out the arbitrary boundaries the racial categories draw in a continuum of human variance, were unsuccessful. The researcher resisted abandoning these racial classifications, as she feared this would also reject statistically proven differences concerning certain parts of the human body. The individual types of the classificatory schema appear here to have become identified with genetic differences, even though there are other ways of determining the latter scientifically (for example, via ancestry classifications or genetic similarity (National Academies of Sciences, Engineering, and Medicine 2023, 197); methodically (for example, via trait differences, so-called clines); or via a separate, specific typification of body parts, completely independent of race or ancestry. Our questions and objections that these racial classifications are too coarse and arbitrary to capture genetic differences were not taken by the researcher as a compelling reason to discard them. Rather, it was—once again—the terminology that worried her, and to which she repeatedly referred. During the exchange, she asked us for suggestions of correct descriptors, pointing out that the terminology problem had been repeatedly discussed at her research institution and that the terms used for these categories had also led to conflicts with students in lectures.

The Caucasian Liver

Racial classifications can also be employed to link regional studies to the global discourse of specialists, thus making them part of a larger puzzle. The case of the Caucasian liver is especially striking here. A medical study involving 2,773 people from the region of Western Pomerania[15] determined the reference values for the liver volume of the sample—presented from the outset in the title, with a racialized description as a "Caucasian population." The use of this term was intended to signal that the significance of the study sample extends far beyond its local survey location.[16] The results of a single, locally defined sample are thus taken here to represent all Caucasians, though the article does not explain whether and how membership of this group was determined, nor who exactly the large group of Caucasians comprises. Yet, it should be noted, the population descriptor "Caucasian" has been widely criticized, both for its vagueness and for its origin in early racial typologies.[17] After using this term to enhance the significance of its sample as a representation of this "race," the study then juxtaposes the Caucasian liver to that of other

15 Against the historical background of the division of Pomerania, it is disconcerting that the study in its title refers to "Pomerania" as the study region and only later clarifies that its own sample is recruited from the Western Pomerania region. Western Pomerania is the much smaller part of Pomerania that lies in present-day Germany. Eastern Pomerania belongs to Poland. See https://mki.wisc.edu/library-archive/external-links/genealogy-and-heritage-societies/pomerania-pommern/.

16 Not for the first time, in fact. Heinemann et al. (1999) already addressed a similar topic with a similar title. The paper discussed here omits all mention of the previous study.

17 Physicians in Germany also point out the history and ambiguity of the term and call for the "Caucasian to finally be buried" (Meißner 2021).

races. Comparing surveys from other parts of the world, namely, Switzerland (identified as another Caucasian sample), China, and Japan, the authors conclude that "racial differences seem likely" (Kromrey et al. 2018, 35). The imprecise assignment of the sample and the identification of differences as racial illustrate the ease with which German research teams, even without international partners, apply racial classifications. Without any apparent reservations, the study uses racial terms to integrate its results into a schema that is assumed to be valid worldwide. In one blow, its recourse to racial classification multiplies the population represented in the study and establishes a relation to other world populations. Having measured not just a West Pomeranian, but the average Caucasian liver amplifies the study's visibility within its discipline and guarantees disciplinary compatibility, even as it also most likely overgeneralizes its local findings (Takezawa et al. 2014, 2). The publication gives no explanation for the observed differences between the races. In this, it reflects a wider tendency: genetic reasons are not ruled out and are usually assumed when race is understood as a global sorting scheme.

Race as a Regional Concept

Besides the version of race as a concept with worldwide validity, there are a number of medical studies that view racial classifications as tied to the social context in which the data was collected. It is not always obvious in such cases whether the authors actually preclude the transfer of the classification from a smaller area, e.g. Germany, to another regional context. However, there are at least two studies in our corpus in which race was understood as a place-based classification and was therefore not applied to the authors' own sample. In both studies, the authors exclude race as a global schema, yet it is still used (otherwise the case would not have been included in our study corpus). The first case has already been presented above. The researcher involved explicitly opposed the option of using racial classifications in Germany. Its use was only legitimate for one survey location where one participating researcher lived, who assured their German colleague that it was normal and uncontroversial to record details of race in that country.

The second study is different in nature. In the case of comparative studies among countries, researchers often face the question of whether and, if so, how they can compare different descent-based classification systems. In German studies and databases, patients or subjects are often grouped according to their "migration background"[18]. This leads to compatibility problems when comparing German and US datasets, because the two classifications—migration background and race/ethnicity—are fundamentally incomparable. Nevertheless, such comparisons are sometimes undertaken. In order to be able to compare the two datasets with regard to their descent-associated differences, an international diabetes research group transformed both classifications into a new binary category "ethnic minority" (DeSalvo et al. 2018). This makes it possible to compare the minorities in one national context with those in another and to show, for instance, whether

18 For an explanation of the concept of "migration background" *Migrationshintergrund*, which is in widespread use in Germany, cf. Will (2019) as well as the chapter by zur Nieden et al. in this volume. A number of major databases, such as the German Cancer Registry have so far avoided gathering any data on people's backgrounds, which means, that no comparisons can be made with other global databases in this respect.

a given disease affects a certain group more on one side of the Atlantic or the other. "Ethnic minority" here takes the form of a boundary object, making different things equal or comparable (Star 2010). Methodologically, this transformation is imprecise—the group categories cannot be easily ethnicized, nor can the groups chosen be generalized to constitute underprivileged minorities. That said, the transformation "works" in the sense that statistical tools, colleagues, and even reviewers make no objection. Such a procedure of transformation has been extended to other datasets from England and Wales, and to Australia, increasing the methodological problems noted above (Craig et al. 2017).

Both examples show, in their own way, that racial classifications do not need to be understood as "global templates," but can retain their local reference and still be effective. Researchers integrate them, but do not adopt them. This may lead to the development of controversial solutions, as researchers seek to harmonize their own datasets with those of others. Beyond this, recognizing a classification as context-bound makes it more difficult to justify differences between its categories in terms of genetic factors. This is still not completely impossible. However, if a classification is reserved for a certain social context but is considered inappropriate for others, its categories are interpreted more as social entities and less as groups with specific natural characteristics.

Race in the Lab: Research and Identity

In a remarkable ethnographic study, Duana Fullwiley (2008) uses the example of two laboratories specializing in medical genetics to show how analyses of ancestry and admixture can lead to the regeneticization of race. One observation she particularly emphasizes is that this is not exclusively the work of white researchers, let alone of committed racists. In the laboratories, she writes, she encountered "minority researchers" taking part in this research for identity-political reasons, "driven not by racist notions of human difference, but by a commitment to reduce health disparities and to include 'their' communities" (Fullwiley 2008, 695). To capture this diversity-sensitive turn towards greater recognition and representation of different social groups in medical research and treatment, which can be observed in the United States since the 1990s, Steven Epstein (2007) coined the term "inclusion-and-difference paradigm."

Likewise, in Fullwiley's study, we find a similar motive in the case of a physician who sees the need to draw attention to a particular physical characteristic of "his" descent group. According to this physician, who claims to draw his insights from experience in medical practice, the bodily proportions assumed in the medical literature generally reflect white, European norms. Since physicians use this information as a guide during surgery, he considers this to be a risk. It would be better, he argues, if surgeons took the patient's descent into account and adjusted their surgical intervention accordingly. To verify his hypothesis of a descent-associated physical difference, the researcher utilized a database containing the body profiles of various patients classified by their race. His analyses confirmed his hypothesis; significant differences between the races were found, ultimately leading him to recommend that operations affecting this part of the body be performed slightly differently depending on the patient's race. Racial differences in physique, in other words, should be taken into account.

It is the combination of two elements that initially puzzled both Fullwiley and us. Not only does it become apparent here that research into physical and/or genetic racial differences can be a matter-of-fact subject of medical research today. The medical study continues to imply that racial typologies correspond to differences found in nature (Reardon 2009, 38–39). When quizzed by the interviewer, the physician conceded that the results show "rough differences using a rough classification" (Dr. Berry, 40, para. 14). And yet he also asserted that no other dataset was available. That said, the real surprise in Fullwiley's article, as in ours, is that these studies are promoted by researchers who refer to their own biographies, drawing on experiences of racialization, as one reason for conducting the study. The researchers report being othered or experiencing everyday racism, and they describe themselves as having a migration background or even: "having a different race, so to speak, born in Germany, born in [city], being socialized here, but every day I am asked by my patients whether I want to go back, and so on, things like that" (Dr. Berry, 40, para. 44). At the same time, however, these researchers reinforce racialized categories by biologizing them. Although, or precisely because, these researchers are regularly reminded of their difference, they are attentive to characteristics of descent groups ("theirs" and others), thereby genetically substantiating racial differences between "themselves," "others," and "Caucasians" in the same way that white researchers have done in the past. Historically, the assertion of natural racial differences for the "other races" has not had the advantage that is now being promised—quite the contrary. The scientific naturalization of racial differences has rather tended to legitimize devaluation, discrimination, and exploitation on the basis of race. It is therefore unsettling when racialized researchers today participate in a controversial scientific endeavor to which "their" races have historically fallen victim. The discomfort this provokes also reflects our own socialization in a country that sees itself as being postracial, as well as the difficulty of drawing a clear distinction between racialization and racism.

Researchers position themselves in different ways in relation to or within a self-applied system of racial classification (figure 3). In retrospect, three modes of affiliation can be identified in the field. Novices and physicians, in particular, tend to understand racial classifications as regional schemas and position themselves as (1) observers of a process of classification carried out by others. They do not assign themselves to a category within this system, but view the classification of human races as an element of a social context to which they do not belong, e.g., Brazilian or US society. Researchers who adopt race as a global observation schema, by contrast, see themselves and others as (2) biological specimens (Hirschauer 2017, 42–45) of a race. They can assign themselves to a category as a matter of fact, but for them personally this remains insignificant and ineffective in terms of their identity. Unlike the racialized researchers, they do not simultaneously see themselves as (3) members of a race, with a racial affiliation of which they are reminded in many situations.

Figure 3: Three modes of self-positioning towards/in the racial classification

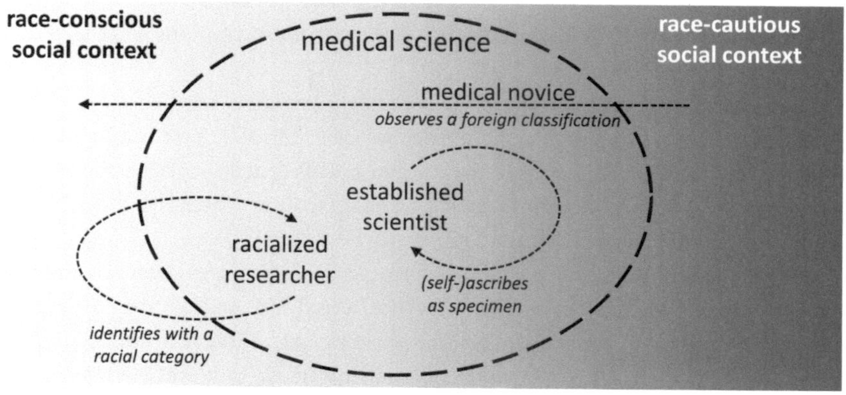

Sociobiographical experiences of being different and academic research interests do not necessarily have to interact as described, however. They need not be related to each other at all. A second researcher with a similar background also spoke in his interview about his experiences of being othered, but did not relate this directly to his research findings. Instead, he emphasized the broad diversity within the research group at his institution. The study he published also examines a certain phenotypic difference between two races but comes to the conclusion that, although research has assumed a difference so far, it cannot be verified. He sums up: "Our mission, rather, is to standardize research protocols, to make them the same worldwide, to make results more comparable—and to say that things are probably not as different as we originally thought. It is, so to speak, a positive side effect of our research to be able to say that these differences that were assumed are perhaps not so relevant after all" (Prof. Garden, 45, para. 54).

Of course, the two cases considered here must be regarded on their own terms. No generalized statements can be derived from them as to the interaction between sociobiographical experiences, the experience of diversity, and the focus of research. Beyond this, however, it is possible to recognize what will be discussed again in detail in the last section. Neither of the two medical researchers who experience being racialized reports concerns in using racial classifications to study phenotypic differences. Neither the classification nor the results obtained by this means are problematic, as their study complies with scientific principles: Scientific standards were adhered to, even enhanced, the research was open-ended, and new knowledge was ultimately obtained. Against such a background of scientific rigor, it no longer matters whether genetic differences could be verified or falsified. While in one instance the therapeutic advantages and in another equality as a "positive side effect" are emphasized as benefits of the study, the question of the scientific appropriateness of the classification disappears.

Crossover Scientists: Science and Society

Alongside the novice and the established researcher, the crossover scientist represents a third ideal type. The name speaks for itself. Interviewees who represent this type (to various degrees) usually have no medical training but occupy the border area between medicine and their own discipline. They include specialists such as epidemiologists and psychologists, and they conduct research on medical issues using the tools and theories of their own disciplines while working in medical science institutions and collaborating with medical researchers.

With regard to their approach to race, crossover scientists are characterized by two particular features. In addition to their (1) decisive rejection of an "essentialist interpretation" of race (Dr. Saleh, 40, para. 74), they (2) outline different dimensions of responsibility that emerge at the intersection of science and society and which also become evident in how they present themselves as scientists (to the subject).

Unlike medical novices and established researchers, crossover scientists emphasize their rejection of a scientific notion of race that describes genetically distinct groups of people. "Human races cannot be substantiated biologically. There's no indication of that." Races are understood as "racialized group[s]" as a historical product of social ascriptions, as a "social construct" (Dr. Fromme, 35, para. 72 and 94). Despite or precisely because of this clear definition of race as a collective idea, crossover scientists show themselves to be particularly challenged by the need to explain the possibility and consequences of any genetic differences between the races. The focus of the debate is less on the medical or scientific value of such a difference. Rather, the social consequences that such a difference would have are negotiated.

> Somehow the question looms over everything: Are we all the same or are there differences? And if there are differences, are they socially constructed or do they perhaps have very small genetic causes at the end of the day? Suppose we were to find out that a drug doesn't work for Asian women. [...] How do you communicate that to the outside world without people getting the idea: "Oh, they've biologically uncovered a human *Rasse* after all," right? (ibid, para. 96)

Although race is understood to be a social construct, it is not ruled out that medically significant differences could have "very small genetic causes." The challenge, however, is not so much the empirical, scientific proof of genetic difference itself, but rather the question of how this can be correctly presented in terms of its significance for the public understanding of race. How can an understanding of race as a social construct be defended—this is the rhetorical question—against society when subtle biological differences threaten to undermine it? Can an understanding of race as a social fiction be maintained in society when medical or human genetic research results claim natural differences, even if only minimal ones? How much nature can the social construct tolerate? The concept of race as a social construct appears fragile if it can be shaken by research results and their incorrect reception. Science runs the risk of uncovering facts that are capable of reviving a sinister myth.

At this point, the weakness of a misinterpreted social constructionism becomes clear. Social constructionism does not determine which constructs are more real than others but rather analyses how social facts operate, how they are built and reproduced, and how they shape and sustain social relationships. Outside of sociology, however, the statement that race is a social construct is often understood to mean that it is not real. It is thus placed in opposition to the natural world—and faced with epistemic strength of science, it can hardly win the contest. On the other hand, it is peculiar that limited statistical significance is ascribed the power to create the public impression that races are proven, but the strong *scientific* reservations against the biological evidence of distinct groups of people are not cited.

Rationalities of Scientific Responsibility

Concern about a revitalization of a supposedly anachronistic understanding of race is typical of the interdisciplinary crossover scientist, whereas medical professionals we interviewed rarely expressed such sentiments. One exception is the study coordinator who claimed that research results need to be screened how they could be interpreted by others inside and outside the scientific world. "In the end, it's about not providing a starting point for misuse of this data. [...] You have to be careful that groups don't use your scientific results differently than you ever thought they would be used, right?" (Dr. Muske, 50, para. 78). In terms of a *reflexive rationality*, scientific responsibility here presupposes that researchers have "the ability to foresee the [social] consequences of their own practice" (Glerup and Horst 2014, 38). In our data sample, the concern about and anticipation of public misreading or deliberate instrumentalization of research results contrasts strongly with the emphasis on science as committed to the principle of a value-free approach. While the crossover scientist underlines the responsibility to foresee and prevent possible negative social consequences of one's own research with race, the other position reminds us of the importance of value-free science. This *demarcation rationality* is based on the assumption that science has its own professional code and structures of oversight, distinct from the rest of society; both enable responsible scientific action (ibid., 37).

The crux of the second position is the argument that an unbiased and objective view is the prerequisite for scientific value neutrality. One interviewee stressed this principle, expressed in different variations and strengths by the vast majority of interviewees, as a particularly important point. For him, the use of racial classifications is "not about any kind of valuation, right? It's about distinguishing between groups. And that could of course also be other groups. But it's not a value judgement. It's clearly about seeing a treatment option or creating an understanding of a disease. Yes, that is perhaps the most important point, do you understand?" (Prof. Kreh, 55, para. 65). Firstly, the quote reveals another rationality that is particularly important in the *medical* field as a means of justifying *scientific* action as responsible. It underlines the beneficial *contribution* of science to the fulfillment of social needs, e.g., the cure or prevention of disease. This rationality here complements that of demarcation. The latter expresses the researcher's conviction that a neutral, scientific use of classifications such as race is possible. While race is controversial and frowned upon in the social context in Germany, "the scientific application [of this or similar classifications] is on the safe and understandable side" (Prof. Hausinger, 60, para. 48).

These rationalities are not mutually exclusive, although they are based on different approaches to dealing responsibly with race. The demarcation rationality relies on basic principles of scientific practice which, if adhered to, guarantee that racial classification can be applied responsibly in science. Science is advised to conduct *value-free* research, to be guided by "disinterested interest" (Bourdieu 1998, 27; my translation) and to not unnecessarily mix and confuse value judgments and factual assessments (Weber 1904, 3); this detachment of research also grounds the right to freedom of research, which in turn benefits the progress of knowledge. "Research must be independent and must be allowed to deal with any topic. Dealing with something does not mean endorsing it—otherwise we would not be able to research many things—but learning to understand something better and finding explanations for it" (Prof. Hausinger, 60, para. 52). If normative value judgments were to set limits on science, many paths would be blocked. For the sake of its own interest in knowledge, science must operate independently of these. Admittedly, the idea that science ultimately benefits society as a whole (Merton 1974, 275–77) is also behind the guiding principle of a disinterested science that is subsequently pursued as purely as possible. However, this idea remains abstract and neither negotiates the concrete social effects of a given piece of research nor uses such effects to guide the direction of future research.

In a sense, the crossover scientist complicates the relationship between science and society, not seeing it as regulated solely by the principle of value freedom. This becomes clear in interactions in which the crossover scientists see themselves as entangled with society. For example, one interviewee toyed with the idea of creating a classification of descent-associated categories adapted to the German population. To do this, he explained, he would first conduct an open survey to find out which categories people in Germany would provide for themselves. "I wouldn't want to come in from the outside and suggest something" (Dr. Grabe, 35, para. 56). In the sense of a regionally specific schema, descent-associated differences are explicitly understood as social differences, which vary from society to society and cannot be presupposed. Another interviewee discusses the possibility of bringing about a deeper involvement of the research participants in the research process, simultaneously emphasizing how difficult it is to implement participatory concepts: "Some communities are also very active and have this slogan 'nothing about us without us.' They demand to be included in research about their group. And I think that's right, but at the same time I cannot implement it completely" (Dr. Grabe, 35, para. 108). In her view, existing research structures make little room for participatory research in medicine; her own academic socialization, the obstinacy and skepticism of research subjects, and the additional costs would make it difficult.

Finally, a heightened awareness of social issues also impacts the interpretation of research results. At the very least, the claim is formulated that the higher disease burden of individual social groups cannot be explained by individual irresponsibility. Health inequalities can no longer be thought of as the result of individual decisions and individualized as unhealthy actions that need to be corrected accordingly. Instead, it is important to recognize that "it is complex social contexts that lead to these differences and that this should also be taken into account when interpreting the results" (Dr. Grabe, 35, para. 42). Race has become one social structure among many and is no longer an individual disposition.

Discourse Limits instead of a Discourse Arena

A discourse arena in which the significance of race or racial classifications in medical science is negotiated can only be recognized to a very limited extent in the available empirical material. Debates on the nature of race or even on the scientific way of dealing with racial classifications are hardly ever held at German research institutes. It is the study's central finding that the German *Rasse* taboo blocks any extensive discussion. It prevents an open and fruitful debate about both the actual scientific importance, and the ontological quality, of race by channeling attention solely to the term and its normative affect. A primary concern of the interviewees is whether race can be used and how it can be correctly labeled. Such considerations are frequently expressed in the interviews, they, however, rarely played an explicit role in the research process. Furthermore, where conflicts do occur in the field, the situation is very simliar: these conflicts are about the term but do not dive deeper into the topic. In such situations, an indignation based on terminology is articulated, yet a thoroughgoing examination of underlying concepts does not take place. What race means, what racial differences are—and what they are not—remains largely unclear.

It is not surprising that medical novices in particular find dealing with racial classifications challenging. Prior to participating in the recent research examined here, and sometimes right up to the interview with us, they had no reason to question the *socially* dominant conviction that the existence of races is *scientifically* untenable, morally reprehensible, and politically dangerous. However, as they become aware of established racial classifications in their own scientific field, these firm convictions are challenged. The manifest contradiction between their habitual race-caution and the exigencies of science generates efforts to reduce the resulting tension via terminological vagueness. Many established researchers are also at odds with the word; at first, it is only appropriated to the extent that its use is permitted when referring to another society.

The novices' discontent with race takes place in their heads; an open debate on how race can or should be thought of in medical-biological terms is only just beginning to take place in Germany (Aikins et al. 2021, Gießelmann and Martin 2023, DeZIM 2023, 109–214). It remains a diffuse both/and category that can be used to suggest social, cultural, biological, and/or genetic reasons for medical findings. Anti-essentialist, gradualist arguments rejecting the idea of races as sharply definable biological and genetic groups are nowhere to be found, although they should be easily accessible to scientific thinking (Morning 2011: 113–117). Rather, one can speak of a great uncertainty about what race actually is.

The situation is different for many established researchers. Assuming that they too once had reservations when they first came into contact with the concept, they have now come to terms with it after several research contacts. They make a sharp distinction between the scientific field and the social context, between scientific and political utilization. At the center of the medical-scientific world, race has a recognized scientific role that is no longer morally irritating, and which is no longer subject to the German *Rasse* taboo. It has become a matter of scientific distinction and habituation. However, the comparison with the interdisciplinary crossover scientist shows that this is not an inevitable process. These researchers are not comfortable with racial classification, as they do not subscribe to the scientific principle of value neutrality, but rather derive their concept of scientific responsibility in exchange with the social environment in which the scientific world is embedded. The crossover scientist remains concerned with the social consequences that medical work using racial classifications may have.

There are, however, two fundamentally different directions that can be taken after this first step. Either the acceptance of race as an established classification in medical science can be followed by a second distinction that delineates between different social contexts and the corresponding descent-related classifications (i.e., race/ethnicity in the United States, migration background in Germany, cultural and ethnic group in Australia, ethnic group in England/Wales etc.) and takes this distinction into account in its own studies. Or race can be universalized and adopted as a global schema for local samples, as well. Both paths bring problems of their own, which could only be discussed in part here. Yet, it is important to recognize that each imposes a certain interpretation that is distinct from the diffuse concept of race held by novices. If we take into account the social context of different classifications, then it is hardly plausible to link genetic differences back to racial classifications, in contrast to cases where the race schema is universalized.

The second important question that this study was looking to answer is that of "why?" Why do German research institutions continue to work with race? We have considered a number of motives over the preceding pages. The use of racial categories is mainly due to international research collaborations in which datasets from race-conscious contexts that permit this classification are studied. If studies additionally contribute a separate dataset from Germany, it is either integrated into the schema as a sample of a white/Caucasian race or a new super-schema is designed that incorporates the native dominant classification (i.e., migration background) and the racial classifications in a modified form (for instance, specifying "ethnic minorities"). Not considering the descent-related classification in view of its fundamental incomparability would have been a third possibility (which we did not discuss here due to our search strategy). Researchers, therefore, face important decisions, especially when working with international datasets. It is of great importance not to leave early researchers alone to manage the issues around race when they are required to conduct research using racial classifications in the context of achieving qualifications (when researching or studying abroad). A thorough understanding of compatibility issues, the conceptual schemas behind descent-related classifications (even that which is designated by "race" is not always the same) and the reasons why they are required to apply it should be ascertained. Otherwise, it is to be feared that the unspoken uncertainties will continue to translate into terminological strategies of concealment.

It is worth mentioning that some national research projects make use of racial classifications where there is no need to do so—where it is neither suggested by a medical database nor expected by an international research cooperation. Even today, race is still used as a tool to enable disciplinary compatibility. It creates links between new research and existing keywords, a body of work, and a set of assumptions within a given discipline, thus increasing visibility and the likelihood of reception by other researchers. A major reason for the use of racial classification systems (even where the sampling location is exclusively Germany) is therefore their global entrenchment and inertia. A system exists in which they act as stable reference categories. In this position, they continue to fuel the myth that races are distinct genetic groups of people. Postcategorical solutions (gradients; or setting the focus on the phenotypic phenomenon of interest) still do not seem to be gaining acceptance.

Finally, racial classifications also arise because in recent years there have been individual and institutional efforts to include subjects and patients of all races in medical and other life science studies. Researchers at German research institutions also frequently work in line with the requirements of the U.S. Food and Drug Administration (FDA) or international medical science journals (although it should at least be possible to fulfill the requirements of journals using the German classification *migration background*). Racialized researchers in Germany also make a conscious effort to take different races into account in their studies. In terms of mere representation, this may initially seem unproblematic. However, such inclusion criteria are always about more than simply employing different social assignments. It is also always about making differences visible and being able to take them seriously in *medical* terms. Such biological racialization runs the risk of once again asserting races as substantial natural entities and setting false medical priorities (Rajagopalan and Fujimura 2012). These concerns must be taken very seriously—and they are likely to lead to proposals that reject the use of racial classifications in life science research in most cases (National Academies of Sciences, Engineering, and Medicine 2023). In Germany, however, the quiet vehemence of the taboo against *Rasse* complicates any discussion about the risks and opportunities of descent-related classifications, though it is urgently necessary. The question of when medical research using racial classifications leads to bad racializations—when it even reinforces old stereotypes and keeps them alive—cannot be answered with silence.

Acknowledgements

I would like to thank Hannah Schönberger for her support in organizing, recording, and transcribing the interviews. I am indebted to Tino Plümecke, Isabelle Bartram, Andrea zur Nieden, all participants in the authors' workshops for this edited volume, and an anonymous reviewer for critical comments on previous versions of this text. Graeme Currie and Michael Taylor helped to significantly improve the linguistic quality of this article. The research was funded by the German Federal Ministry of Education and Research (BMBF, grant number 01GP1790).

References

Aikins, Muna AnNisa, Teresa Bremberger, Joshua Kwesi Aikins, Daniel Gyamerah, and Deniz Yıldırım-Caliman. 2021. *Afrozensus 2020: Perspektiven, Anti-Schwarze Rassismuserfahrungen und Engagement Schwarzer, afrikanischer und afrodiasporischer Menschen in Deutschland*. Berlin.

Ankerst, Donna P., Johanna Straubinger, Katharina Selig et al. 2018. "A Contemporary Prostate Biopsy Risk Calculator Based on Multiple Heterogeneous Cohorts." *European Urology* 74 (2): 197–203. DOI: https://doi.org/10.1016/j.eururo.2018.05.003.

Bartram, Isabelle, Laura Schnieder, Nils Ellebrecht, Florian Ruland, Tino Plümecke, and Andrea zur Nieden. 2023. "Categorizing People in the German Life Sciences: A Systematic Literature Review of Classifications of Human Diversity." *Discover Social Science and Health* 3 (1). DOI: https://doi.org/10.1007/s44155-023-00033-5.

Becker, Howard S. 1966. *Outsiders. Studies in the Sociology of Deviance*. London: Free Press of Glencoe.

Bourdieu, Pierre. 1998. *Vom Gebrauch der Wissenschaft: Für eine klinische Soziologie des wissenschaftlichen Feldes*. Konstanz: UVK.

Bourdieu, Pierre. 2002. *Distinction: A Social Critique of the Judgement of Taste*. Cambridge, Mass. Harvard Univ. Press.

Bourdieu, Pierre. 2018. *Classification Struggles: Lectures at the Collège De France (1981–1982)*. Cambridge, UK: Polity.

Bowker, Geoffrey C., and Susan Leigh Star. 1999. *Sorting Things Out: Classification and Its Consequences*. Cambridge: The MIT Press.

Braun, Lundy. 2014. *Breathing Race into the Machine: The Surprising Career of the Spirometer from Plantation to Genetics*. Minneapolis, Minn. University of Minnesota Press.

Brubaker, Rogers. 2002. "Ethnicity without Groups." *European Journal of Sociology* 43 (2): 163–189.

Chin, Rita. 2017. "Thinking Difference in Postwar Germany: Some Epistemological Obstacles Around 'Race'." In *Migration, Memory, and Diversity: Germany from 1945 to the Present*, edited by Cornelia Wilhelm, 206–230. New York: Berghahn Books.

Clarke, Adele E. 1990. "Controversy and the Development of Reproductive Sciences." *Social Problems* 37 (1): 18–37. DOI: https://doi.org/10.2307/800792.

Clarke, Adele E., Carrie Friese, and Rachel Washburn. 2018. *Situational Analysis: Grounded Theory After the Interpretive Turn*. Second edition. Los Angeles: Sage.

Craig, Maria E., Nicole Prinz, Claire T. Boyle, et al. 2017. "Prevalence of Celiac Disease in 52,721 Youth with Type 1 Diabetes: International Comparison Across Three Continents." *Diabetes Care* 40 (8): 1034–1040. DOI: https://doi.org/10.2337/dc16-2508.

Corbin, Juliet, and Anselm Strauss. 2008. *Basics of qualitative research. Techniques and procedures for developing grounded theory*, 3. ed. Los Angeles: Sage.

DeSalvo, Daniel J., Kellee M. Miller, Julia M. Hermann, et al. 2018. "Continuous Glucose Monitoring and Glycemic Control Among Youth with Type 1 Diabetes: International Comparison from the T1D Exchange and DPV Initiative." *Pediatric Diabetes* 19 (7): 1271–1275. DOI: https://doi.org/10.1111/pedi.12711.

DeZIM (Deutsches Zentrum für Integrations- und Migrationsforschung). 2023. *Rassismus und seine Symptome. Bericht des Nationalen Diskriminierungs- und Rassismusmonitors*. Berlin. Accessed Januar 27, 2025. https://www.dezim-institut.de/fileadmin/user_upload/Demo_FIS/publikation_pdf/FA-5824.pdf.

Durkheim, Emile. 1982. *The Rules of Sociological Method*. New York, N.Y.: The Free Press.

Ellebrecht, Nils, Hannah Schönberger, Isabelle Bartram, Tino Plümecke, and Andrea zur Nieden. 2023. "Tabu, Problem oder Standard? Die Ethnisierung von race-Klassifikationen in der deutschen Medizin- und Gesundheitsforschung." In *Polarisierte Welten. Verhandlungen des 41. Kongresses der Deutschen Gesellschaft für Soziologie in Bielefeld 2022*, edited by Paula-Irene Villa Braslavsky. Accessed Januar 27, 2025. https://publikationen.soziologie.de/index.php/kongressband_2022/article/view/1638/1873.

Epstein, Steven. 2007. *Inclusion: The Politics of Difference in Medical Research*. Chicago: University of Chicago Press.

Festinger, Leon. 1957. *A Theory of Cognitive Dissonance*. Stanford, Calif.: Stanford Univ. Press.

Fischer, Martin S., Uwe Hoßfeld, Johannes Krause, and Stefan Richter. 2019. "Jena Declaration. The concept of race is the result of racism, not its prerequisite." Accessed Januar 27, 2025. https://www.uni-jena.de/en/190910-jenaererklaerung-en.

Fullwiley, Duana. 2008. "The Biologistical Construction of Race: 'Admixture' Technology and the New Genetic Medicine." *Social Studies of Science* 38 (5): 695–735. DOI: https://doi.org/10.1177/0306312708090796.

Gibson, Barry, and Olga Boiko. 2012. "Luhmann's Social Systems Theory, Health and Illness." In *Contemporary Theorists for Medical Sociology*, edited by Graham Scambler, 49–70. New York: Routledge.

Gieryn, Thomas F. 1983. "Boundary-Work and the Demarcation of Science from Non-Science: Strains and Interests in Professional Ideologies of Scientists." *American Sociological Review* 48 (6): 781–795.

Gießelmann, Kathrin, and Mirjam Martin. 2023. "Medizinische Algorithmen und Risikoscores: Die Rolle von ‚race' überdenken." *Deutsches Ärzteblatt* 120 (45): A-1903 / B-1616.

Glerup, Cecilie, and Maja Horst. 2014. "Mapping 'social responsibility' in science." *Journal of Responsible Innovation* 1 (1): 31–50. DOI: https://doi.org/10.1080/23299460.2014.882077.

Hardtke, S., R. Rocco, J. Ogata, et al. 2018. "Risk factors and seroprevalence of hepatitis E evaluated in frozen-serum samples (2002–2003) of pregnant women compared with female blood donors in a Southern region of Brazil." *Journal of Medical Virology* 90 (12): 1856–1862. DOI: https://doi.org/10.1002/jmv.25274.

Harmon-Jones, Eddie, and Judson Mills. 2019. "An Introduction to Cognitive Dissonance Theory and an Overview of Current Perspectives on the Theory." In *Cognitive Dissonance: Reexamining a Pivotal Theory in Psychology*, edited by Eddie Harmon-Jones, 3–24. Washington: American Psychological Association.

Heinemann, A., F. Wischhusen, K. Püschel, and X. Rogiers. 1999. "Standard Liver Volume in the Caucasian Population." *Liver Transplantation and Surgery* 5 (5): 366–368. DOI: https://doi.org/10.1002/lt.500050516.

Heintz, Bettina. 2024. "Tiere als 'lebende Artefakte'. Die Erfindung des Rassehundes als kulturelle Kategorie und als materielles Objekt." *Zeitschrift für Soziologie* 53 (3): 223–242. DOI: https://doi.org/10.1515/zfsoz-2024-2023.

Hirschauer, Stefan. 2017. "Humandifferenzierung. Modi und Grade sozialer Zugehörigkeit." In Un/doing differences: Praktiken der Humandifferenzierung, edited by Stefan Hirschauer, 29–54. Weilerswist: Velbrück Wissenschaft.

Hirschauer, Stefan. 2023. "Telling People Apart: Outline of a Theory of Human Differentiation." *Sociological Theory* 41 (4): 352–376. DOI: https://doi.org/10.1177/07352751231206411

Hirschman, Albert O. 1970. *Exit, Voice, and Loyalty: Responses to Decline in Firms, Organizations, and States.* Cambridge: Harvard University Press.

Hofmann, F.C., J. Neumann, and U. Heilmeier et al. 2018. "Conservatively treated knee injury is associated with knee cartilage matrix degeneration measured with MRI-based T2 relaxation times: data from the osteoarthritis initiative." *Skeletal Radiology* 47 (1): 93–106. DOI: https://doi.org/10.1007/s00256-017-2759-6.

Hochman, Adam. 2021. "Janus-Faced Race: Is Race Biological, Social, or Mythical?" *American Journal of Physical Anthropology* 175 (2): 453–464. DOI: https://doi.org/10.1002/ajpa.24169.

Jugert, Philipp, Marie J. Kaiser, Francesca Ialuna, and Sauro Civitillo. 2022. "Researching race-ethnicity in race-mute Europe." *Infant and Child Development* 31 (1). DOI: https://doi.org/10.1002/icd.2260.

Kelle, Udo, and Susann Kluge. 2010. *Vom Einzelfall zum Typus: Fallvergleich und Fallkontrastierung in der qualitativen Sozialforschung*. Wiesbaden: VS.

Krieger, Nancy, Jarvis T. Chen, Pamela D. Waterman, David H. Rehkopf, and S. V. Subramanian. 2005. "Painting a Truer Picture of US Socioeconomic and Racial/Ethnic Health Inequalities: The Public Health Disparities Geocoding Project." *American Journal of Public Health* 95 (2): 312–323. DOI: https://doi.org/10.2105/AJPH.2003.032482.

Knorr-Cetina, Karin. 1989. "Spielarten des Konstruktivismus: Einige Notizen und Anmerkungen." *Soziale Welt* 40 (1/2): 86–96.

Kromrey, M. L., T. Ittermann, C. v. Wahsen, et al. 2018. "Reference Values of Liver Volume in Caucasian Population and Factors Influencing Liver Size." *European Journal of Radiology* 106: 32–37. DOI: https://doi.org/10.1016/j.ejrad.2018.07.005.

Latour, Bruno. 2013. *An inquiry into modes of existence*. Cambridge: Harvard University Press.

Lett, Elle, Emmanuella Asabor, Sourik Beltrán, Ashley Michelle Cannon, and Onyebuchi A. Arah. 2022. "Conceptualizing, Contextualizing, and Operationalizing Race in Quantitative Health Sciences Research." *Annals of Family Medicine* 20 (2): 157–163. DOI: https://doi.org/10.1370/afm.2792.

Lipphardt, Veronika. 2009. "Von Der »Europäischen Rasse« zu den »Europiden«. Wissen um die biologische Beschaffenheit des Europäers in Sach- Und Lehrbüchern, 1950–1989." In *Der Europäer – Ein Konstrukt: Wissensbestände, Diskurse, Praktiken*, edited by Lorraine Bluche, Veronika Lipphardt, and Kiran K. Patel. Göttingen: Wallstein.

Livingstone, Frank B. 1962. "On the Non-Existence of Human Races." *Current Anthropology* 3 (3): 279.

Luhmann, Niklas. 1990. "Der Medizinische Code." In *Konstruktivistische Perspektiven*, edited by Niklas Luhmann, 183–195. Opladen: Westdeutscher Verlag.

Marchart, Oliver. 2018. *Das unmögliche Objekt: Eine postfundamentalistische Theorie der Gesellschaft*. Berlin: Suhrkamp.

M'charek, Amade. 2023. "Curious about race: Generous methods and modes of knowing in practice." *Social Studies of Science* 53 (6), 826–849. DOI: https://doi.org/10.1177/03063127231201178.

Meißner, Thomas. 2021. "Lasst uns den „Kaukasier" endlich begraben: Rassismus in klinischer Forschung." *HNO-Nachrichten* 51 (2): 10–13. DOI: https://doi.org/10.1007/s00060-021-7496-8.

Merton, Robert K. 1974. *The Sociology of Science: Theoretical and Empirical Investigations*. Chicago: Univ. of Chicago Pr.

Mol, Annemarie. 2007. *The Body Multiple: Ontology in Medical Practice*. Durham: Duke University Press.

Morning, Ann. 2011. *The Nature of Race. How Scientists Think and Teach about Human Difference*. Berkely: University of California Press.

Morning, Ann, and Marcello Maneri. 2022. *An Ugly Word: Rethinking Race in Italy and the United States*. Chicago: Russell Sage Foundation.

Mulinari, Shai, and Anna Bredström. 2024a. "'Black Race', 'Schwarze Hautfarbe', 'Origine Africaine', or 'Etnia Nera'? The Absent Presence of Race in European Pharmaceutical Regulation." *BioSocieties* 19: 19–39. DOI: https://doi.org/10.1057/s41292-022-00291-7.

Mulinari, Shai, and Anna Bredström. 2024b. "Race in Clinical Trials in Sweden: How Regulatory and Medical Standards in Clinical Research Trump the Post-Racial Discourse." *Sociology of Health & Illness* 46 (2): 315–332. DOI: https://doi.org/10.1111/1467-9566.13709.

National Academies of Sciences, Engineering, and Medicine. 2023. *Using Population Descriptors in Genetics and Genomics Research: A New Framework for an Evolving Field*. Washington, DC: The National Academies Press. DOI: https://doi.org/10.17226/26902.

Rajagopalan, Ramya, and Joan H. Fujimura. 2012. "Will Personalized Medicine Challenge or Reify Categories of Race and Ethnicity?" *The virtual mentor* 14 (8): 657–663. DOI: https://doi.org/10.1001/virtualmentor.2012.14.8.msoc1-1208.

Reardon, Jenny. 2009. *Race to the Finish: Identity and Governance in an Age of Genomics*. Princeton: Princeton University Press.

Reeves, Frank. 1983. *British Racial Discourse: A Study of British Political Discourse About Race and Race-Related Matters*. Cambridge, New York: Cambridge University Press.

Root, Michael. 2003. "The Use of Race in Medicine as a Proxy for Genetic Differences." *Philosophy of Science* 70 (5): 1173–1183. DOI: https://doi.org/10.1086/377398.

Rose, Geoffrey. 2001. "Sick individuals and sick populations." *International Journal of Epidemiology* 30 (3): 427–432. DOI: https://doi.org/10.1093/ije/30.3.427.

Schenk, Alina, Saioa López, Maik Kschischo, and Nicholas McGranahan. 2020. Germline ancestry influences the evolutionary disease course in lung adenocarcinomas. *Evolutionary Applications* 13 (7): 1550–1557. DOI: https://doi.org/10.1111/eva.12964.

Schenk, Liane. 2016. "Migrationssensible Gesundheitsforschung. Theoretische und empirische Forschungsergebnisse." Habilitation, Medizinische Fakultät, Charité.

Star, Susan Leigh. 2010. "This Is Not a Boundary Object: Reflections on the Origin of a Concept." *Science, Technology, & Human Values* 35 (5): 601–617. DOI: https://doi.org/10.1177/0162243910377624.

Star, Susan Leigh, and Anselm Strauss. 1999. "Layers of Silence, Arenas of Voice: The Ecology of Visible and Invisible Work." *Computer Supported Cooperative Work (CSCW)* 8 (1–2): 9–30. DOI: https://doi.org/10.1023/A:1008651105359.

Strauss, Anselm L. 1978. "A Social World Perspective." *Studies in Symbolic Interaction* (1): 119–128.

Strauss, Anselm L. 1987. *Qualitative Analysis for Social Scientists*. Cambridge: Cambridge University Press.

Suda, Akira, Andreas Seitz, Yuji Odaka, et al. 2021. "Assessment of coronary vasomotor responses to acetylcholine in German and Japanese patients with epicardial coronary spasm—more similarities than differences?" *Heart Vessels* 36 (3): 337–344. DOI: https://doi.org/10.1007/s00380-020-01707-9.

Takezawa, Yasuko, Kazuto Kato, Hiroki Oota, et al. 2014. Human genetic research, race, ethnicity and the labeling of populations: recommendations based on an interdisciplinary workshop in Japan. *BMC Med Ethics* 15, 33. DOI: https://doi.org/10.1186/1472-6939-15-33.

Thomas, William I., and Dorothy S. Thomas. 1928. *The Child in America: Behavior Problems and Programs*. New York, NY: Knopf.

Trong, Philip Dao, Christopher Beynon, Andreas Unterberg, Till Schneider, and Jessica Jesser. 2018. "Racial Differences in the Anatomy of the Posterior Fossa: Neurosurgical Considerations." *World Neurosurgery* 117: e571–e574. DOI: https://doi.org/10.1016/j.wneu.2018.06.089.

Uhlig, Annemarie, Johannes Uhlig, Lutz Trojan, and Hyun S Kim. 2020. "Stereotactic Body Radiotherapy for Stage I Renal Cell Carcinoma: National Treatment Trends and Outcomes Compared to Partial Nephrectomy and Thermal Ablation." *Journal of Vascular and Interventional Radiology* 31 (4): 564–571. DOI: https://doi.org/10.1016/j.jvir.2019.11.009.

Wacquant, Loïc. 1997. "For an analytic of racial domination." *Political Power and Social Theory* 11 (1): 221–234.

Wacquant, Loïc. 2022. "Resolving the Trouble with 'Race'." *New Left Review* (133): 67–88.

Weber, Max. 1904. "Die 'Objektivität' sozialwissenschaftlicher und sozialpolitischer Erkenntnis." *Archiv für Sozialwissenschaft und Sozialpolitik* 19 (1): 22–87.

Weber, Max. 1972. *Wirtschaft und Gesellschaft. Grundriß der verstehenden Soziologie*. Tübingen: Mohr.

Will, Anne-Kathrin. 2019. "The German Statistical' Category 'Migration Background': Historical Roots, Revisions and Shortcomings." *Ethnicities* 19 (3): 535–557. DOI: https://doi.org/10.1177/1468796819833437.

Yudell, Michael, Dorothy Roberts, Rob DeSalle, and Sarah Tishkoff. 2016. "Taking Race Out of Human Genetics." *Science* 351 (6273): 564–565. DOI: https://doi.org/10.1126/science.aac4951.

zur Nieden, Andrea. 2014. "The Presence of the Past. 'Ashkenazi BRCA Mutations' and Transnational Differences in Categories of 'Race' and 'Ethnicity': The German Case." In *Breast Cancer Gene Research and Medical Practices: Transnational Perspectives in the Time of BRCA*, edited by Galen Joseph, Jessica Mozersky, Andrea zur Nieden, and Sonja Palfner, 17–34. London: Routledge.

zur Nieden, Andrea. 2025. "Persistent Ideas: Race, Ethnicity, and Migration Background in German Epidemiological and Pharmaceutical Research." In *Behemoth. A Journal of Social Disorder* 18 (1): 39–55. DOI: https://doi.org/10.6094/behemoth.2025.18.1.1119.

Identifying the Asian Body
Ancestry and Belonging in Forensic Anthropology

Trudi Buck and Yulia Egorova

Abstract *Forensic anthropology is characterized by the identification of deceased individuals through the creation of a biological profile from the skeletal remains. This profile includes the age-at-death, sex, stature, and ancestry of the skeleton. Ancestry, one way of describing human morphological variation, is often seen as being used as a proxy for social race, and the use of the trifecta of continental groups, Africa, Asia, and Europe, as the main stays of identification is still prevalent. The chapter attempts to critically explore the use of ancestry-related classificatory systems deployed in forensic anthropology by focusing on osteology collections. In doing so it engages two separate sets of problematics. First, it highlights deficiencies in the representation of Asian populations in forensic databases. Second, it addresses ethical challenges of teaching with collections of human remains derived from postcolonial contexts without alienating students of ethnically minoritized backgrounds.*

Introduction

In this chapter we will address the topic of the diverse conceptualizations of human classifications in the life sciences by focusing on the problematics of the classification of human remains by ancestry in osteology collections. The authors are a forensic anthropologist (Trudi Buck) and a social anthropologist (Yulia Egorova), both based in an integrated department of anthropology where teaching and research are conducted in different fields of anthropological inquiry. The paper is an attempt to critically explore the use of ancestry-related classificatory systems deployed in forensic anthropology, as well as to start reflecting on the ethical challenges of teaching about ancestry classification with the use of osteology collections.

Our analysis in the paper will be two-pronged. Firstly, we will discuss the geographical biases inherent in the development of existing techniques for the estimation of ancestry and the mismatches of terminology used in forensic practice in different parts of the world. Secondly, we will reflect on the ethical challenges of teaching the available methods within forensic anthropology and working with human remains in classrooms with diverse student audiences. In such an academic setting, talking about the ancestry of individuals whose remains are represented in the collection and the provenance of these remains may result in subjecting students of minoritized backgrounds to potentially alienating discourses of othering.

In starting this discussion, we will be contributing to the body of literature in social studies of science that has focused on the history and sociology of the development of ancestry classification methods in the life sciences (e.g. Abu El-Haj 2012; Egorova 2013; Lipphardt 2014; Lipphardt et al. 2021; Liu 2010; Reardon 2005; Schwartz-Marin and Restrepo 2013; Sommer 2016), and to the growing research in critical heritage studies which has discussed the problematic of the material representations of marginalized groups' histories (e.g. Burch-Brown 2022; Mookherjee 2022), colonial legacies of museums (Hicks 2020), and, more broadly, ways to address what Sharon Macdonald has described as "difficult heritage" (2009). It will also contribute to recent and ongoing conversations in biological anthropology about the situated nature of scientific knowledge production in this area (Athreya 2019).[1]

Ancestry Classification in Forensic Anthropology

The main aim of forensic anthropology is to achieve positive identification of deceased unknown individuals for medicolegal purposes. This is conducted through the creation of a biological profile from their skeletal remains, including the estimation of biological sex, age-at-death, stature, and geographic ancestry. Ancestry, one way of describing human morphological variation or population affinity, can be perceived as a proxy for social race in forensic identification and on these grounds has been the center of debates as to whether or not forensic anthropologists should abolish the practice of ancestry estimation altogether (Bethard and DiGangi 2020; DiGangi and Bethard 2021; Stull et al. 2020).

One of the main areas of contestation around ancestry classifications is the historical use of the trifecta of continental groups, Africa, Asia, and Europe, to classify individuals, which still forms the basis of the main estimation methods today (Hefner 2018; Bethard and DiGangi 2020; Go, Tallman, and Kim 2019). These groupings are built upon the work of early physical anthropologists such as Aleš Hrdlička, although the aims of such work included the construction of racial hierarchies rather than identification (Blakey 1996; Caspari 2009). As with much methodology and theory in forensic anthropology, the techniques used for the estimation of ancestry originate in US academic institutions, and thus the terminology employed follows this geographic bias, with the earliest cranial discrimination method attempted by Giles and Elliot (1962) between "Whites," "Blacks," and "Indians."[2] The techniques that have subsequently followed this were also largely based on anatomical collections derived from US population groups (Albanese et al.

1 We borrow the term "situated knowledge" from Haraway (1988).

2 See Pilloud et al. (2021) for a discussion regarding terminology used to describe human morphological variation in forensic anthropology.

2022). In the Forensic Anthropology Data Bank[3], originally developed by The University of Tennessee, Knoxville, in 1986, for example, the contemporary "Asian" subset consists exclusively of individuals of Chinese, Japanese, and Vietnamese descent. In contrast with these classifications, in the United Kingdom the term Asian would be understood to include individuals of South Asian origin. This geographic area, including India and Pakistan, is not well represented in the methods used to estimate ancestry.

At the center of the production of scientific knowledge in this discipline, there is, therefore, a bias often unacknowledged in the literature.[4] This bias is still included within the methods highlighted for ancestry estimation in introductory textbooks used for undergraduate teaching (e.g. Christensen and Passalacqua 2018; Byers 2024). The three-group model that forms the basis of historical research on ancestry in forensic anthropology thus has larger implications for the epistemology of ancestry studies and the discipline globally.[5] Acknowledgment should be given, but often is not, to the arbitrary and historical nature of these classifications of skeletal variation, which may lead to over-simplistic assumptions of morphological homogeneity.[6] The Macromorphoscopic Databank, for example, is designed to provide "large and appropriate reference samples" for developing identification standards in forensic anthropological ancestry estimation (Hefner 2018, 994). This ancestry estimation methodology uses morphological features such as the shape of the nasal aperture or the width between the orbits as classificatory variables (Hefner 2018). Identification starts at the large scale, using the three-group model of European, African, and Asian, then moves through tiers to finer levels of classification that end with geographic origin (Hefner 2018). The "Asian" group encompasses individuals as varied and disparate in geographical terms as Indigenous Americans, Thais, Australians, or Maori (e.g., Atkinson and Tallman 2020; Plemons et al. 2018). Out of thirty-five potential areas of "geographic origin" (Hefner 2018, 997) within the Asian group, only one of these is from South Asia, Sri Lanka. The Sri Lankan population data in the Macromorphoscopic databank are listed under the second tier of "ancestry (geographic)" as being of Pacific Island origin, together with populations of a wide range of areas including, for instance, Australia, Indonesia, Malaysia and Samoa (Hefner 2018, 997).

3 www.statsmachine.net/software/Fordisc/support/Help/Fordisc3_Help.pdf, last accessed 5 December 2024.

4 Research has begun to address the issue of people with Latin American and Filipino origins within the United States (e.g. Go et al. 2019, Hughes et al. 2018).

5 Note that in this paper we are only specifically addressing the use of ancestry estimation in forensic anthropology, which is a specialized application of anthropological knowledge to questions about medico-legal significance. The wider discipline of biological anthropology across the globe has differing histories of how race has been studied.

6 For further discussion of assumptions of morphological homogeneity in inequality of identification in Black Caribbean populations, see Dwyer et al. (2023).

Perpetuating Lack of Representation

The lack of representation of South Asian individuals in ancestry estimation is due to much research relying on the same anatomical collections and databases in the creation of specific methodologies without addition of new data from wider geographic areas, even when this is potentially available. Craniometric data from over 2000 individuals from thirty global populations is freely available to researchers based upon data collected by W. W. Howells between 1965 and 1980.[7] This dataset forms the basis of many ancestry estimation studies (e.g., Algee-Hewitt 2016; Dong et al. 2021; Navega et al. 2015). However, in this dataset, as well, the only extensive South Asian representatives are peoples of the Andaman Islands (Howells 1995). Howells acknowledges that due to limitations of collected material there are "spacious interstices" for which there is no information included, notably South Asia and Latin America (Howells 1995, 3). While the lack of individuals from South America in forensic anthropological databases is being addressed (see for example Spradley 2021), the lack of South Asian data continues to fall into this "lacuna" of knowledge (Howells 1995). One possible explanation for the underrepresentation of South Asian individuals is the historical and racial biases in the development of biological anthropology that are inextricably linked with the colonial ideologies of their time. In early typological studies (e.g., Hooton 1946; Coon and Hunt 1965), people of India were referred to as "Caucasoid" alongside European populations. While these typological studies are no longer considered acceptable in science, their legacy continues within the uncritical use of classifications they established.

Rick Smith and Deborah Bolnick remind us that biological anthropology, which has been instrumental in producing scientific knowledge about humans and nonhumans, has been historically shaped by majoritarian and settler colonial agendas. These epistemological biases created a flawed and incomplete scientific corpus about human variation and have continued to live on in some of the contemporary knowledge production practices, leading to the marginalization and exclusion of minoritized bodies and voices (Smith and Bolnick 2019, 465). For instance, they cite the example of genetic studies of Indigenous groups in North America that have for a long time been influenced by settler colonial preoccupation with the notion of biological purity, which did not take into consideration the social, political, and legal factors that determine belonging in these communities (Smith and Bolnick 2019, 465; TallBear 2013).

Addressing the broader context of biological and evolutionary anthropology, Sheela Athreya points out that Eurocentric biases in the models of human evolution continue to inform knowledge production in these fields, and "the seeming near universality of observations and assumptions upon which we rest our evolutionary models actually reflects the lack of diversity in paleoanthropology and the clear power structure that privileges Western knowledge" (Athreya 2019, 473).

The creation of biological profiles, including the estimation of ancestry, is a form of narrative, a literal interpretation of the trope of "reading the body." The exclusion of the South Asian body from this narrative raises questions about whose voices are heard or bodies seen, and contrarily who are being diminished. In a call for a reflexive analysis of

7 https://web.utk.edu/ auerbach/HOWL.htm, last accessed 5 December 2024.

such praxis, Sabrina Agarwal describes this as a form of biopower and highlights the need for the transformation and decolonization of the discipline (Agarwal 2024a; idem 2024b). Recognition of the United States as the "gatekeepers" of the state of the art of ancestry estimation and forensic skeletal identification is a step towards changing the status quo, which is required if human variation is to be classified in a truly global, meaningful, and inclusive manner.

Western bodies and Western scientific knowledge are being prioritized in the development of methodologies, but at the same time students are expected to learn from the bodies of the poor and from unclaimed bodies as subjects/objects that are not represented within the techniques of analysis that they study. It is not even possible to estimate the "ancestry" of such an individual from an anatomical teaching collection because the "test subject" data does not exist in the statistical programs available for use. This highlights the inequalities and lack of diversity encapsulated in the discipline. For the teachers of forensic anthropology, this presents a challenge of explaining accepted epistemological knowledge as it exists in the textbooks set as reading against the reality that is not present in the established literature.

The lack of the representation of the South Asian body in the modern forensic classifications of ancestry leaves South Asian populations outside of the forensic anthropological context as it currently stands. The obvious dangers of this academic structure are inherent in the possible misclassification of an individual within a forensic context who is not represented within the extant biological data. This has been already discussed in detail in a theoretical context, for example regarding individuals from Brazil (Jacometti et al. 2023; Fernandes et al. 2021) and Filipino individuals (Go et al. 2019) who cannot be correctly identified by the existing techniques such as AncesTrees and Fordisc. Lorenzo Franceschetti (2023) highlights the more practical application and ongoing dangers of the lack of appropriate reference collections in published methods. His research into the attempts to identify 150 of the 500 people killed in the Lampedusa shipwreck showcases where anthropological methods fail to meet the diversity of individuals that attempted to travel to Europe. As we discuss in the final section, just as dangerous is how these inherent biases are built into the epistemology of the discipline being transmitted, often uncritically, to practitioners and students, including people of color who are excluded from the fundamental analyses.[8]

Identity and Belonging in the Teaching of Forensic Anthropology

When it comes to the specifics of teaching forensic anthropology, Amelia Hubbard (2021) warns of the dangers of hidden curricula, and of unconscious and implicit bias when teaching forensic anthropology—concerns centering on the lack of critical reflection. Such hidden curricula are present in the skeletal remains of individuals that are widely used for teaching in global university classrooms, such as in one of the authors' own laboratories at Durham University in England. Indeed, the pedagogical context of forensic anthropology is that while the South Asian body is excluded from accepted methods

8 Sabrina Agarwal (2024a) refers to this as the legacy and disposability of the brown body.

of ancestry estimation, it exists as a subject/object for osteological teaching, as many universities have anatomical teaching collections that have their origins in India. Agarwal notes that colonial India offered unique opportunities for the collection of human remains and that after the end of the British rule, too, export of human bodies out of India continued in complex ways (Agarwal 2024a; idem 2024b). Indeed, at its peak the once legal "Calcutta Bone Trade" was exporting over 50,000 bodies a year to universities and medical students around the world (Carney 2011). To use a concept introduced by Lawrence Cohen, populations and communities who were the source of these remains became "bioavailable" (Cohen 2004)[9], as they did not have the means to resist the methods of acquisition of these skeletons, which did not provide channels of consent for the individuals involved, leading to the bodies of the poor often being collected by illegal means and/or without the knowledge of their families. The Indian Government criminalized the trade in 1985 but there is evidence that it continues in the present, albeit having gone underground. The earliest trade in human remains for medical teaching also ran out of France, Russia, Germany, and China though India became the main supplier from the mid-1930s (Jones 2023, 611).

Acknowledging the difficult truths regarding the existence of these remains in the undergraduate laboratory calls on us to reflect not only on the theoretical development of forensic anthropology but also on how we transmit this situated knowledge in the classroom. In the cases of using human remains in teaching laboratories like ours, we suggest that what needs to be assessed is not only the politics of ancestry identification practices, but also the ethics of presenting findings and hypotheses about the provenance of human remains collections that the laboratory holds. For instance, in a situation where students from the Global South or other parts of the world, where human remains may have originated from, constitute a minority in the classroom (and in the general population), accounts highlighting the violent histories that may have led to the generation and acquisition of these remains may be experienced by these students as traumatizing and alienating, even if these accounts are informed by liberal humanistic politics. In this respect, teaching forensic anthropology shares the problematics of the ethics of presenting material pertaining to the histories of minoritized individuals and groups with teaching in social anthropology and other social sciences and humanities.

Moreover, the very mention of the Indian origin of human remains collections is very likely to be traumatizing for students from the Global South. A poignant example is provided by Agarwal, a Canadian bioarchaeologist of South Asian descent, who recounts her experience in an undergraduate human osteology class. Upon learning of the Indian provenance of the skeletons used in her studies, Agarwal found herself identifying with the subjects presented in the collection more than with her predominantly White student cohort (Agarwal 2024b, 3). For such an experience, Watkins offers a theoretical framework for understanding anatomical collections, conceptualizing them as representations of the "anthropological other" (Watkins 2018). She demonstrates how these remains, integral to both teaching and research, are positioned within normative

9 As Miriam Ticktin (2011, 155) explains, "in Lawrence Cohen's (2004) terms, people must be understood as differently 'bioavailable,' a term he uses to describe the likelihood for a person or population that its tissues may be disaggregated and transferred to some other entity or process."

scientific practices. Rachel Watkins contends that although such collections are foundational to anthropology, the lack of adequate social and historical contextualization obscures their relevance to broader debates about race, science, and inequality. In line with these critiques, we argue that the relevance of smaller-scale anatomical teaching collections, curated within university settings, is similarly obscured in these debates. While usually the "White body" is seen as standard while the "non-White body" is constructed as marked with difference, in this case, on the contrary, human remains from South Asia are generalized as the standard for all humans, but this can have negative implications for the identification of individuals of South Asian ancestry.

It is essential to understand these assemblages within their social and historical context, recognizing the ways in which they shape the perspectives of both researcher and students studying the remains. Within the UK, it is crucial to acknowledge not only the social and historical context of the anatomical skeletons but also the structural violence that their existence and use represent. Furthermore, the underrepresentation of the South Asian bodies in the praxis of forensic anthropology highlights a broader gap in the discipline. Addressing these issues requires ongoing reflection on the sensitivities associated with teaching about the historical and contemporary experiences of marginalization and inequality.

References

Abu El-Haj, Nadia. 2012. *The Genealogical Science: The Search for Jewish Origins and the Politics of Epistemology.* Chicago: University of Chicago Press.

Agarwal, Sabrina C. 2024a. "The Disposability and Inclusion of Brown Bodies." *American Journal of Biological Anthropology* e25003. DOI: https://doi.org/10.1002/ajpa.25003.

Agarwal, Sabrina C. 2024b. "The Bioethics of Skeletal Anatomy Collections from India." *Nature Communications* 15 (1): 1692. DOI: https://doi.org/10.1038/s41467-024-45738-6.

Albanese, John, Abegail Dagdag, Carmen Skalic, Stephanie Osley, and Hugo Cardoso. 2022. "The Fall and Rise of Identified Reference Collection: It Is Possible and Necessary to Transition from a Typological Conceptualization of Variation to Effective Utilization of Collections." *Forensic Sciences* 2 (2): 438–454. DOI: https://doi.org/10.3390/forensicsci2020033.

Algee-Hewitt, Bridget F. B. 2016. "Population Inference from Contemporary American Craniometrics." *American Journal of Physical Anthropology* 160 (4): 604–624. DOI: https://doi.org/10.1002/ajpa.22959.

Athreya, Sheela. 2019. "'But You're Not a Real Minority': The Marginalization of Asian Voices in Paleoanthropology." *American Anthropologist* 121 (2): 472–474. DOI: https://doi.org/10.1111/aman.13216.

Atkinson, Megan L., and Sean D. Tallman. 2020. "Nonmetric Cranial Trait Variation and Ancestry Estimation in Asian and Asian-Derived Groups." *Journal of Forensic Sciences* 65 (3): 692–706. DOI: https://doi.org/10.1111/1556-4029.14234.

Bethard, Jonathan D., and Elizabeth A. DiGangi. 2020. "Letter to the Editor: Moving Beyond a Lost Cause: Forensic Anthropology and Ancestry Estimates in the United

States." *Journal of Forensic Sciences* 65 (5): 1791. DOI: https://doi.org/10.1111/1556-4029.14513.

Blakey, Michael L. 1996. "Skull Doctors Revisited: Intrinsic Social and Political Bias in the History of American Physical Anthropology: With Special Reference to the Work of Aleš Hrdlička." In *Race and Other Misadventures: Essays in Honor of Ashley Montagu in His Ninetieth Year*, edited by Larry T. Reynolds and Leonard Lieberman, 64–95. New York: General Hall.

Burch-Brown, Joanna. 2022. "Should Slavery's Statues be Preserved? On Transitional Justice and Contested Heritage." *Journal of Applied Philosophy* 39 (5): 807–824. DOI: https://doi.org/10.1111/japp.12485.

Byers, Steven N. 2011. *Forensic Anthropology Laboratory Manual*. 3rd edition. London: Pearson.

Carney, Scott. 2011. *The Red Market: On the Trail of the World's Organ Brokers, Bone Thieves, Blood Farmers, and Child Traffickers*. New York: William Morrow.

Caspari, Rachel. 2009. "1918: Three Perspectives on Race and Human Variation." *American Journal of Physical Anthropology* 139 (1): 5–15. DOI: https://doi.org/10.1002/ajpa.20975.

Christensen, Angi, M., and Nicholas V. Passalacqua. 2018. *A Laboratory Manual for Forensic Anthropology*. London: Academic Press.

Cohen, Lawrence. 2004. "Operability: Surgery at the Margin of the State." In *Anthropology In the Margins of the State, edited by Veena Das and Deborah Poole*, 165–190. Santa Fe: School of American Research Press.

Coon, Carleton S., and Edward E. Hunt, Jr. 1965. *The Living Races of Man*. New York: Knopf.

DiGangi, Elizabeth A., and Jonathan D. Bethard. 2021. "Uncloaking a Lost Cause: Decolonizing Ancestry Estimation in the United States." *American Journal of Physical Anthropology* 175 (2): 422–436. DOI: https://doi.org/10.1002/ajpa.24212.

Dong, Yibo, Andrew Gao, Ian Hou, et al. 2021. "A Deep Learning Model for Ancestry Estimation with Craniometric Measurements." In *Proceedings: 2021 IEEE International Conference on Bioinformatics and Biomedicine*, edited by Yufei Huang, Lukasz Kurgan, Feng Luo, et al., 3350–3357. Piscataway: IEEE. DOI: https://doi.org/10.1109/BIBM52615.2021.9669742.

Dwyer, Isis, Delande Justinvil, and Andreana Cunningham. 2023. "Caribbeanist Casualties: Interrogating the Application of Structural Vulnerability to Forensic Anthropology." *Forensic Science International*: Synergy 6: 100327. DOI: https://doi.org/10.1016/j.fsisyn.2023.100327.

Egorova, Yulia. 2013. "The Substance That Empowers? DNA in South Asia." *Contemporary South Asia* 21 (3): 291–303.

Fernandes, Larissa Chaves Cardoso, Maria Izabel Cardoso Bento, Patrícia Moreira Rabello, et al. 2021. "Analysis of the Accuracy of AncesTrees Software in Ancestry Estimation in Brazilian Identified Sample." *Advances in Anthropology* 11 (2): 163–178. DOI: https://doi.org/10.4236/aa.2021.112011.

Franceschetti, Lorenzo. 2023. *Estimating the Biogeographical Origin of the Unidentified Bodies of the Shipwreck of April 18th, 2015 in the Mediterranean: Comparison of Genetic and Anthropological Assessments*. PhD dissertation, Università degli studi di Milano.

Giles, Eugene, and Orville Elliot. 1962. "Race Identification from Cranial Measurements." *Journal of Forensic Sciences* 7 (2): 147–157.

Go, Matthew C., Ansley R. Jones, Bridget F. B. Algee-Hewitt, Beatrix Dudzik, and Cris E. Hughes. 2019. "Classification Trends Among Contemporary Filipino Crania Using Fordisc 3.1." *Forensic Anthropology* 2 (4): 293–303. DOI: https://doi.org/10.5744/fa.2019.1005.

Go, Matthew C., Sean D. Tallman, and Jieun Kim. 2019. "Advances in Forensic Anthropological Research in East and Southeast Asia." *Forensic Anthropology* 2 (4): 197–203. DOI: https://doi.org/10.5744/fa.2019.1023.

Haraway, Donna. 1988. "Situated Knowledges: The Science Question in Feminism and the Privilege of Partial Perspective." *Feminist Studies* 14 (3): 575–599.

Hefner, Joseph T. 2018. "The Macromorphoscopic Databank." *American Journal of Physical Anthropology* 166 (4): 994–1004. DOI: https://doi.org/10.1002/ajpa.23492.

Hicks, Dan. 2020. *The Brutish Museums: The Benin Bronzes, Colonial Violence and Cultural Restitution*. London: Pluto Press.

Hooton, Ernest A. 1946. *Up from the Ape*. New York: The Macmillan Company.

Howells, William White. 1995. *Who's Who in Skulls: Ethnic Identification of Crania from Measurements*. Cambridge, MA: Harvard University Press.

Hubbard, Amelia R. 2021. "Explanation of Ancestry Estimation in Forensic Anthropology Textbooks: Lessons for the Undergraduate Classroom." *Forensic Anthropology* 4 (4): 341–359. DOI: https://doi.org/10.5744/bi.2021.0006.

Hughes, Cris E., Bridget F. B. Algee-Hewitt, and Lyle W. Konigsberg. 2018. "Population Identifiability from Forensic Genetic Markers: Ancestry Variation in Latin America." *Human Biology* 90 (3): 161–175. DOI: https://doi.org/10.13110/humanbiology.90.3.03.

Jacometti, Victor, Marco Aurelio Guimarães, Luis Otávio Carvalho de Moraes, Sérgio Ricardo Marques, Eugénia Cunha, and Ricardo Henrique Alves da Silva. 2023. "Ancestry Estimation in Forensic Anthropology: Accuracy of the AncesTrees Software in a Brazilian Sample." *Forensic Sciences Research* 8 (3): 202–210. DOI: https://doi.org/10.1093/fsr/owad030.

Jones, David G. 2023. "Anatomists' Uses of Human Skeletons: Ethical Issues Associated with the India Bone Trade and Anonymized Archival Collections." *Anatomical Sciences Education* 16 (4): 610–617. DOI: https://doi.org/10.1002/ase.2280.

Lipphardt, Veronika. 2014. "'Geographical Distribution Patterns of Various Genes': Genetic Studies of Human Variation After 1945." *Studies in History and Philosophy of Biological and Biomedical Sciences* 47 (A): 50–61. DOI: https://doi.org/10.1016/j.shpsc.2014.05.006.

Lipphardt, Veronika, Mihai Surdu, Nils Ellebrecht, Peter Pfaffelhuber, Matthias Wienroth, and Gudrun A. Rappold. 2021. "Europe's Roma People Are Vulnerable to Poor Practice in Genetics." *Nature* 599 (7885): 368–371. DOI: https://doi.org/10.1038/d41586-021-03416-3.

Liu, Jennifer A. 2010. "Making Taiwanese (Stem Cells): Identity, Genetics, and Hybridity." In *Asian Biotech: Ethics and Communities of Fate*, edited by Aihwa Ong and Nancy N. Chen, 239–262. Durham: Duke University Press.

Macdonald, Sharon. 2009. *Difficult Heritage: Negotiating the Nazi Past in Nuremberg and Beyond*. New York: Routledge.

Mookherjee, Nayanika. 2022. "Irreconcilable Times." *Journal of the Royal Anthropological Institute* 28 (S1): 153–178. DOI: https://doi.org/10.1111/1467-9655.13760.

Navega, David, Catarina Coelho, Ricardo Vicente, Maria Teresa Ferreira, Sofia Wasterlain, and Eugénia Cunha. 2015. "AncesTrees: Ancestry Estimation with Randomized Decision Trees." *International Journal of Legal Medicine* 129 (5): 1145–1153. DOI: https://doi.org/10.1007/s00414-014-1050-9.

Pilloud, Marin A., Cassie E. Skipper, SaMoura L. Horsley, et al. 2021. "Terminology Used to Describe Human Variation in Forensic Anthropology." *Forensic Anthropology* 4 (4): 257–282. DOI: https://doi.org/10.5744/fa.2021.0001.

Plemons, Amber M., Joseph T. Hefner, and Kelly R. Kamnikar. 2018. "Refining Asian Ancestry Classifications via Cranial Macromorphoscopic Traits." *American Journal of Physical Anthropology* 165 (S66): 210.

Reardon, Jenny. 2005. Race to the Finish: Identity and Governance in an Age of Genomics. Princeton: Princeton University Press.

Schwartz-Marín, Ernesto and Eduardo Restrepo. 2013. "Biocoloniality, Governance, and the Protection of 'Genetic Identities' in Mexico and Colombia." *Sociology* 47 (5): 993–1010. DOI: https://doi.org/10.1177/0038038513494506.

Smith, Rick W. A., and Deborah A. Bolnick. 2019. "Situating Science: Doing Biological Anthropology as a View from Somewhere." *American Anthropologist* 121 (2): 465–467. DOI: https://doi.org/10.1111/aman.13213.

Sommer, Marianne. 2016. *History Within: The Science, Culture, and Politics of Bones, Organisms, and Molecules*. Chicago: University of Chicago Press.

Spradley, M. Katherine. 2021. "Use of Craniometric Data to Facilitate Migrant Identifications at the United States/Mexico Border." *American Journal of Physical Anthropology* 175 (2): 486–496. DOI: https://doi.org/10.1002/ajpa.24241.

Stull, Kyra E., Eric J. Bartelink, Alexandra R. Klales, et al. 2020. "Commentary On: Bethard JD, DiGangi EA. Letter to the Editor: Moving Beyond a Lost Cause: Forensic Anthropology and Ancestry Estimates in the United States." *Journal of Forensic Sciences* 66 (1): 417–420. DOI: https://doi.org/10.1111/1556-4029.14616.

TallBear, Kim. 2013. *Native American DNA: Tribal Belonging and the False Promise of Genetic Science*. Minneapolis: University of Minnesota Press.

Ticktin, Miriam. 2011. "How Biology Travels: A Humanitarian Trip." *Body & Society* 17 (2–3): 139–158. DOI: https://doi.org/10.1177/1357034X11400764.

Watkins, Rachel. 2018. "Anatomical Collections as the Anthropological Other: Some Considerations." In *Bioarchaeological Analyses and Bodies: New Ways of Knowing Anatomical and Archaeological Skeletal Collections*, edited by Pamela K. Stone, 27–47. Cham: Springer.

Postracial Performative Identifications and Subjectifications

"Who Is the Most European of Us All?"

Occidentalism, European Identity, and Counterknowledge on Race and Genomics

Claude-Olivier Doron

Abstract *This article examines how various online communities with racialist or identitarian agendas, such as White nationalists or European preservationists, engage with and produce knowledge about human genomics and biogeographical ancestry. It begins by showing how these communities have developed and disseminated a form of "counterknowledge" online, focusing on human biodiversity, race, and genomics, in opposition to what they perceive as "mainstream science." It emphasizes that this counterknowledge cannot be dismissed as mere "ideology" or "junk science," highlighting its complex and ambivalent relationship with mainstream population genetics. It characterizes this counterknowledge as a form of decentralized citizen science that fosters participatory research programs on the internet, often exploring the deep genomic history of specific ethnopolitical groups. The second section examines how the genetic identity of Europe is a contested topic within these online communities. Rather than broadly referring to "Whiteness" or "European descent," these discussions often emphasize distinctions within "Whiteness," particularly between "Mediterraneans" and "Nordicists." The third section investigates how notions of European DNA acquire a "social life" on these forums and how individual identities are negotiated in these spaces. The article underscores that the genomic identities discussed on these platforms are not entirely novel biosocialities; instead, they are deeply rooted in older frameworks for constructing and defining identity, such as race, blood, territory, and ancestry.*

Introduction

In this article, I examine how various online communities with racialist or identitarian agendas, such as White nationalists or European preservationists, engage with and produce knowledge about human genomics and biogeographical ancestry.[1] During the

1 It is difficult to find the appropriate label to describe these communities. While some explicitly claim they are White supremacists or "European preservationists," many others prefer to define themselves as "Nordicists" and "Mediterraneans." It would be inappropriate to picture them

early years of the internet's development, enthusiastic naiveté abounded. Some scholars sincerely believed that the internet would enable a global and fluid world and put an end to racial identifications and racist claims, leading to the formation of radically new biosocialities.[2] Since then, however, a very different, bleak reality has appeared. Scholars are now focusing mainly on the fact that the internet and social media are actually a locus of intense work reelaborating racial identities and hierarchies. Many groups with racist or racialist agendas have used the internet as a strategic field to promote their claims.[3] I focus here on a specific dimension of this phenomenon, which was still largely ignored when I launched this research in 2013.[4] White nationalists, self-described European preservationists, and other racist communities have largely appropriated the theme of human biological diversity on the internet. They have harnessed the growing knowledge about genomic variations and biogeographical ancestry to claim that race is a biological reality, reinforce their own racial identities, and, more broadly, promote genetic determinism and racial hierarchies. Until very recently, scholars have focused largely on the uses that Afro-American, Jewish, Native American, Irish, and other minority groups had made of DNA ancestry testing, showing how DNA had been integrated into a positive (albeit highly problematic) quest for social justice, reparation, and heritage (Nash 2008 and Nash 2015; Abu El-Haj 2012; Bliss 2012; TallBear 2013; Nelson 2016; Abel 2021). They have also analyzed how genomics has been integrated, through national genome programs, into national identity building or genomic sovereignty (Wade et al. 2014; Reardon 2019). While these are all very important points to study, we should not neglect other

as forming homogeneous groups. On some forums, such as anthroscape or forum.biodiversity, one finds not only individuals clearly driven by ideological and political agendas, but also people merely interested in genetic ancestry and in the deep history of their regions, without necessarily sharing the same racialist or identitarian agenda. The Apricity, Occidental Enclave or Stormfront are more explicitly White supremacist and racist.

2 The concept of biosociality was initially coined by Paul Rabinow in his seminal article "Artificiality and Enlightenment: From Sociobiology to Biosociality" (Rabinow 1996), to describe how natural or biological material (such as genes) can be invested and worked on through sociocultural practices, in such a way that it may lead to the formation of radically "new group and individual identities." This concept has since been widely used, especially by Hacking, Novas, Rose, and others, to contrast the old ways of thinking about race, ethnicity, or biomedical entities, with more creative and plastic identities articulating biological materials to define new collective identities and forms of subjectivities.

3 On this renewed interest in race and racism on internet since the mid-2000s, see, among others, Daniels 2009; Daniels 2013; Chow-White 2008; Nakamura, and Chow-White 2012.

4 The situation has changed after Donald Trump's first election, which showed the strength of alt-right communities online in the United States, and after the growth of explicitly racist political agendas in numerous countries, which were largely promoted through internet and social networks. In 2016, the Jewish New York-based journal The Forward published an article on "Human Biodiversity: The Pseudoscientific Racism of the Alt-Right," rightly insisting on the fact that the keywords "human biodiversity" were systematically associated with far-right websites on the Internet. An article in Vox reiterated the same point in 2017. Between 2019 and 2021, Aaron Panofsky and his colleagues published some results of their important studies on US White supremacists (Panofsky and Donovan 2019; Panofsky, Dasgupta, and Iturragia 2021). These studies, which mainly focus on US White nationalists, converge largely with my own, which was more interested in European preservationists. On the US White supremacists' agenda, see also Stern 2019.

political and social uses relating to neither discriminated minorities nor national programs. One of the first empirical surveys on bloggers who revealed their DNA ancestry tests emphasized that a vast majority identified as "White, European and European American" (88 percent) (Wagner and Weiss 2012).[5] This information contrasted with the scarcity of studies focusing on the uses of DNA ancestry tests by this kind of population in scholarship. I examine some of these uses here, which are clearly connected to alt-right and European ethnonationalist agendas. I am certainly not claiming that these uses represent the majority of the uses of DNA ancestry tests by this population; many individuals are only interested in their genealogy or the deep history of their country. It is, however, worth noting that these uses are far from marginal. In 2019, the forum The Apricity claimed to have 20,000 members and some 180,000 visits per month; until it was closed down in 2020, Forumbiodiversity claimed to have 12,000 members and 290,000 visits per month; and Dieneke's anthropology blog counted approximately 1.3 million page views since its foundation in 2005, as well 1,500 followers, mainly from the United States and Europe.[6] This is not a huge community, but it does not address a small audience, either. Moreover, one has to keep in mind that, on the internet and social media, the main issue is *visibility*, and in terms of visibility, alt-right and European ethnonationalists strategically dominate the field of human biological variations. The term "human biological diversity" has been largely monopolized by these groups on internet, and an online search for information on this theme immediately leads to their websites.[7] Clearly, they are pushing an ideological struggle on this issue, enabled by their clever use of various internet tools: page ranking, cloaked websites, blogs presenting themselves as purely informative, specific links or information excerpts, online communities and participatory programs, etc. They are thus producing something that I describe as a "counterknowledge" on race and genomics.

---

5 As the authors noted, "the demographic characteristics of our sample are … worth noting: there is no clear explanation why our sample predominantly comprised White, European or European-American males. This is intriguing since many scholars assume that the industry is predominately marketing its products to other demographics, e.g. Blacks or Afro-Americans and Native Americans" (Wagner and Weiss 2012, 53).

6 The internet is a living archive that evolves rapidly. I did this research mainly between 2013 and 2015 and its results were still valid as a description of the present in 2020. Since then, a large part of the landscape I am describing here has, however, suffered a major transformation, probably due to the rise of violence on the part of White supremacists and its potential political and juridical consequences. Forums such as Anthroscape and Forumbiodiversity have been closed and are now history.

7 This was true until 2020. See, for instance, the website Humanbiologicaldiversity.com. Quite interestingly, one of the first results that a search on Google now yields (in 2023) is an article by Jon Entine and Patrick Whittle disclaiming that, as "many people believe" due to recent press articles, "human biodiversity" is an alt-right code for embracing racism. This article has been published on the Genetic Literacy Project, a program run by the same Jon Entine whose moto is "science not ideology," though it is in fact mainly run by nonscientists and clearly designed to promote genetic reductionism and genetic engineering. It is worth noting that Entine was one of the members of the Human Biodiversity Institute founded by Steve Sailer, which mainly featured heroic figures of the counterknowledge on race, genetics, and intelligence, such as Rushton, Bayley, Murray, Pinker, or Entine himself.

A Counterknowledge on Race and Genomics: An Alternative Truth about Racial Reality

The Main Features of This Counterknowledge

I am using "counterknowledge" in the sense that Michel Foucault uses it in *Il faut défendre la société* (*Society Must be Defended*), where he describes the discourse on the struggle of races promoted by French nobility in the seventeenth century as a "counterknowledge" directed against the juridical and administrative apparatus of absolute monarchy. As Foucault showed, these counterknowledges proposed alternative narratives, based on a mix of erudition (a form of expert knowledge) and the revalorization of forgotten, disqualified or lay knowledge. They were weapons in a struggle against the centralizing and homogenizing power of the monarchy (Foucault 2003, 129). I do not want to push the analogy too far, particularly because Foucault rightly praised the richness of this counterknowledge, which is absolutely not the case with that which I study here. However, I believe that the concept of counterknowledge in Foucault's sense is pertinent to capture some dimensions of how genomics is used by these online communities. It helps us to avoid dichotomist oppositions, such as the one between science and ideology (or "counterknowledge" mainly reduced to "nonscience"), and to describe more adequately some of the operations involved in this case. I will successively emphasize two points.

First, such counterknowledge has a *strategic* dimension. As we will see, it is clearly directed against certain enemies and narratives considered to have dominated the field of human genetics from the 1970–1980s. Rather than treating this counterknowledge as being merely junk science, we must note the fact that those who produce it select and emphasize some widely admitted scientific results and techniques, but turn them into weapons to contradict what they describe as "official science." They cleverly play on various ambivalences in dominant scientific discourses on race, genomics, and human variations, which means that the best strategy to fight such counterknowledge cannot be to reassert only "true" science against its ideological misuses. It is also necessary to clarify some ambivalences of core concepts (such as heritability, ancestral populations, etc.), categories (such as racial and ethnic labels) and techniques (such as admixture studies or uses of PCA) from human genomics. It is also important to develop more suitable ways of communicating and teaching these ideas so that the public at large, and geneticists themselves, embrace a more critical and reflexive perspective on genomic narratives and genomic data.

Second, the development of such a counterknowledge is strongly linked to a broader movement of decentralization in knowledge production. In that respect, the internet and social media have played a decisive part. Foucault's concept of "counterknowledge" is very useful in this context because he rightly draws a historical contrast between the building of counterknowledges, on the one hand, and the processes of "disciplining knowledges" through central institutions, control of procedures, and so on, on the other, that occurred during the late seventeenth to early eighteenth centuries (Foucault 2003, 178). What is described, today, as a "post-truth era"[8] may more adequately be seen as a moment in which the main institutions that regulated and controlled the production and circulation of knowledges are experiencing a multifaceted crisis. In particular, they have faced widespread criticism concerning their legitimacy, their connections with political or economic interests, and the ways in which their modalities for the production of knowledge have tended to exclude or neglect "lay expert knowledge" and many other forms of knowledge. One consequence of this crisis is that it allows various groups and individuals to participate actively in the production and circulation of knowledges, and thus to secure access to new forms and levels of visibility through internet and social media. As we will see, this is a striking feature of the counterknowledge of genomics and human diversity: it is strongly linked to the participative and citizen sciences and to the blurring of the divide between professionals and amateurs.[9] But, as in many other cases of participative movements, beyond the so called symmetry and democratization of participative knowledge, it will prove easy here to identify distinct political strategies and establish that certain groups acquire more visibility than others.

A Counterknowledge and Its Enemies

As noted above, such counterknowledge on race and genomics has a clear strategic and polemic dimension: it is directed against four main enemies.

1. Emphasizing genetic diversity between human groups versus homogeneity of humankind

First, this counterknowledge is directed against those who, inspired by a narrative stemming from antiracist population genetics of the early 1970s, insisted both on the remarkable genetic *homogeneity* of humankind and on the fact that variations between individuals (or "interindividual variations") are far more important than differences between populations (or "interpopulation differences"). The promoters of such counterknowledge insist, on the contrary, on human biological *diversity* and on the existing gaps between populations, in particular between what they call "racial groups." In that respect, they strategically use the numerous biomedical studies that have focused on genomic diversity since 2000 and on several statistical devices that help to visualize separate racial clusters. It is important to keep in mind that the interpretation of genomic data depends heavily on

8 On this topic, see Sismondo 2017 and McIntyre 2018. As we will see, it would be totally wrong to describe the counterknowledge I am referring to as a kind of "science denial." On the contrary, its proponents generally claim to be fighting for "real science." The question here is not the repudiation of "truth," or even of "scientific truth," but rather a struggle over what constitutes scientific truth as far as genomic human diversity is concerned.

9 On the various meanings and practices covered by the label "citizen sciences" and the problem of public participation in science's production, see Strasser et al. 2019. To use Strasser's typology, the counterknowledge on genomics goes from mere self-reporting (i.e., exposing and giving access to one's own genetic results) to computing and analyzing (i.e., developing software to analyze genomic data) and even "making" science (through the building of ethnonational programs on genomic diversity).

the dimension that is emphasized, and the purposes for doing so. Lewontin's groundbreaking 1972 article emphasized that the percentage of genetic variations covered by interpopulation differences, and more specifically by taxonomic categories such as races or ethnic groups, was very poor (15 percent) compared to the percentage covered by interindividual variations (85 percent).[10] This led him to conclude that these taxonomical categories were not relevant as far as genetic information was concerned. Significantly, even in Lewontin's mind, this did not imply that one could not use such a small number of genetic variations for various practical aims, such as, for instance, tracing ancestry (Gannett 2021). Lewontin—and later the Human Genome program—emphasized the unique homogeneity of the human species, insisting on the fact that about 99.8–99.9 percent of the human genome is similar. On the other hand, the remaining 0.1–0.2 percent still represent millions of differences in human DNA, and biomedical and human genomics studies from the 2000s have focused mainly on these differences either for medical and pharmaceutical purposes, or to develop various applications, such as forensic, commercial, or political identity uses. These investigations primarily looked at supposedly significant patterns of variations between races or biogeographical ancestry groups. They did so not for taxonomical reasons, but for many other purposes, such as tracing individual biogeographical ancestries, breaking down a given population into ancestry groups, associating genomic variations with an augmentation of risk diseases, and so on.

The promoters of new forms of online racialism only needed to select some of these studies to advance their own agenda. The emphasis on genomic differences between races or ethnicities, which became so fashionable in biomedical research after 2002, has actually found a ready welcome on White supremacist forums. Beginning in 2005, for instance, the main such forum in the United States, Stormfront, opened with the thread: "Evidence that Racial Groupings Match Real Genetic Profiles." This thread aimed to collect links to scientific studies that supposedly established such a correspondence. It included numerous biomedical studies linking risk factors to race. The thread grew considerably over the following years and its contributors grew more confident as they accumulated more "proof" from scientific publications. When Ancestry Informative Markers (AIMs) were first used for forensic purposes in 2005[11], one of the contributors claimed: "I saw this today! The Boasian anthropology 'no such thing as race' crowd are taking some major hits these days."[12] In 2006, another contributor, referring to the developing research on haplotypes,[13] concluded: "I think the genographic findings will

10 "The mean proportion of the total species diversity that is contained within populations is 85.4% ... Less than 15% of all human genetic diversity is accounted for by differences between human groups! Moreover, the difference between populations within a race accounts for an additional 8.3%, so that only 6.3% is accounted for by racial classification" (Lewontin 1972, 396).

11 Ancestry Informative Markers are a set of selected punctual variations on the genome whose proportions vary strongly according to populations and that are used to identify the different components of the biogeographic ancestry of an individual or a group.

12 By "Boasian anthropology," these groups generally mean a cultural anthropology which supposedly comes, in the United States, from the anthropologist Franz Boas and emphasizes the influence of culture and education rather than focusing on biological and genetic determinations.

13 Haplotype is a group of genomic variations that are linked and tend to be inherited together. Haplotypes on Y-DNA or Mt-DNA are generally used as a "signature" to link one paternal or maternal

play right into our hands. For example, to be a native European, your DNA must display certain Y and mtDNA markers."[14]

2. Proving that "race is a biological reality" and how former racial categories match genomic studies

This brings us directly to the second enemy of the advocates for this counterknowledge: the "liberal creationists" or "cultural Marxists"[15], as they label them, who argue that race is a social construct. Proponents of this counterknowledge contradict this claim by affirming that race is a "biological reality."

Figure 1: The "war" against cultural Marxists and liberal creationists

The Cultural Marxist War against Darwinism

Creationists: evolution is a social construct, not biologically real.
Liberal Creationists: race is a social construct, not biologically real.
Charles Darwin: I'm not a creationist; I actually wrote: "There is, however, no doubt that various races, when carefully compared and measured, differ much from each other...."

Source: http://www.humanbiologicaldiversity.com, last accessed march 2017.

They try to prove that the most advanced genomics confirm the reality of common sense and of older racial categories. Here again, they cleverly play with certain ambiguities in the arguments of various geneticists concerning the overlap between genetic ancestry and racial categories (Bliss 2012). The proponents of this counterknowledge thus welcomed AIMs as a way to construct clusters that correspond broadly to traditional racial categories ("Caucasoid," "Negroid," etc.) and they seized upon a quite famous study by Rosenberg & al. (2002), along with others studies, to translate AIMs into these categories (Figure 2).

line of one individual ancestry to an ethnic group where these haplotypes are supposedly statistically more frequent.

14 All these quotations are taken from the thread "Evidence That Racial Groupings Match Real Genetic Profiles" on https://www.stormfront.org/forum/t182050/, last accessed 12 June 2014.

15 "Cultural Marxism" is a concept that was initially coined by German far right groups during the Weimar Republic years.

Figure 2: AIMs as supposedly corresponding to more traditional racial groups. In this illustration, the blog "racialreality," operated by an Italian blogger, wants to show "clearly the divisions and admixture between the six main racial groups: Capoid (red), Negroid (orange), Caucasoid (blue), Americoid (green), Mongoloid (yellow) and Australoid (purple)." The aim is to demonstrate that phenotypes and more ancient categories from the first half of the twentieth century (here from Carleton S. Coon's physical anthropology) largely correspond to the results of more recent AIMs and admixture studies.

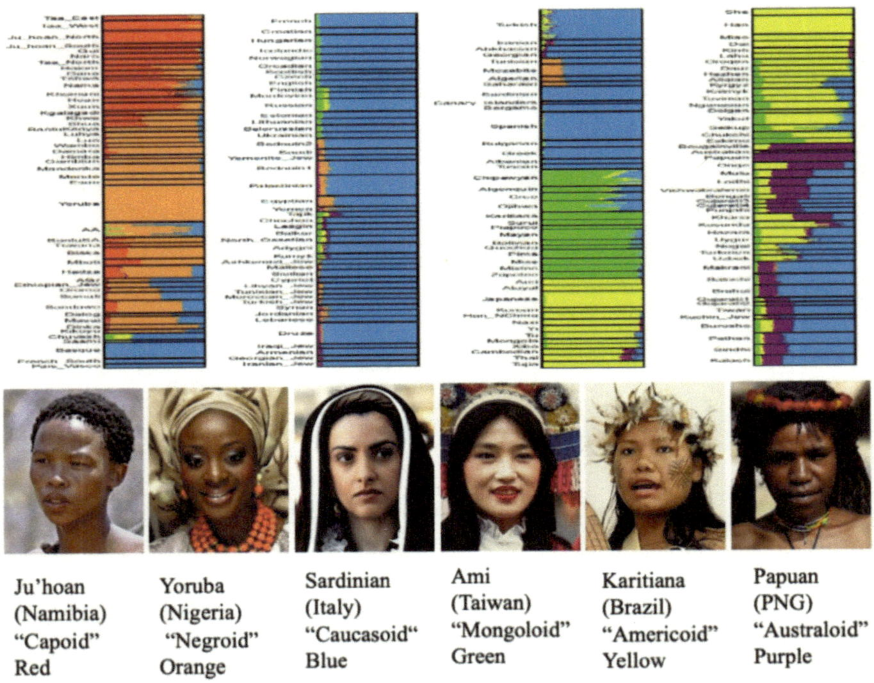

Source: http://racialreality.blogspot.com/2013/12/global-admixture-analysis-at-k6.html, last accessed 9 April 2024.

One of the main problems in human genomics is the vast amount of data it has accumulated, which requires different visualization techniques to be understood. These techniques have numerous limits and biases, which are generally not made explicit (Ekhaik, 2022). They give the false impression of a self-evident representation of reality, immediately accessible to the mind. This is particularly true for studies using what is known as "Principal Component Analysis" (PCA) to visualize the distribution of genomic variations into different clusters. Some of the main actors in the counterknowledge on race and genomics are perfectly aware of the pitfalls of PCA studies, but they nevertheless use them to illustrate supposed correspondences between ancient racial categories and genomic "data." It is worth noting that this line of argument is not exceptional, for one can easily find scientific publications or interpretations by authors inspired, for instance, by the new realist trend of the philosophy of race that draw the same conclusions from the same so-called "data" (Spencer 2015).

Figure 3: Racial groups supposedly confirmed by PCA studies on dienekes.blogspot.com. The genome blogger Dieneke comments on the above graph, taken from Nassir et al. 2009, as follows: "In PC1 vs PC2 (top left), the five major races (Caucasoids, East Asian Mongoloids, Amerindians, Australoids, Negroids) are clearly separable." This is a typical PCA visualization of genomic variations. In Dieneke's interpretation, the predominantly purple cluster on the top left of graph A supposedly represents "Negroids" (though it actually reflects a few individuals classified as "South-Africans"); the top right cluster of widely mixed colors, "Caucasoids"; the predominantly light blue cluster, "Australoids" (who are described as being better identified by graph B as a neatly separated cluster); the predominantly red/yellow cluster, supposedly "East Asians"; and the predominantly green cluster, supposedly "Amerindians." Noteworthy is how problematic the varying labels applied to these populations are, as well as the fact that only a few individuals are taken to represent an entire population.

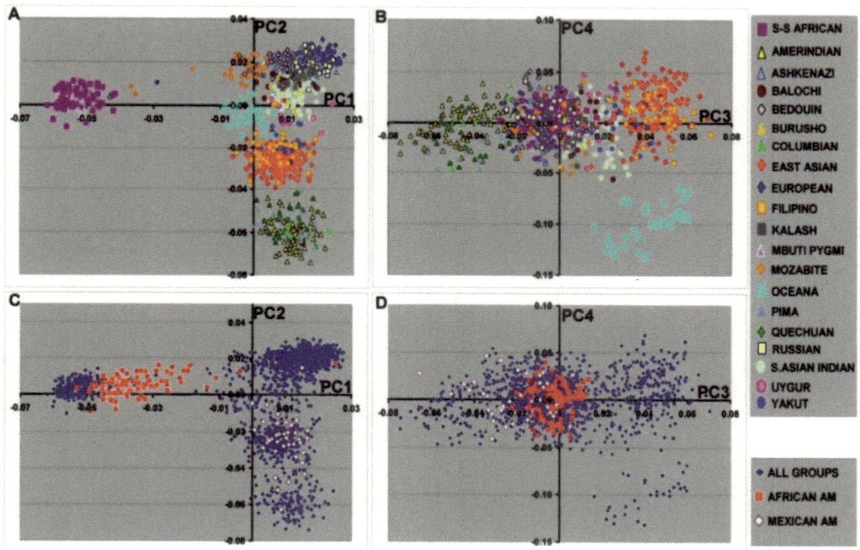

Source: http://dienekes.blogspot.com/2009/07/ancestry-informative-marker-set-for.html, last accessed 9 April 2024.

They also focus on haplotypes, which they always describe as "*racial markers*": "I think Negroid Y-DNA is E1b1a or something of that nature. J1+2 and R1b/a are all Caucasoid/Middle Eastern and are not found in sub-Saharan Africa, with the exception of a special branch of R1a or b and J1 in some parts of Ethiopia where they are mixed race (J1 was given to them by Caucasoid Mediterraneans on the paternal side)."[16]

This same desire to establish overlaps between the racial categories of traditional physical anthropology and the results of genomics lies at the heart of various projects, such as the Society for Nordish Physical Anthropology, founded in 1999 and hosted on the website *The Apricity*, or numerous blogs that focus on "human biodiversity." One such blog, *racialreality.altervista.org*, explicitly aims to show that Carleton Coon's classification

16 https://www.stormfront.org/forum/t182050/, last accessed 12 June 2014.

of races is not "outdated"; it adopts his major taxonomic categories by relating them to studies of human population genetics and embodying them in portrait-types (Figure 2).

Figure 4: Y-DNA haplotypes European distribution. Here, as in many other examples one can find on the web, haplotypes are directly connected to so-called ethnic or racial groups. It is worth noting, again, that this is not peculiar to this kind of website. One finds similar operations, for instance, in various DNA ancestry testing companies.

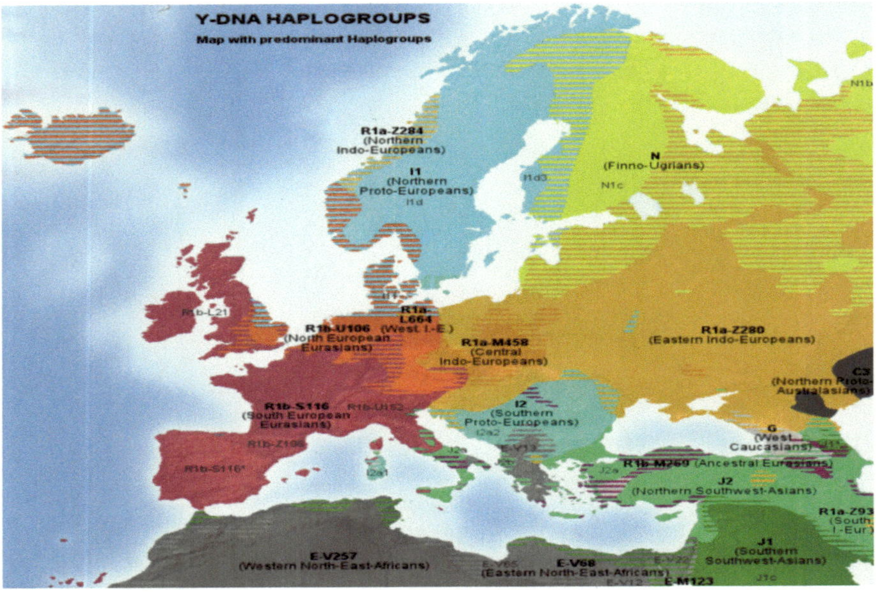

Initially published on Eupedia.com, it can now be found everywhere on the web: for instance, https://www.reddit.com/r/MapPorn/comments/302202/genetic_map_of_europe_with_dna_haplogroups_743_764/ (last accessed 9 April 2024), sometimes with notes such as "if European borders were drawn by DNA instead of ethnicity."

3. Genetic barriers and genetic determinism

Third, arguing against those who celebrate or, at least, insist on admixture and fluidity as the fundamental patterns of human population dynamics, and on the clinal dimension of human variations, these websites and forums generally emphasize biogeographical barriers and stable genetic identities. This is a point I will explore in more detail in the second part of this article, when I examine their approach to the issue of European genetic identity. But it is important to stress the fact that, here again, these sources are able to cleverly rely on some recurring ambiguities in human population genetics, particularly the fact that, for methodological purposes, geneticists tend to build stable and distinct groups and exclude the most recent migration waves from their studies. For methodological reasons, many human population studies tend to include only individuals whose four grandparents lived within a defined perimeter of the locality being studied. The famous study published in 2008 under the title "Genes Mirror Geography within Europe," for instance, which claimed to describe the distribution of genetic variations in Europe, systematically excluded from its studies those whose "putative geographic origin was from outside Europe" or were "not European," according their grandparents' origins (Novembre et al. 2008).

One can understand the scientific motivations for such exclusions, which are actually quite common in mainstream human population genetics.[17] Nevertheless, methodologically it mirrors the political logic of Far Right movements in Europe, who tend to consider the last waves of migrants as not European, even if they were born in Europe and are citizens of a European country. It is therefore not surprising that a self-confessed Nordicist, such as the genome blogger running the Fennoscandia Biogeographic Project, which aims to neatly map and differentiate Fennoscandia (e.g. Scandinavia, Denmark, and Finland) as a genographic entity, can proceed in exactly the same way. As he explains, his project proceeds in two phases. In its first phase,

> the project is open for participants of close to 100 percent Fennoscandian origin. Individuals of mixed Norwegian, Swedish, Finn or Saami origin are acceptable. Individuals of only partly Fennoscandian origins are not accepted in this project, nor are people of non-Fennoscandian origin [Then, in the second phase] we look at the analysis possibilities of differentiating Scandinavians (Swedes, Norwegians, Danes). This time all individuals who showed considerable Saami (both North-Saami or South-Saami) or Finnish admixture in the previous analysis were removed from further analysis to keep any outside influence reduced to a minimum and lumped into the "other" group containing the rest of the world.[18]

Finally, this counterknowledge—or at least some of it—is also directed against those who, after the Bell Curve controversy[19] and the marginalization of racist psychologists, claim that the majority of genomic variations that are used to trace ancestry have no impact on the phenotype and, in particular, on behavioral and cognitive traits. I will not focus specifically on this question in this article, but most of the actors we are studying generally insist on the fact that race matters, that genetic ancestry is a fundamental

---

17 Selecting only individuals whose grandparents are known to be born in a restricted area around the place under study has been usual in many population genetics studies from the 1950s, in particular for European territories. It has been justified by different arguments. Due to the important movements of populations (internal migrations, immigration from non-European countries, etc.) that happened from the 1950s to the 1960s, describing the distribution, structure, and dynamics of genetic variation in different regions of Europe supposedly required selecting individuals whose family was known to have lived in such a region before the 1950s. It is also described as a way to build a reference panel and avoid false associations when trying to identify correlations between the distribution of genetic variation and risks of some diseases.

18 http://fennoscandia.blogspot.com/2011/03/?m=0, last accessed 9 April 2024.

19 *The Bell Curve* was a book published in 1994 by psychologist Richard Herrnstein and political scientist Charles Murray. Among other themes, they claimed to identify some correlations between IQ average and racial groups, insisting that these IQ differences were largely determined by genetics and suggesting that misguided public policies and immigration had caused the United States to suffer a degradation of average IQ. The book sparked a controversy among North American psychologists, as well as strong criticism from other scientists opposed to sociobiology and genetic reductionism. It is now a widely shared reference in White supremacist groups.

component of individual and collective identities, and that it determines many mental or behavioral abilities.[20]

Beyond the Science and Ideology Divide

1. A counterknowledge that cannot be reduced to mere denial of science

As argued above, it would be misleading to dismiss such counterknowledge as mere "junk science" or an "ideology" that can easily be opposed to "science" and "scientific facts." This is why I am referring here not to "counterknowledge," defined by Damian Thompson as "misinformation packaged to look like fact" (Thompson 2008, 1), referring to conspiracy theories and so-called "alternative facts," but rather to Foucault's more fertile concept. It would certainly be comfortable and reassuring to believe that this counterknowledge has no connection to scientific research and science. But in fact, their relationship is far more complex. First, as noted above, the contributors to this counterknowledge often exploit ambiguities inherent in the scientific discourse on human population genetics, race, and ancestry. These ambiguities enable them to selectively use scientific studies, fitting them into their ideological framework without distorting the studies themselves. To debunk this counterknowledge, one in fact needs to criticize some of the concepts and ways of reasoning and of interpreting data that are common in mainstream human population genetics. Second, some of these contributors have truly expert knowledge on genomics that helps them to select and interpret the articles they use, to choose the right software and ancestry tests to maximize their Whiteness or their European ancestry, or even to design programs that fit their political agendas. Some are scholars trained in population genetics, well aware of academic debates, and two of them were even celebrated by *Nature* in 2010, in an article titled "the rise of the genome bloggers." This article portrayed two central figures of these online communities: Dienekes Pontikos and David Wesolowski, the latter known as "Polaco" (Callaway 2010, 880). Ironically (the author of the article *in Nature* did not know their ideological positions), they represented the two trends dividing European identity discussions on the web: Dienekes can be described as representing the "Mediterranean" point of view, while "Polaco" is one of the main figures referred to by the "Nordicists" and was an administrator of the forum forumbiodiversity. In Nature's article, the population geneticists Doron Behar praised them, claiming "they are not amateurs. They are far from being amateurs ... I cannot stress enough the level of appreciation I have for their efforts." The fact is that many of them are perfectly aware of the most technical aspects of population genetics.

Such counterknowledge is located in a hybrid space between science and folk knowledge. On the one hand, it tends to praise common sense and the supposed self-evidence of the existence of races. Such self-evidence is said to be denied by "mainstream scientists" only because of ideological and political biases. On the other hand, many of them are truly experts on population genetics and claim to have a scientific ethos. In their view, "official science", as they would call it, is biased with unscientific and ideological positions. They distinguish between "scientific facts", which they present as corresponding

20 They should consequently be located in an older history that goes back to the debates from the 1970s to the 1980s on sociobiology, genetic determinism, and IQ.

to common sense and being congruent with the existence of races, and the way scholars present these facts under the sway of "political correctness" or unscientific, ideological reasons. In this respect, the proponents of this counterknowledge skillfully play with the ambiguities of scholars, opposing their off-stage versus on-stage discourses.[21] Following a very common strategy of alt-right activists, they oppose mainstream media and official science in the name of a supposedly concealed truth, a denied reality that people are well aware of but that officials try to conceal. In so doing, they turn the demand for "free speech" (vs. "political correctness") and of "true science" (vs. "ideology") against their opponents (Titley 2020). The moto of the Genetic Literacy Project is a fine example of this strategy: "Science, not ideology." This is why they eagerly spread declarations of renowned scientists or journalists (such as David Reich, James Watson, or Nicholas Wade) that are dissonant with so-called "political correctness." For instance, in France, Far-Right networks have widely promoted extracts of David Reich's work. In 2018, the cloaked website operated by Far-Right identitarians, Breizh-Info, noted: "according to a respected American geneticist, genetic differences between races are real." A positive review of David Reich's work has also been published on the Illiade Institute for "the deep European memory." This Institute "refuses the great replacement and calls for the defense of our civilization." It features all the greatest names of the French Far Right, including Renaud Camus, Marion Maréchal, Alain de Benoist, Jean-Yves Le Galloux, and others. This institute also offers training courses at the European Awakening School, with its own ENA-type "promotions" with evocative names: promotion Jünger, Dominique Venner, Tolkien, and so on. This example shows another dimension of such counterknowledge: its efforts to duplicate official science and bear the external signs of a respectable knowledge. Some of these online communities produce their own online journals, such as *Mankind Quarterly* or the platform *Openpsych*. They have their own so-called scientific institute, with affiliated researchers (the Ulster Institute for Social Research, created by Richard Lynn). They organize series of conferences (such as the London Conference on Intelligence, at the University College of London, from 2014 to 2017).

2. A peculiar kind of participative knowledge

Yet the most interesting side of this counterknowledge is elsewhere, in the fact that it exploits all the resources internet provides for the development of new forms of participative knowledge or "citizen science." The internet has paved the way for a radical decentralization of the production of knowledge on genomics. People can edit their own blogs and publish posts without any oversight or gatekeepers and, most interestingly, they can even build their own participatory research programs online. I want to emphasize this point, which is probably the most original feature of such counterknowledge. Genomic data can easily be translated into informatics data that can be exchanged, transferred and accumulated online. Today, many clients of direct-to-consumer DNA testing companies use these services not only to obtain interpretations and narratives about their ancestry. They use them, rather, as mere extractors and providers of raw genomic data that

21 On these ambiguities, see, for instance, Bliss 2012, 112–135, who discusses in particular the positions of Rosenberg, Pritchard, and Reich.

they can then introduce themselves online, directly, into software such as GEDMATCH, GENOtation, or INTERPRETOME (Figure 5).

Figure 5: An INTERPRETOME result. This graph shows how a personal result (the black square) introduced into Interpretome is located in supposedly national clusters of DNA. It is worth noting the choice of picturing DNA data with national flags. Such pictures can then be displayed in forums to show one's personal DNA identity as, in this case, in the Occidental Enclave.

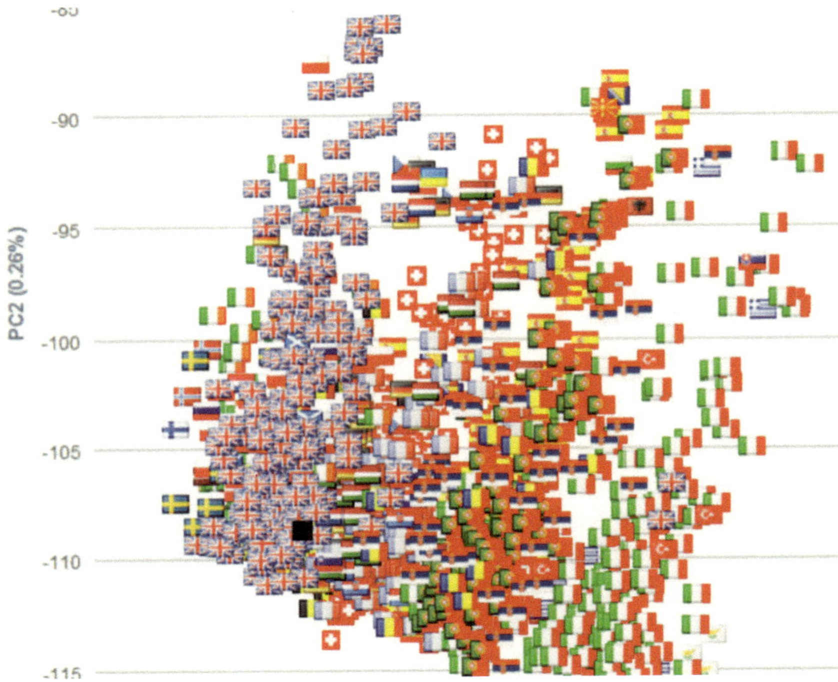

Source: http://www.occidentalenclave.org/viewtopic.php?f=28&t=3208, last accessed 9 April 2024.

Such users of these tests can also send these data to "genome bloggers," who may incorporate the data into their own research projects, advancing more or less explicit political agendas. Generally, these online projects are designed to identify the genomic bases of a region (such as Fennoscandia, Austrasia, or Armorica) or of an ancient political entity (such as the Duchy of Lithuania). Some also aim to determine the most important components of European identity, as we will see later. The following box contains an indicative list and brief description of some of these programs.

Table 1: Description of the main genome bloggers' sites and programs, and of their (nordicist or mediterranean) agendas

I Mediterraneans ("Meds")
Dodecad Ancestry Project (Dieneke) – "To provide detailed ancestry analysis, primarily for Eurasian individuals" – "To build samples of individuals for regions of the world (e.g. Greeks, people from Anatolia and Caucasus, Balkans, Southern Italians, etc.) currently under-represented in publicly available datasets" – To establish the importance of the Med component in European identity **Italianthro.blogspot** – "This blog [is] a collection of material focusing on the biological and social anthropology of Italy and Italian people around the world ... An underlying aim will be to refute falsehoods about Italians spread by Afrocentrists, Nordicists and Northern Italian supremacists"
II. Nordicists
Eurogenes (Polaco) – To provide different admixture tests (K 9, 10, 11, 12, 13 + Hunter gatherer vs. Farmer) that focus on European populations and give detailed individual admixture analysis – To collect more data on the genetic components of Europe populations (specifically the Baltics and northeast Europe), or on similar supposed populations, such as Polish genes – To define the main components of European identity (hunter-gatherers from the Baltics) **Magnus Ducatus Lituaniae Project** – "BGA analysis project for the territories of former Grand Duchy of Lithuania" **Fennoscandia Biographic Project (Anders Palsen)** – To map the genetic relationship and origin of the current populations in Fennoscandia: Norwegians, Swedes, Finns and Saami – Then to "look at the analysis possibilities of differentiating Scandinavians (Swedes, Norwegians, Danes). This time all individuals who show considerable Saami (both North-Saami like or South-Saami like) or Finnish admixture in the previous analysis were removed from further analysis to keep any outside influence reduced to a minimum, and were lumped into the 'other' group containing the rest of the world"

These programs generally operate through an exchange of services between the participants, who provide the data, and the designers of the projects—not so different from the way more official research programs work to attract participants.[22] It is important to keep in mind that, contrary to DNA ancestry services specifically designed for Afro-descendants or other minority communities, there is no company offering specific and detailed information for people claiming their "White," "European" identity or wanting to explore their regional roots in detail. Therefore, many of them are not satisfied with the information provided by generalist DNA testing companies. To obtain more information and tests for their uses, they may turn to these "genome bloggers" who provide this information in exchange for their data. "Polaco," in particular, has designed a specific test ("Eurogenes") which is very popular among these clients.

22 To enroll participants, the CANDELA program, for instance, led by a consortium of various universities in South America, proposed personalized ancestry tests for its contributors (Abel 2021, 72).

To illustrate how these programs work, we can pick the more neutral project called "Explore Your DNA/France Project"[23]. Noting that "in the world of ethnicity research by DNA, France is still at the end of the pack," "a group of enthusiasts," who describe themselves as a "team of specialists … made up of enlightened amateurs" (originating mainly from a group that investigated the region of "Brittany") proposes to "dissect" participants' "raw data" and to give them "more precise" results "than what your DNA genealogy site gives." To do so, they rely on "a revolutionary tool [that] has appeared on the scene of ethnicity research, created by David Wesolowski [a.k.a Polaco]": the calculator Eurogenes. They ask participants to send their "raw data" to Polaco, as well as twelve dollars to cover the expenses of the tests, allowing them to be enrolled in the program and receive the information they seek. For this group, the project offers a means to progressively build detailed maps of biogeographical ancestry components from France and Brittany. It is worth noting that, in this case as in many others, the analogy with Foucault's analysis of "counterknowledge" is striking: by articulating erudition, amateurism, and localism, these counterknowledges largely continue the trends encountered in the formation of various antiquarian societies in Europe from the eighteenth century onward.[24]

This localism, focusing on specific local roots and identities, does not imply that the various groups building this counterknowledge are isolated and trapped in their own local identity; quite the contrary. The internet is also a highly effective way to build connected transnational communities through links, networks, and forums. This is an important point, because the internet helps to build racial, transnational communities based on supposed shared ancestry, a shared sense of belonging, and shared fears of endangered identities supposedly needing to be preserved against common enemies (migrants, Muslims, liberal politics).[25] Such communities can be based either on the claim of a common "Whiteness" and shared European identity (in forums such as Stormfront, The Apricity, The Occidental Enclave) or on more dividing affinities within this "Whiteness," such as the very important line opposing Mediterraneans ("Meds") versus Nordicists, which actually structures the landscape of many blogs and forums. This is what we will examine now.

"Who Is the Most European of Us All?" European Genetic Identities and the Division of Whiteness

In this section of my chapter, I investigate more closely how certain members are discussing the issue of European genetic identity in these online communities. On the one

---

23 https://www.exploreyourdna.com/index.aspx, last accessed 9 April 2024.

24 On the importance of local antiquarian societies all over Europe in the building of racialized regional identities, see for instance Thiesse (2001), Sweet (2004), Manias (2013).

25 It is important to keep in mind that various mass killings over the past decade in the United States and New Zealand were carried out by young White nationalists who were part of this online community and believed in the "great replacement" theory. The "Manifesto" of Payton Gendron, for instance, who shot ten people in Buffalo in 2022 and claimed he was "simply a White man seeking to protect and serve my community, my culture and my race," was full of references picked up from these online forums.

hand, Europe appears on these forums as a central point of focus, where it is described as the "homeland" from where Whites or Indo-Europeans originated. "Europe" in this sense is taken to define a shared ancestry and a common origin, which is supposedly endangered by immigration and liberalism. This is why some of those on these forums explicitly describe themselves as "European preservationists." They embrace the language of biological conservation and picture themselves as attempting to investigate and preserve the vanishing genetic human diversity of Europe. Yet, surprisingly, on forums such as Anthroscape, Forumbiodiversity, The Apricity, or The Occidental Enclave, as well as on many blogs, the main issue appears not to be Europe as a shared genetic identity or Whiteness as a common status.[26] Quite the contrary, as we will see, the main issue is rather distinguishing and hierarchizing different groups within Whiteness and separating various unequal components of a supposed European identity. In doing so, many of them explicitly connect their own way of identifying with the mid-nineteenth-century opposition between Nordics and Mediterraneans. Others seem to be more preoccupied with more regional identities, which they see as suffering from political and cultural erasures (Brittany, Duchy of Lithuania, etc.).

A shared sense of community? Europe as a common endangered homeland

It is useful to bear in mind two important principles that one finds, for instance, in the US alt-right leader Richard Spencer's Charlottesville statement in 2017, titled "What It Means to Be Alt-Right?": "1. Race is real. Race matters. Race is the foundation of identity. 'White' is a shorthand for a world-wide constellation of peoples ... derived from the Indo-European race ... European refers to a core stock ... from which related cultures and shared civilization sprang. 2. The ethno-state: [the State is an] existential entity, [the] physical manifestation of people's being, order and will to survive."[27] According to Spencer, "racially or ethnically defined states are legitimate and necessary."[28] Here, the counterknowledge of genomics is often taken to provide supposedly "scientific" underpinnings to these kinds of claims.

As Spencer's quote clearly indicates, Europe occupies a specific place in these narratives: it is pictured as the ancestral homeland of the Indo-Europeans, as the place from which White peoples originated, along with their civilization and culture. Moreover, the narratives of European preservationists generally describe the biological and cultural identity of such a homeland as endangered. For those living outside Europe, to put forward their percentages of European ancestry is also a way to claim their Whiteness. This is why defining European genetic identity is so central in all these blogs and forums. In a thread on forumbiodiversity titled "Which Are the Purest Races/Ethnicities?," David Wesolowski reformulated the question as follows: "who is the most European of us all?"

---

26 These findings contrast with the existing literature, which has focused more on Stormfront and other predominantly North American White supremacist communities.

27 https://altright.com/2017/08/11/what-it-means-to-be-alt-right/, last accessed 2 February 2018.

28 Ibid.

Arguing against someone who claimed that Europe was a mere geographical entity, he then affirmed: "as you can see from my map, it's also a genetic concept."[29]

Figure 6: Europe as a genetic concept according to David Wesolowski (Polaco).

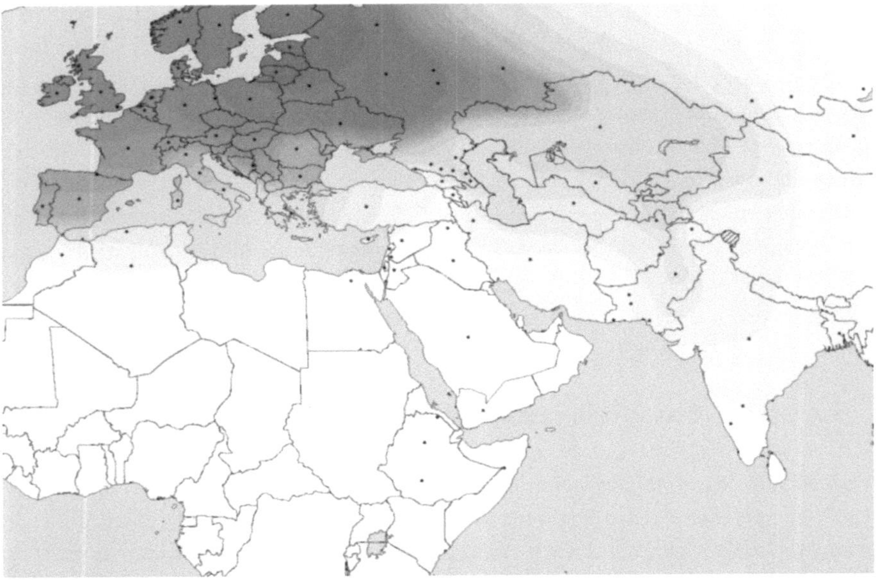

Source: https://www.forumbiodiversity.com, last accessed 9 April 2024.

Wesolowski's map (Figure 6) supposedly shows that "the European component correlates precisely with the indigenous Upper Paleolithic/Mesolithic hunter-gatherer ancestry ... based on ancient genomes."[30] As we have seen, Wesolowski is also the designer of the Eurogenes admixture test, which is the favorite test for White supremacists and other European ethnonationalists, because it maximizes information on their European ancestries and minimizes their potential non-European admixture. This test also helps them to know to which side of "Whiteness" they belong: Nordicist or Mediterranean.

The latter question seems to be the most important one; for many of the users of these forums, the main problem is not, as we will see, to prove their Whiteness but rather to prove that they belong to the right side of the Whiteness: the *most European*. It is worth noting, in this respect, that another very fashionable test proposed by Polaco is an admixture test evaluating the proportions of Mesolithic Hunter Gatherer vs. Neolithic Farmer in one's ancestry. To naïve scholars, this test may seem to be a mere recreational tool or a way to define a new form of "biosociality," but Polaco's claim, which connects explicitly European genetic identity with Mesolithic hunter gatherer ancestry, reveals it to be a strategic test for determining one's genetic European identity.

29 https://forumbiodiversity.com, last accessed 8 January 2018.

30 Ibid.

Figure 7: Self portrait of a Nordicist according to Eurogenes. "Rambo Von München," who presents herself as "female, nationality: American; meta-ethnicity: German and colonial American; ethnicity: German; political orientation: conservative preservationist; mtDNA: H1," uses this picture showing her Eurogenes results as an identity marker on the forum Occidentalenclave.org. She very clearly emphasizes her two main components, "North European" and Mediterranean, and which of the two is the most important for her.

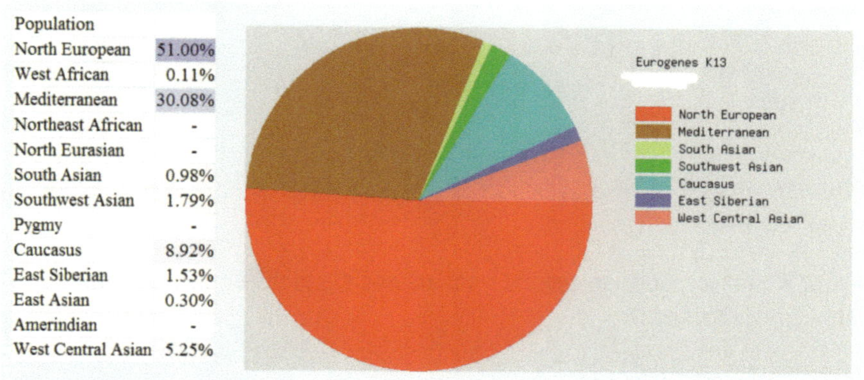

Source: https://www.occidentalenclave.org, last accessed 9 April 2024.

Genetically defining such a European identity first requires to establish the contours of Europe's frontiers and emphasizing its biogeographical barriers. We can take one example of that operation from a blogger who is located on the fringes of Europe and could therefore be suspected of being admixed with non-Europeans, as part of the great debate between Meds and Nordicists.[31] The Norwegian Anders Palsen, founder of the Fennoscandia Biographic Project, stresses that it is of the utmost importance to distinguish Scandinavian people genetically from Mongoloid people and to show, in particular, that Finns and Saami are "genetically Caucasoid or European." For that purpose, he employs a combination of various PCA studies to distinguish Europeans from Africans and Asians (Figs. 8 and 9).

These two illustrations take a global perspective to emphasize the fact that, at a broader scale, Finns/Sammi cluster systematically with "Europeans" and are neatly differentiated either from Africans (Figure 8) or from East Asians/Siberians (Figure 9).

31 Meds and Nordicists tend to fight over the percentage of non-European admixture in their respective genome.

Figure 8: PCA1—Brown: Finns/Sammi; Blue: Africans

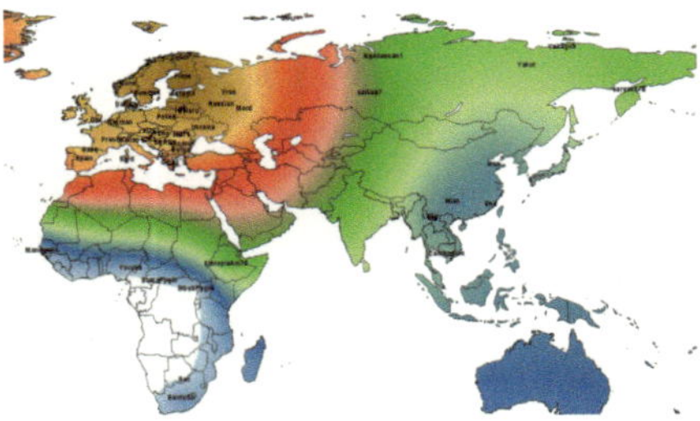

Source: http://fennoscandia.blogspot.com/2014/01/is-there-east-asian-influence-in.html, last accessed 9 April 2024.

Figure 9: PCA2—Brown: Africans; Blue: East Asians/Siberians.

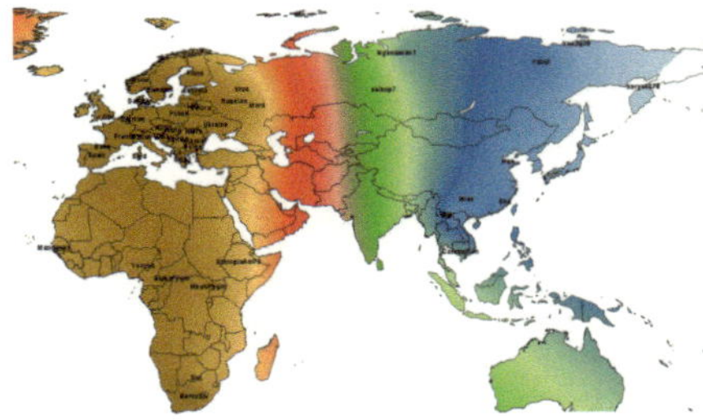

Source: http://fennoscandia.blogspot.com/2014/01/is-there-east-asian-influence-in.html, last accessed 9 April 2024.

Beyond Whiteness: building genetic frontiers and hierarchies within European identity

Figure 10: Zooming into Europe to define internal hierarchies. Here, Anders Palsen uses the same operations to show how people from "Fennoscandian" descent are those with the least African, East Asian, and Siberian influences in Europe, while "Meds" are supposedly deeply infused with either African or East Asian influences.

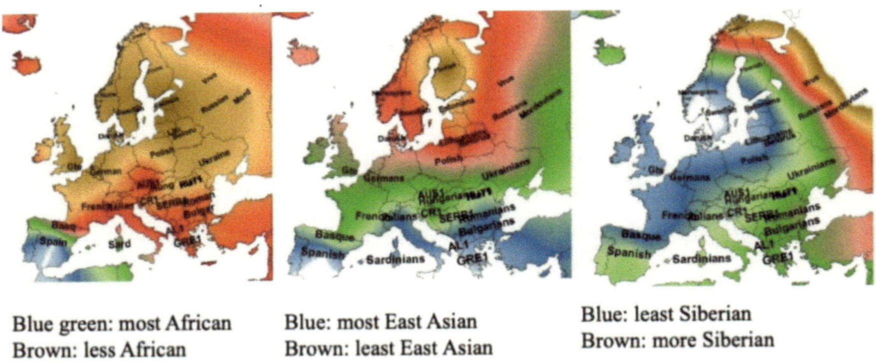

Source: http://fennoscandia.blogspot.com/2014/01/is-there-east-asian-influence-in.html, last accessed 9 April 2024.

As we have seen, genomics can be used in various participatory projects (such as Fennoscandia) to determine the peculiar genetic make-up of some specific European regions, reinforcing them as ethnopolitical stable entities. One should always keep in mind that, in Europe, Whiteness is far from being homogenous. It is a divided field, pervaded by tensions between transnational or regional identities. "Whiteness" may be a relevant concept to describe the political stance of some White supremacists in the United States, but it is not very effective as far as European identitarians are concerned. Of course, they all share a European preservationist agenda, and they always exclude the last waves of migrants from European genetic make-up. However, the true purpose of many of blogs and forums is to highlight the numerous distinctions and differences within so-called "Whiteness," especially the supposed great divide between Nordicists and Mediterraneans. Admixture with non-Europeans and migrations can be instrumental in these debates. Palsen, for instance, zooms inside Europe and uses his components to define a gradient showing who is the most European of all (unsurprisingly, the Scandinavians) and who is the most "infected" with non-European admixture (unsurprisingly, the Meds) (Figure 10).

The great divide between "Meds" and "Nordicists" structures and polarizes the whole field of Occidentalism and European preservationism on the internet. One can trace it through blogs' constellations and preferential links, which define differentiated and polarized networks of websites. One can find it in the opposition between forums: anthroscape, for instance, was mainly Med (created by the Italian blogger "racialreality"), while forum.biodiversity and the Apricity were more neatly Nordicist. Ironically, we even find this distinction in the two genome bloggers who were celebrated in *Nature*: David Wesolowski (Polaco) is a Nordicist of a peculiar sort who, unlike some other Nordicists, insists on European Baltic and northeastern roots, rather than on Germanic or Scandinavian ones. He claims that the fundamental component of European genetic identity date from the Upper Paleolithic/Mesolithic hunter gatherers. The other prominent genome blogger, Dienekes (Pontikos), can be associated with a "Med" agenda. He wants to ensure better representation of peoples from southeastern Europe (Greeks, Italians, etc.) and from the Middle East in genomic databases, and he emphasizes the importance of Neolithic migrations from the Caucasus and Anatolia to southern Europe as the core component of European identity. This opposition between Nordicists and Meds dates back to nineteenth-century racial science, but today it also mirrors political oppositions inside Europe, particularly in light of the opposition of Greek or Italian nationalists to Germanic influences in Europe, which some, in these forums, define as a "Nordicist imperialism." Such a differentiation also appears clearly in the ways individuals present themselves on the forums and how they exhibit their pedigrees through admixture tests.

This leads us to another point. Genomics can be used not only to define ethnopolitical identities, to differentiate Europeans from others, and to create divisions inside Whiteness. Genomics can also be used—albeit *not always*—to hierarchize races or ethnic groups according to a narrative of genetic determinism that correlates genomic clusters with psychological and cognitive abilities, criminal or behavioral tendencies, or entrepreneurial skills. In this case, the counterknowledge on race and genomics comes together with the widely developed counterknowledge on racial psychology and behavioral genetics. This field has been largely marginalized in academia since the 2000s[32], but it is thriving online. It can be illustrated, in particular, by the Ulster Institute and figures such as Emil Kierkegaard or Razib Khan. I will examine only two examples to illustrate these connections. The first is the French blog intelligence.humaine, which is praised by the most far-right movements in France and has been involved in an online controversy with the son of Carla Bruni-Sarkozy (the wife of the former French president). This blog relies mainly on Richard Lynn's works and states that "the genetic bases of intellectual

32 On the history of this field of research, see Panosky 2014.

differences between nations are due to the racial identity of the population." It correlates the racial composition of the world's populations (according to their genetic admixture), their mean IQ, and their average salary.[33] The second example comes from the site Eupedia.com, created by Maciamo Hay (whose alias is Satyavrata, the ancestor of the Aryans), who was highly active on Academia.edu, publishing a cartography of haplotype distribution in Europe.[34] Eupedia (Figure 11) is a cloaked website that presents itself as an informative guide to European countries but focuses mainly on northwestern Europe (countries with Frankish or Germanic influences). It is largely devoted to human population genetics and to European peoples' cultural traits, supposedly determined by genetics.

Figure 11: Eupedia.com and European genomics

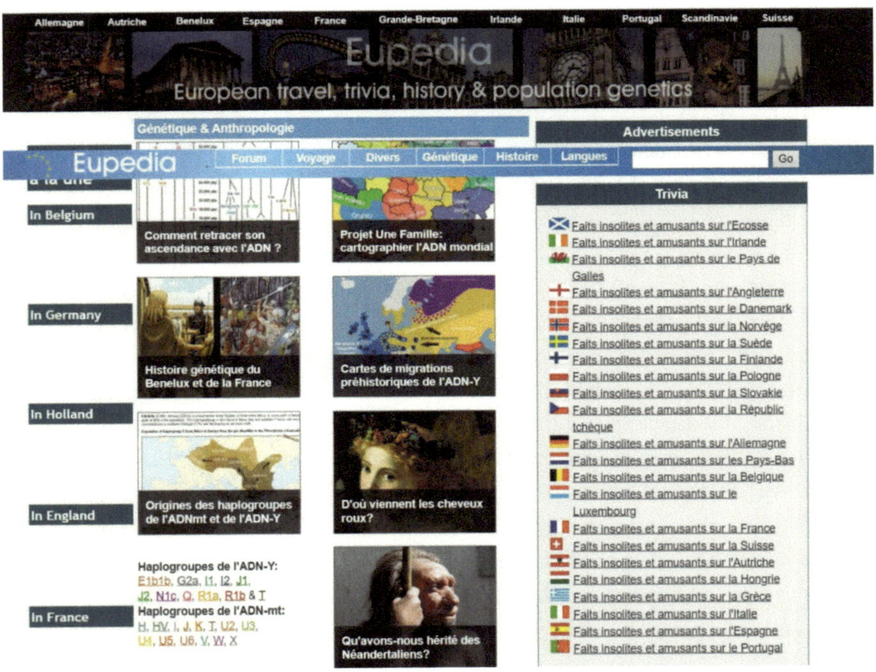

Source: https://www.eupedia.com/forum/threads/are-some-countries-doomed-to-high-unemploy ment -due-to-their-genetic-pool.26930/, last accessed 9 April 2024.

Maciamo Hay promotes an internal hierarchy inside Europe, based on genetics. He has developed an "entrepreneurialism index," for instance, based on the relation between individualism and uncertainty avoidance, to classify European countries and explain their rate of unemployment. He correlates his index with the genetic pool of

33 https://intelligence-humaine.com/qi-par-pays-et-economie/, last accessed 9 April 2024.

34 Maciamo Hay describes himself as "a Belgian-born researcher in genetics, as well as a techno-progressive futurist, an historian and a travel writer" (https://www.vitamodularis.org/about, last accessed 9 April 2024). He is a promoter of transhumanism.

these countries, convinced that "individualism is a trait shared by ethnically Celtic and Germanic countries" and that "entrepreneurialism, like individualism and uncertainty avoidance, is deeply rooted in one's genes." He argues that the Italians, Greeks and Spanish have significant rates of unemployment precisely because of their gene pool. Such articulation between economic neoliberalism, transhumanism, and racial, genetic determinism is far from being rare on internet and warrants more scrutiny.

The Social Life of European DNA:[35] How New "Biosocialities" Incorporate Old Racial Identities

Negotiating biological identities online: a new biosociality?

I want to conclude by detailing how such European DNA acquires a social life in these online communities. These forums cover a wide range of activities, from quite serious and scientifically informed discussions concerning European haplotypes, biases of genetic genealogy software, and so on, to more unconventional activities. People may ask others to classify individuals (celebrities, anonymous, friends, oneself, etc.), for instance, based upon a photo. This strange practice oscillates between a kind of folk taxonomy and a practical exercise of racial prejudice that is an occasion for the worst racist commentaries.

I do explore these activities here, focusing rather on how individuals on these forums present and negotiate their racial and ethnic identities through genomics. This question has been widely studied in the recent literature on DNA ancestry tests, through interviews and other methodology, primarily to determine the extent to which DNA results may modify and destabilize the ways clients perceive and live their racial or ethnic identities, and how they effectively negotiate the tensions between the results they obtain from these tests and their expectations. Scholars' answers are, generally, quite disappointing. They tend to conclude that individual attitudes towards DNA results vary widely, from deep disorientation to mere confirmation of initial self-identifications, with many people selecting what they want to retain (or not) from their DNA tests (e.g., Nelson 2016, Abel 2021, Panosky and Donovan 2019). Scholars tend to emphasize (and rightly so) the fact that clients are generally not passive in the way they relate to their tests. They may buy multiple tests, and they may selectively affiliate and retain some elements from their results and discuss the validity and pertinence of others. It is worth noting that such an active relation to genealogy and ancestry is not at all new, and that genealogy has always been selective, emphasizing strategically such or such ancestry and obscuring others. It has nothing to do with a "new" relation to nature and biology, which should be distinguished from older, more essentialist ways of conceiving this relation, such as those claims sometimes found in Rabinow- or Rose-style narratives on biosociality.[36]

35 The title paraphrases Nelson 2016

36 These narratives, which sharply contrast new "biosocialities" from "old" ways of relating to biology, nature, race, and identity, are generally based on a partial and in parts grossly exaggerated vision of how racial identities, ancestry, and heredity have actually functioned in the past, and on a largely

Figure 12: Two exercises (among many others) of folk taxonomy or racial prejudices. The second picture showed the photo of an individual. I purposely altered it to avoid any identification.[37]

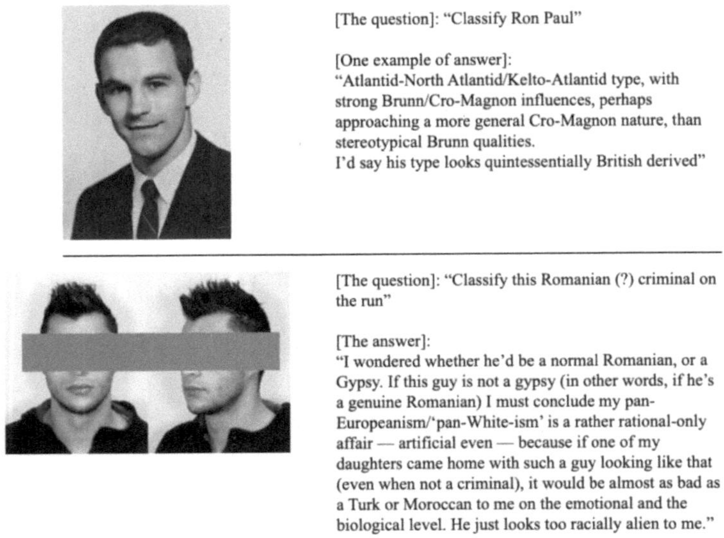

Source: https://www.theapricity.com/forum/forumdisplay.php?25-Taxonomy (last accessed 9 April 2024).

It may be true, as Rabinow and Rose have argued, that genomics can sometimes produce radically new biosocialities and imply new forms of relations to oneself and others. On these forums, for instance, almost everyone identifies themselves indicating, among other information, their haplotypes on Y-DNA and Mt-DNA, which they may even integrate into their avatars or pseudonym. They also publish their personal admixture results and discuss others' tests, integrating these results in their signatures (Figure 13).

---

fantasized dichotomy between "nature" and "culture" that was supposedly dominant in the past and is "dissolving" through the "transformations of the category of life" (Gibbon and Novas 2008, 3).

37 At the source one can find more than 5000 examples of this kind. Ron Paul is a key libertarian political figure in the United States. As there are many links between libertarian ideology and the groups I am analyzing, one should not be surprised to see his "type" discussed on the forum.

Figure 13: Two examples of signatures integrating Y-DNA, Mt-DNA and Eurogenes admixture results.

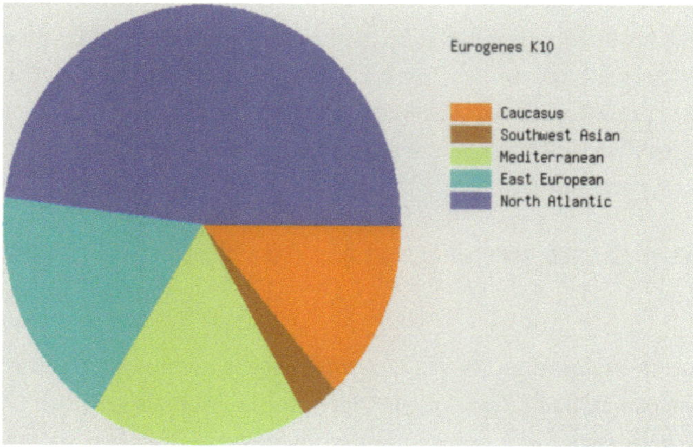

Source: http://www.occidentalenclave.org, last accessed 9 April 2024.

Figure 14: Locating oneself in a national cluster. Angharad's 23andMe results and his comments: "Well I am 100 percent European. Unsurprisingly, I cluster with Northern Europeans."

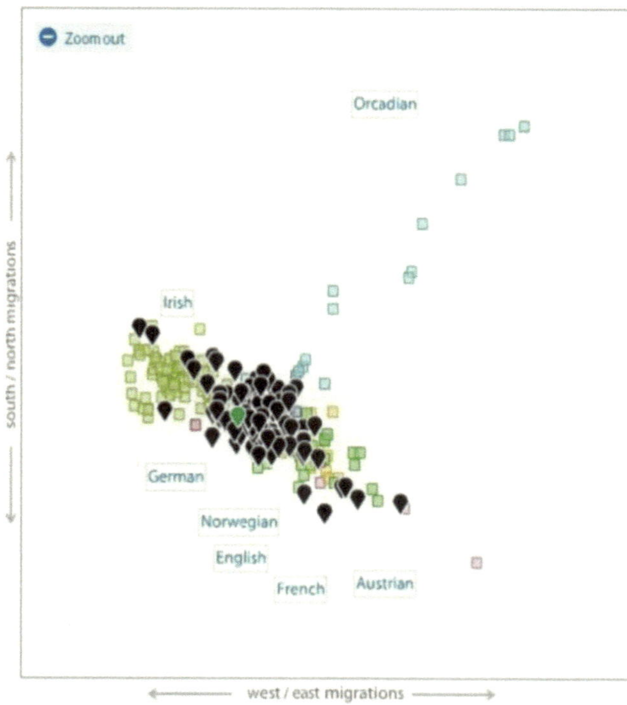

http://www.occidentalenclave.org, last accessed 9 April 2024.

They may even use maps obtained from genealogical testing services or certain blog projects, locating them individually within more broader entities or clusters (ethnic groups, nations, etc.) (Figure 14).

Here, DNA acquires a true social life. It helps to claim identities through various markers that are exhibited online. These identities are negotiated and discussed with other members of the forum. DNA serves to prove membership of larger communities, to demonstrate relative purity, and, sometimes, even to facilitate dating (some websites can have a dating section).

"The tradition of all dead generations weighs like a nightmare on the brains of the living."

In most of the cases, however, we are very far from the dream of an ex nihilo creation of brand new biosocial and fluid identities. First, the individuals generally pick their genealogy tests with a specific purpose in mind, namely, to reinforce and confirm their feeling of belonging to a preestablished community. The strategies they follow may differ slightly, depending on whether they are from the United States or from Europe. American citizens seem to be more eager to prove their European roots in general and to minimize their African or Native American ancestry.[38] In the case of Europeans (but also many Americans or citizens from other countries, such as Australians), the question is generally more complicated, because they also want to assert a position between Nordicists and Meds. Pallantides (an English member), for instance, emphasizes very clearly his "North Atlantic" component (Figure 15).

Figure 15: Claiming a Nordicist identity

Pallantides: "professional racial anthropologist"; English; ethnicity: Atlantean:

60.00% North Atlantic [highlighted in bold by Pallantides]

29.16% East European

6.70% Mediterranean

1.82% Siberian

1.82% Caucasus

0.46% East Asian

Source: http://www.occidentalenclave.org, last accessed 9 April 2024.

The vast majority of them are very well aware that race, nowadays, is a matter of statistics and admixture, but they also know how to select their genealogical tests in such a way that they can minimize their non-European components. They thus come as close as possible to a claim one finds everywhere on these forums: "I'm 100 percent European";

38 Some may be looking for a kind of "diploma of Whiteness," like one Brazilian White supremacist mentioned by Abel 2021, 81, who stated he wanted "a legal certificate" to prove he was 100 percent European. These individuals are, however, generally aware that race is a matter of percentages rather than absolute purity.

"I'm exceptionally White. No surprise there." They also tend to favor tests that help them to emphasize in detail certain European ancestries. In that respect, White supremacists and European ethnonationalists cannot resort to specialized ancestry testing companies on the web. People looking for their African roots can turn to numerous companies, such as African Ancestry or LivingDna, which employ specific databases to provide details about various African origins and "tribes." This kind of service does not exist for European identitarians or White nationalists,[39] which is partly why they resort to Eurogenes and other tests developed online by genome bloggers, which maximizes their information on European DNA.

The integration of Y-DNA or Mt-DNA haplotypes in signatures and profiles is also far from being neutral. Its main function is not to claim any new biosociality but rather to prove belonging to very traditional ethnic groups that tend to be valued (Germans, Vikings, Saxons, ancient Greeks). If these claims connect to any kind of "biosociality," then it is to old biosocial entities from nineteenth-century racial science. One should analyze these DNA identifiers in the more general context of avatars or alias, which work as a single, connected system of identification markers. All the elements comprising these avatars send a clear message. First, ethnicity and meta-ethnicity are always indicated, metaethnicity referring to what used to be called "historical races" (such as Slavs, Germans, Celts, etc.) or to broader races such as Caucasians and other identities (Neanderthal). Phenotypes are usually mentioned, as well (e.g., Alpinids, Barbarians). Aliases regularly refer to a mythology (e.g., Loki, Angharad, Satyavrata) and a romantic-racial imaginary that would sound quite familiar to any historian of the nineteenth century. The same goes for the visual elements of these avatars, which are often quite explicit and may sometimes look very offending to racialized groups. The haplogroups themselves are far from being neutral: haplogroups are considered to be markers of certain ethnocultural identities. Classical Greeks, for instance, are supposedly J2, E1b1b or R1b; Vikings, I1. One finds threads discussing "Which Y haplogroup is most superior?," where R1b, supposedly a marker of Indo-European ancestry, is generally considered as the best, with the following arguments: "R1 has conquered most parts of Europe," "everyone in Europe has R1 ancestors!! It's the greatest of them all."

On the other hand, beyond such echoes from nineteenth-century racial science, many of these genealogical uses of genomics allude to what is actually the core of the concept of race: its genealogical dimension, inscribed in the "blood." If the more recent confusion between race and color has partly obscured this relationship in such a way that some scholars even believe it is possible to distinguish "race" from "ancestry," the history of the concept of race tells a very different story,[40] and these European or American na-

39 At least, they do not seem to consider some existing companies, such as Igenea, to provide services they need for their interests. On Igenea, see Sommer 2012, 115–140.

40 It is not the place to develop this point extensively. Historically, the genealogical dimension of the concept of race is clear and can be traced in any fields where it has been used from the fifteenth century onward. "Race" systematically referred to ancestry and to certain lineages in which physical and moral traits were transmitted over generations (see, for instance, Doron and Haddad 2021). Even in the specific field of natural history—and then, in physical anthropology and biology—the concept of race was introduced and used to describe inherited varieties and genealogically grounded groups within a species (Doron 2012; ibid. 2016). Skin color was one characteris-

tionalists seem to understand it quite well. More specifically, their genealogical practices and references allude directly to the old background of the concept of race referring to nobility and quality of blood. Such a background never disappeared from certain racist doctrines of the nineteenth and twentieth centuries (especially the Nazis' interpretation of race). This link is perfectly illustrated on humanbiologicaldiversity.com. Surprisingly, in a glossary that is supposed to explain technical terms to understand biological human diversity (alleles, bottleneck effect, etc.), one finds the following definitions:

> *Aristocracy*: Historically, aristocrats originated as a military caste and have tended to be fairer-skinned and taller than commoners (e.g. the "fair princess"), both in Europe and North Asia. Aristocrats also have reproduced at a higher rate than commoners. According to Guillaume Faye, aristocrats are those who defend their people before their own interests. An aristocracy has a sense of history and blood lineage, seeing itself as biologically representative of the people it serves.
>
> *Blue Blood*: The fair-skinned upper-class / aristocracy (e.g. a fair princess). During the Medieval period in Southern Europe, one's skin was supposed to be fair enough to see the blue veins (hence 'blue blood'), thus distinguishing fair-skinned Europeans from the duskier Moors and others.

The fact is that these forums are saturated with references to nobility, as many members choose titles such as Duke, Prince, King, and so forth as aliases (Figure 16). An example is motto of the Skadi forum, which was described as "a free speech forum for people of Germanic heritage," dedicated to "the preservation of our Germanic heritage" and the "development of all-Germanic consciousness": "Are you noble enough to join?"

Figure 16: Two examples of aliases with nobiliary references.

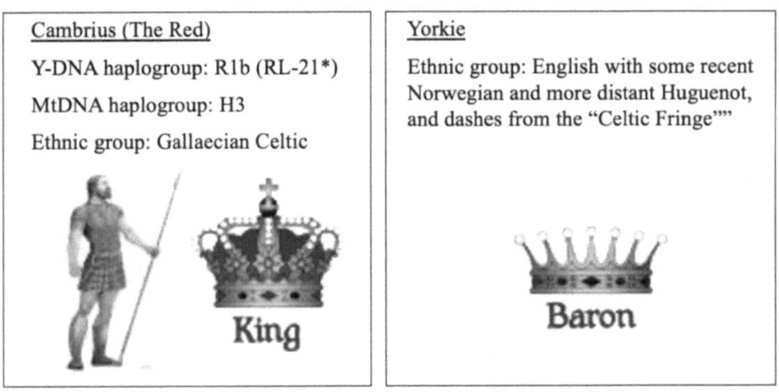

Source: https://www.eupedia.com, last accessed 9 April 2024

---

tic, among many, that was supposed to be hereditary and could serve to define these groups. All this was still true in the 1930 to 1950s, when genetics began to be presented as the best way to define scientifically "race" and identify "racial relations" by scholars such as Laurence Snyder, William Boyd, or Ronald Fisher, precisely because genetics gave access to supposedly strictly inherited traits that are stable and insensitive to environment. See for instance Boyd and Asimov (1958).

Many contributors insist on the fact that their ancestry is related to the history of noble, conquering, and "smart" peoples and the very use of admixture proportions clearly echoes "quarters of nobility." An individual's proportion of "noble blood" is quantified through the various contributions of their ancestors, in terms of ethnic or racial components. Just as the genealogical knowledge of nobility has often proved quite plastic and easy to manipulate in order to emphasize a particular ancestry, depending on the political or symbolical stakes[41], so too does a similar game operate in these peculiar communities.

Conclusion

This article had three main objectives. The first was to explore how various online communities that share European identitarian and racialist agendas use genomic data. It is important to emphasize one point in this respect. Even though these online communities can be described as developing a new kind of decentralized, participative counterknowledge on race and genomics, it would be misleading to believe we can dismiss it as mere junk science or an ideology opposed to the pure science of human population genetics. Fighting such a counterknowledge and its extrapolations supposes a drastic epistemological and historical critique of some of the core categories and techniques of mainstream human population genetics itself. In its ways of producing, processing, and reasoning from genomic data, the discipline more or less consciously incorporates problematic assumptions, ambiguities, and biases that can then be exploited for political and ideological motives. The second objective was to examine how these online communities focus, in particular, on European genetic identities and how they define them. One important element here is to emphasize that, in most of these cases, such identity debates relate not to Whiteness, but rather to structural divisions within Europe, which refer to more ancient distinctions in European racial science, such as that between Nordicists and Mediterraneans. Third, I aimed to explore how, in these online communities, European DNA acquires a social life. In this respect, it is important to note that we are very far from the dream of radically new biosocialities celebrated by various scholars. Even in the cases that may look as creative as new biosocialities, when individuals exhibit their Y-DNA or mt-DNA results and include their admixture tests in their signatures, they actually define their identities according to older categories and ways of reasoning. I conclude by paraphrasing Karl Marx (1852, 176):

> men make their own history, but they do not make it as they please; they [make it] under circumstances existing already, given and transmitted from the past. The tradition of all dead generations weighs like a nightmare on the brains of the living. And just as they seem to be occupied with ... creating something that did not exist before ... they anxiously conjure up the spirits of the past to their service, borrowing from them

41 This is a common place well known among the historians of nobility and genealogy. See Jettot and Lezowski (2016), Pietri and Butaud (2006), Rouchon (2014).

names ... and costumes in order to present this new scene in world history in time-honored disguise and borrowed language.

This conclusion actually holds true both for the most advanced population genomics and for the so-called new biosocialities that supposedly emerged from it. The only way to completely eliminate the spirit of the past is to investigate, without complacency, how it is continually reinscribed in scientific practices and in common methods of identification and categorization.

References

Abel, Sarah. 2021. *Permanent Markers: Race, Ancestry & the Body after the Genome*. Chapel Hill: University of North Carolina Press.

Abu El-Haj, Nadia. 2012. *The Genealogical Science: The Search for Jewish Origins and the Politics of Epistemology*. Chicago: University of Chicago Press.

Bliss, Catherine. 2012. *Race Decoded: The Genomic Fight for Social Justice*. Stanford: Stanford University Press.

Boyd, William C., and Isaac Asimov. 1958. *Races and People*. London: Abelard-Schuman.

Buteau, Germain, and Pietri, Valérie. 2006. *Les enjeux de la généalogie (XIIe-XVIIIe siècles): pouvoir et identité*. Paris: Autrement.

Callaway, Ewen. 2010. "The Rise of the Genome Bloggers." *Nature* 468 (7326): 880–881.

Chow-White, Peter A. 2008. "The Informationalization of Race: Communication Technologies and the Human Genome in the Digital Age." *International Journal of Communication* 2: 1168–1194.

Daniels, Jessie. 2009. *Cyber Racism: White Supremacy Online and the New Attack on Civil Rights*. Lanham: Rowman & Littlefield.

Daniels, Jessie. 2013. "Race and Racism in Internet Studies: A Review and Critique." *New Media & Society* 15 (5): 695–719.

Doron, Claude-Olivier. 2012. "Race and Genealogy: Buffon and the Formation of the Concept of 'Race.'" *Humana Mente. Journal of Philosophical Studies* 5 (22): 75–109.

Doron, Claude-Olivier. 2016. *L'homme altéré: Races et dégénérescence (XVIIe–XIXe siècles)*. Ceyzérieu: Champ Vallon.

Doron, Claude-Olivier, and Elie Haddad. 2021. "Race et histoire à l'époque moderne." *Revue d'histoire moderne & contemporaine* 68 (2): 7–34 and 68 (3): 7–36.

Elhaik, Eran. 2022. "Principal Component Analyses (PCA)-based Findings in Population Genetics Studies are highly biased and must be reevaluated." *Nature. Scientific Reports* 12 (14683)

Foucault, Michel. 2003. "Society Must Be Defended": *Lectures at the Collège de France, 1975–76*. New York: Picador.

Gannett, Lisa. 2021. "Human Genetic Diversity: Fact and Fallacy." In *Remapping Race in a Global Context*, edited by Ludovica Lorusso and Rasmus Winther, 58–63. New York: Routledge.

Gibbon, Sahra, and Carlos Novas, eds. 2008. *Biosocialities, Genetics and the Social Sciences: Making Biologies and Identities*. New York: Routledge.

Jettot, Stéphane, and Lezowski, Marie, eds. 2016. *L'entreprise généalogique. Pratiques sociales et imaginaires en Europe (XVe-XXe siècle)*. Berlin: Peter Lang.

Lewontin, Richard C. 1972. "The Apportionment of Human Diversity." In *Evolutionary Biology*, vol. 6, edited by Theodosius Dobzhansky, Max K. Hecht, and William C. Steere, 381–398. New York: Appleton-Century-Crofts.

Manias, Chris. 2013. *Race, Science, and the Nation. Reconstructing the Ancient Past in Britain, France and Germany*. New York: Routledge

Marx, Karl. 1852. "Le 18 brumaire de Louis Bonaparte." In *Les luttes de classes en France, suivi de La Constitution de la République française adoptée le 4 novembre 1848, suivi de Le 18 Brumaire de Louis Bonaparte*, edited by Maximilien Rubel, 175–314. Paris: Gallimard (2002).

McIntyre, Lee. 2018. *Post-Truth*. Boston: MIT Press.

Nakamura, Lisa, and Peter A. Chow-White, eds. 2012. *Race after the Internet*. New York: Routledge.

Nash, Catherine. 2008. *Of Irish Descent: Origin Stories, Genealogy, & the Politics of Belonging*. Syracuse: Syracuse University Press.

Nash, Catherine. 2015. *Genetic Geographies: The Trouble with Ancestry*. Minneapolis: University of Minnesota Press.

Nelson, Alondra. 2016. *The Social Life of DNA: Race, Reparations, and Reconciliation after the Genome*. Boston: Beacon Press.

Novembre, John, Toby Johnson, Katarzyna Bryc, et al. 2008. "Genes Mirror Geography within Europe." *Nature* 456 (7218): 98–101.

Panofsky, Aaron, and Joan Donovan. 2019. "Genetic Ancestry Testing among White Nationalists: From Identity Repair to Citizen Science." *Social Studies of Science* 49 (5): 653–681.

Panofsky, Aaron, Kushan Dasgupta, and Nicole Iturriaga. 2021. "How White Nationalists Mobilize Genetics: From Genetic Ancestry and Human Biodiversity to Counterscience and Metapolitics." *American Journal of Physical Anthropology* 175 (2): 387–398.

Panofsky, Aaron. 2014. *Misbehaving Science: Controversy and the Development of Behavior Genetics*. Chicago: University of Chicago Press.

Rabinow, Paul. 1996. "Chapter 5: Artificiality and Enlightenment: From Sociobiology to Biosociality." In *Essays on the Anthropology of Reason*, edited by Paul Rabinow. Princeton: Princeton University Press.

Reardon, Jenny. 2019. *The Postgenomic Condition: Ethics, Justice & Knowledge after the Genome*. Chicago: University of Chicago Press.

Rosenberg, Noah A., Jonathan K. Pritchard, James L. Weber, et al. 2002. "Genetic Structure of Human Populations." *Science* 298 (5602): 2381–2385.

Rouchon, Olivier, ed. 2014. *L'opération généalogique: Cultures et pratiques européennes, XVe-XVIIIe siècle*. Rennes: Presses Universitaires de Rennes

Stern, Alexandra Minna. 2019. *Proud Boys and the White Ethnostate: How the Alt-Right Is Warping the American Imagination*. Boston: Beacon Press.

Sismondo, Sergio. 2017. "Post-truth?" *Social Studies of Science* 47 (1): 3–6.

Sommer, Marianne. 2012. "'Do You Have Celtic, Jewish or Germanic Roots?' Applied Swiss History before and after DNA." In *Identity Politics and New Genetics: Re/Creating Categories of Difference and Belonging*, edited by Katharina Schramm, David Skinner, and Richard Rottenburg. 116–140. New York: Berghahn Books.

Spencer, Quayshawn. 2015. "Philosophy of Race Meets Population Genetics." *Studies in History and Philosophy of Biological and Biomedical Science* 52: 46–55.

Strasser, Bruno J., Jérôme Baudry, Dana Mahr, Gabriela Sanchez, and Elise Tancoigne. 2019. "'Citizen Science'? Rethinking Science and Public Participation." *Science & Technology Studies* 32 (2): 52–76.

Sweet, Rosemary. 2004. *Antiquaries: The Discovery of the Past in Eighteenth-Century England*. London: Hambledon and London

TallBear, Kim. 2013. *Native American DNA: Tribal Belonging and the False Promise of Genetic Science.* Minneapolis: University of Minnesota Press.

Thiesse, Anne-Marie. 2001. *La création des identités nationales.* Paris: Seuil

Titley, Gavan. 2020. *Is Free Speech Racist?* Cambridge: Polity.

Thompson, Damian. 2008. *Counterknowledge: How We Surrendered to Conspiracy Theories, Quack Medicine, Bogus Science and Fake History*. London: Atlantic Books.

Wade, Peter, Carlos López Beltrán, Eduardo Restrepo, and Ricardo Ventura Santos, eds. 2014. *Mestizo Genomics: Race Mixture, Nation, and Science in Latin America*. Durham: Duke University Press.

Wagner, Jennifer K, and Kenneth M. Weiss. 2012. "Attitudes on DNA Ancestry Tests." *Human Genetics* 131 (1): 41–56.

***Otherness* in Focus**

Visualizations of Genetic Distances in DNA Studies of Roma

Veronika Lipphardt and Mihai Surdu

Abstract *Our chapter examines how genetic studies postulate and stabilize the existence of homogeneous, easy-to-study groups using labels, images, and sampling decisions. Through two case studies, we show that all these three elements are required to make the truth claim as strong as possible. We begin with a brief review of the literature on visualizations from science and technology studies (STS), art history, image studies, and sociology. Then, in the first case study we give an overview of visualizations in genetic studies of Roma showing how over the years, labels in visualizations have become increasingly abstract and stand in for ever-larger truth claims. We argue that visualizations in DNA studies of Roma contribute to a leap of faith at an ontological level: they stretch argumentative claims from small, unrepresentative samples to entire populations or even continental groups, from familial disease mutations to ethnic disease mutations, and from historical migration routes to narratives of origins and ancestry. In the second case study, we analyze a PCA (principal component analysis) visualization from a genetic study of Roma to problematize the performativity of techno-scientific images in DNA studies of Roma in stabilizing truth claims about the Indian origin of the Roma and their difference from "Europeans." The two-dimensional PCA graph analyzed and the order of data plotting work together to obscure, albeit not quite successfully, the juxtaposition between data of Roma and "Europeans." We demonstrate that the racialization of Roma in the DNA papers we have analyzed occurs in the interplay of sampling, labelling, and visualization. We conclude that visualizations of human diversity are part of simplification work which often conceals biased sampling, the arbitrariness of population labelling and the manipulation of data in the laboratory.*

Introduction

Geneticists who study human population genetics sometimes discuss population labels: if a conventional, widely used label is perceived as insulting to some of the members of the population, a renaming may be considered appropriate. This is the case for "Roma," a name that has been predominantly used in population genetics for roughly thirty years instead of "Gypsies." In 2010, the leading journal of forensic geneticists decided to completely abandon the term "Gypsies" and replace it with "Roma" (Parson and Roewer 2010).

However, both terms fail to describe a population one could study as such. "Roma" does not denote a group bound together by language, culture, tradition, religion, geography, occupation, physical appearance, lifestyle, or self-identification (see, e.g., Surdu and Kovats 2015; Surdu 2016). For many decades, "Gypsies" and "Roma" referred to the label applied to people by others, that is, by non-Roma. Even individuals who identify as Roma would probably not agree to be classified under one population label given the enormous variety of subgroups with distinct identities included under this label. Yet the reason why some geneticists have been eager to study "Roma" as a population for decades was the assumption that they had been genetically isolated from their European surroundings. As they were believed to have come to Europe from India, the idea was that they would display a large proportion of Indian, i.e., non-European, genetic ancestry.

Genetic literature rarely questions the label "Roma," or whether it denotes a population that can be easily traced and sampled for genetic study. In this paper, we would like to examine three components to demonstrate how genetic studies postulates and stabilizes the existence of a homogeneous, easy-to-study group by means of labels, images, and sampling decisions. While population labels and visual elements are displayed in scientific images to strengthen truth claims about Roma, these images rarely explain the sampling strategies used; rather, these strategies need to be extracted through an in-depth reading of the publication. Yet, as we argue, without specific sampling strategies, these images could never be designed in such a convincing way. All three elements are required to make the truth claim as strong as possible.

We examine two case studies to show how labels stabilize conceptions of Roma groupness in visualizations. In the first case, in which we also give an overview of visualizations in genetic studies of Roma, we demonstrate how over the years, labels in visualizations have become increasingly abstract and stand in for ever-larger truth claims. Second, we analyze a PCA (principal component analysis) visualization from a genetic study in which "Roma" are one of the represented populations and demonstrate how the take-home message extracted from the picture, as it is highlighted in conclusions, title, and abstract, misrepresents the findings.

These two cases are merely two instances of many pictures found in DNA studies on Roma, which result from a vast repertoire of entangled methods, techniques, and visualizations: e.g., linkage maps of chromosomes, single-nucleotide polymorphism (SNP) maps, pedigree structure graphs, phylogenetic trees, electropherograms, allele group frequencies graphs, nonmetric multidimensional scaling (NMDS) graphs, homozygosity maps, scatter plots of short tandem repeat (STR) and SNP profiles, genetic distances maps, and diagrams of median-joining networks. All of these graphics demonstrate the increased diversification in visual strategies initiated at the turn of the twentieth century by physical anthropologists and human geneticists to lend credibility to clustering humans into racial or continental groups (Lipphardt 2015; Lipphardt and Sommer 2015; Sommer 2015).

The publications including the images of our case studies are but a few out of 450 peer-reviewed publications we have collected in a database of genetic studies of Roma, all published between 1921 and 2021. About two-thirds of the articles in this database are DNA-based studies published after 1990 that contain a vast register of visualizations. For analyzing our data material, we depend upon long-standing collaboration and experience with an interdisciplinary group including researchers from the life sciences, mathematics, medicine, and the humanities. This group shares an interest in the genetic data of vulnerable communities, especially of Roma people. In our group, the most fervently discussed topics were visualizations and sampling strategies in DNA studies of Roma. The discussions about epistemic, ethical, and social aspects of DNA studies with Roma have resulted in coauthored publications (Lipphardt, Rappold, and Surdu 2021a; Lipphardt et al. 2021; Lipphardt, Rappold, and Surdu 2021b). The analytical work carried out in our meetings has also been insightful for framing the argument of this article.

Analyzing genetic studies with Roma subjects means taking a perspective from within studies targeting a vulnerable population. While some life scientists argue that the inclusion of minorities and underrepresented groups in global health data collection could allow for a reduction in health inequalities (e.g., Hindorff et al. 2017; Ben-Eghan et al. 2020; Sirugo, Williams, and Tishkoff 2019), scientists also acknowledge the problematic implications of employing ethnic and racializing group labels in medical studies and data collections (e.g., Curtis and Balloux 2020; Ben-Eghan et al. 2020; Lee et al. 2008). Our focus on DNA studies of Roma also has an ELSI (ethical, legal and social implication) dimension which we have addressed in other publications. Roma have a troubled history with human genetics and remain vulnerable today to unethical practices in genetics (Lipphardt, Rappold and Surdu 2021b; Lipphardt et al. 2021). Poor ethical practices also contribute to their misrepresentation in genetic studies (Lipphardt, Rappold, and Surdu 2021a). Finally, our database includes a large enough number of cases to enable an analysis of the style of technoscientific imagery (Bredekamp 2015). For our second case study, we have selected a PCA visualization, as this has become a common form of graphic representation in DNA studies of Roma in the last decade; besides, the article presenting the graph under discussion is a good example of sampling issues we encountered in DNA studies of Roma (Lipphardt, Rappold, and Surdu 2021a).

In what follows, we begin with a cursory review of the literature on visualizations from science and technology studies (STS), art history, image studies, and sociology. Next, we provide a short overview of Roma visualizations in genetic and DNA studies, which includes our first case study, and then continue with an in-depth case study. At the end of the paper, we summarize our argument and draw conclusions.

Visualizations in Science

As Bredekamp et al. (2021[2010], XII) put it, "images ... do not merely represent, but veritably *construct*, do not simply illustrate, but actively *bring forth*, that which they show". In scientific work, images are arguments of a *framing* strategy as their producers decide on what aspects are worthy of making visible and what others to leave out of the picture (Lynch and Woolgar, 1990, 2014, Lipphardt and Sommer, 2015). A framing strategy is most obvious in the production of photographs[1], but all types of scientific visualizations use selective depiction to concentrate on what they intend to show.

The role of visualizations in scientific knowledge production can hardly be overestimated, as they are not just illustrations of aspects the text conveys. Many scientists expect visualizations to represent the overall argument of the text so that reading the full text becomes unnecessary. In this case, the captions and inscriptions of the visualizations (in our case, population labels) are crucial for meaning making. In a private conversation with a scientist, he conceded: "The text is just an appendix in our publications. What the people are *reading* in publications in our field are the graphs". This statement validates Goffman's claim that "the picture itself is designed to tell its little story without much textual assistance" (1979, 26).

What is more, images are not just informative but can have aesthetical appeal and can potentially invite the reader to imbue the image with additional layers of meaning. For example, the Icelandic neurologist Kari Stefansson, founder of the biopharmaceutical company deCODE genetics, commented on the deCODE Map of the Human Genome: "This map is to me a thing of beauty. We are looking in quite high definition at the ingenious processes driving the generation of human diversity" (deCODE 2010).

Visualizations in expert publications can also conceal a textual, methodological, or argumentative deficit by making a plea for a specific issue or epistemic object (Surdu 2016). Definitional problems may be resolved through visual stimuli. For example, in twentieth-century physical anthropology, "Gypsies" were pictured as nomads—nomadism being considered the epitome of the group—through images of tents, horses, and traditional clothes. In more recent policy publications, "Roma" —by definition considered as poor—are depicted with symbols of poverty in the background, such as garbage dumps, ruined houses, and unclothed children (Surdu 2016).

In human genetic diversity research, visualizations can also be understood as *obligatory passage points* (Callon 1984) for establishing truth claims on genetic relationships among populations or as a battleground for contestation. Elhaik (2022, 14683) provides

1 What Barthes (1984) has suggested for the functions of photography, namely "to inform, to represent, to surprise, to cause, to signify, to provoke desire" (28), seems to also hold true for scientific visualizations. Far from being an arbitrary choice, the selection of scientific visualizations, especially those used in an ubiquitous way, can indicate the habitus of a scientific field. On the one hand, the image is not merely an expression of the researcher's individual ideas, intentions, or point of view, but, as Bourdieu et al. (1990) argue, also a representation of collective modalities of understanding. On the other hand, what the public is not ideologically prepared to see remains invisible (Sontag 1977, 18).
In science, as elsewhere, photography was considered the paragon of objectivity due to the noninterference of human action and subjectivity: "essentially an act of non-intervention" (Sontag 1977, 11), "an assertion of neutrality" (Sekula 1984, 6), "mechanical objectivity" (Daston and Galison 1992, 82); "the mechanical is here [in photography] a guarantee of objectivity" (Barthes 1980, 278). Yet the objectivity and truth attributed to photography are more of a myth (Goffman 1979; Sekula 1984; Barthes 1980; Bourdieu 1990). In Goffman's words (1979, 20) the myth of photographic objectivity consists in "the very general tendency to confuse realness with representativeness and ideographic with nomothetic validity." The above observations apply not only to photographs but to a wide range of scientific visualizations.

a discussion of recent controversies on human genetics visualizations using PCA scatterplots and the method behind them. The author argues that PCA and its visualizations outcomes in human population genetics are "highly biased" due to "two fallacies: *cherry-picking* or *circular reasoning* (i.e., 'exploration'), the screening and selecting PCA scatterplots that fit preconceived hypotheses while ignoring the other plots" (emphasis in original), thus contesting thousands of findings in genetic studies on human diversity. Lending credibility to visualizations of genetic distances is also part of simplification work resulting from data filtering in scientific practices, as we will demonstrate in our case study (for the general argument see Star, 1983; for the case of data filtering in DNA studies of Roma see Lipphardt, Rappold, and Surdu, 2021a). Filtering out data (through sampling strategies and by excluding "admixture" in the lab) has the effect of making the images' arguments more appealing, convincing, striking, and visible.

This additional layer of simplification resulting from data filtering adds to the work of simplification (Lynch, 1985) and vivification (Lynch, 1988) already embedded in schematic visualizations of which genetic distances are part. Scientific visualizations, particularly relational graphs presenting statistical data, have the role of reducing the complexity of large data sets and of making data easy and quickly readable. Their role is to bring order to an otherwise messy reality (Lynch 1985, 1988). William Playfair, the political economist credited as the founder of statistical graphics, states that his graphs make statistical "studies more clear, and retained more easily by memory" (Playfair 1801, 15, cited in Tufte 2001, 45). Since they are easy to recall by memory, statistical graphics may be also memorable, meaning they may exert a considerable, long-lasting influence on the viewers' perceptions and beliefs. Images are powerful; they exert a fascination on their viewers and can hold them captive (Bredekamp 2021[2010]).

In fact, as Latour (1986) states, inscriptions (including graphs, maps and diagrammatic images) serve a political function in the sociotechnical domain insofar as they aim at "*mobilization*": the aim of winning over as many allies as possible to gain domination over a field. The circulation of inscriptions establishes authorship and originality. Visualizations are part and parcel of persuasion strategies as they are capable of convincing by showing rather than by saying or writing (Latour 1986, 14). Staging, "dramatic visual effects" and oversimplification of what is supposed to be represented are part of the key repertoire of scientific visualizations (Latour 1986, 14).

STS scholars hold that scientific visualizations are intimately related to the truth claims they come with; sometimes they provide *evidence* for invisible phenomena or relationships (Bredekamp 2019; 2021[2010]. These scholars acknowledge that scientific visualizations are not neutral or value-free in the sense of a mechanistic illustration of "data" but are deliberate choices informed by and dependent on scientific methodologies, instruments, theories, concepts, laboratory and field research practices and the assumptions embedded in them (Lynch 1985, 1987; Werner 2015; Bredekamp, Dünkel, and Schneider 2015). Moreover, scientific visualizations are not idiosyncratic, since they derive from and contribute to disciplinary pictorial traditions (Lipphardt, Sommer, and Werner 2015; Bredekamp 2015; Bruhn 2015) and conventions (Bredekamp, Dünkel, and Schneider, 2015).

From an STS perspective, the interest resides in the production and use of scientific visualizations and how they come into being rather than in their correspondence with the objects, phenomena, or relationships depicted by them (Lynch 1988, 2014; Daston 2014; Woolgar 2014).

Various scientific *practices* such as marking specimens, amplifying, creating graphical space and labeling can make objects, relations, and phenomena visible and confer them reality (Hacking 1983; Lynch 1985, 2014). As Hacking (1983) explains (with an example from small particle physics), marking and naming are crucial evidentiary scientific practices: "*if you can spray them then they are real*" (ibid., 23; emphasis in the original). Labeling, which makes visible things that otherwise are invisible or absent (Latour 1986), is also a key operation for the truth claims of scientific visualizations: "Even in scientific illustrations it is the caption which determines the truth of the picture" (Gombrich 1961, 67–68, cited in Goffman 1979, 14). The effects of labeling are ever more important when they apply to human beings, as they contribute to "making up people" (Hacking 1999[1986]).

Our analysis applies the above insights from STS, genetics, and art history to explore the collective apprehension of human genetic diversity resulting from scientific visualizations of genetic distance in DNA studies. By problematizing visualizations of groupness in DNA studies—which are unavoidably racializing in the case of Roma—we attend to the following questions: how are the assumed genetic distances *made*, and how are they made visible? What ontological entities and relationships populate these visualizations? What functions do visualizations of genetic distances have in the epistemic process? How do visualization practices (graphs, diagrams, tables) concur to generate population boundaries? Which aesthetic conventions are used to visualize genetic distances and what role do they play in creating boundaries?

An Overview of Scientific Visualizations in DNA Studies of Roma

Since 2010, population genetic papers on Roma have included technoscientific images in a wide variety of genres. In some cases (e.g., Moorjani et al. 2013; Mendizabal et al. 2012), they occupy half or more of the printed space of the paper. There are photographs, pedigree charts, phylogenetic trees, tables, maps, and PCA scatterplots.

In the illustrations, population labels are used in generalizing ways. For example, pedigree charts appear frequently in medical genetic studies of Roma. One of their functions is to display the analysis units (often extended families) as inbreeding and consanguineous. But many studies using pedigree charts make a great leap by applying labels such as "inbreeding" and "consanguineous" collectively to the Roma population (for some more recent examples, see Gabrikova et al. 2013; García-Magariños et al. 2015; Mašindová et al. 2015; Morar et al. 2011).

Another example, consisting of three publications on the same topic, shows how generalizations are built and transferred in the interplay of sampling, labeling, the production of images and their circulations. In the paper "The Molecular Genetic Basis of Glanzmann's Thrombasthenia in a Gypsy Population in France: Identification of a New Mutation on the Alpha IIb Gene" in the journal *Blood* (Schlegel et al. 1995, 978, figure1), the authors include a pedigree chart titled "Generation tree of gypsy family 1" and refer to a number of forty-two obviously related individuals. Nevertheless, the title of the paper refers to "Thrombasthenia in a Gypsy Population in France" (ibid., 977). The study reports about the discovery of a novel mutation which is called a "Gypsy mutation" (Schlegel et al. 1995, 980, figure 5).

This "Gypsy mutation" pictured by Schlegel et al. (1995) is confirmed and repictured in a genealogical pedigree of a different family by Ruan et al. (1998); the authors name it "gypsy Glanzmann's mutation" in the caption of an image (ibid., 133, figure 3).[2] They call the disease mutation an "abnormality" in the "gene of a European gypsy tribe" (Ruan et al. 1998, 129). Both papers thus take an extended family to stand in for populations that are deemed foreign in their context: "Gypsies" in France, or "Gypsies" in Europe. This exemplifies how a truth claim can be enlarged and extended beyond its actual reach by making a label ever more abstract.

The same disease mutation is discussed in a paper in 2011, where it is called "French Gypsy mutation" (Fiore et al. 2011). The paper's aim is to provide "insights into demographic and population history" (ibid., 983) based on the "French Gypsy mutation." Here, the mutation, which is marked by the label "Gypsy," is no longer the epistemic object of interest; rather, it is an epistemic instrument to find out about another object of interest, namely, the population history. Consequently, the label is black-boxed together with the mutation, and it is taken for granted that it can stand in for the history of a population. "Dating French Gypsy mutation in thrombasthenia" appears as the heading of a pedigree chart presented by Fiore et al. (ibid., 984, figure 2) whose caption contains the label "French Gypsy mutation."

Finally, figure 3 (Fiore et al. 2011, 986) titled "Origin of the French Gypsy Mutation: Origin of Gypsies," illustrates "the migration of Gypsies in France" (984) with colored arrows. In the color-coding for figure 3, pink stands for "European migration to Balkans" (the route from "North India" to "Byzantium empire" [sic]), blue for "migration in Germany" (the route from "Byzantium empire" to "Germany"), green for "introduction of the French Gypsy mutation in France" (the route from "Germany to France") and red (used for three concentric circles and three arrows positioned over "France") for "the founder effect." Superimposed on the physical maps depicting the migration route are color-coded rectangles with time intervals: a pink rectangle with the text "11th century," a blue rectangle with the text "15th century" and a green rectangle with the text "17th–18th." However, the chart only shows the data of one family. It is used to make plausible just one story of how the disease mutation might have reached France and does not provide any context regarding other cases, or possible factors, contributing to the accumulation.

In many papers, visual strategies and labels are used to demonstrate genetic distances between Roma and non-Roma. In these cases, the non-Roma counterpart is given a label that reveals how foreign and distant the Roma are deemed to be. For example, in two tables of a medical paper, the authors use the labels "Gypsies" and "Whites" (Milanov, Topalov, and Kmetski 1999). Table 2 (ibid., 220) reports the prevalence of multiple sclerosis in the city of Sofia ("Gypsies" – n=1 and "Whites" – n=31) and in Samkov ("Gypsies"

2 The name Glanzmann is associated with the disease mutation because the Swiss pediatrician Eduard Glanzmann in 1918 identified (in children from a village in Swiss Alps) and described the rare disease which now carries his name: Glanzmann's Thrombasthenia. See https://www.hog.org/handbook/section/1/glanzmanns-thrombasthenia.

– n=2 and "Whites" – n=15). The title of the paper is "Prevalence of Multiple Sclerosis in Gypsies and Bulgarians," even though the case numbers say little about any population. The biomedical paper Ramal et al. (2004) also compares "Gypsies" and "WCM, white Caucasian Mediterranean" in Tables 3 and 4 (ibid., 937).

Other papers (e.g., Castella et al. 2011, 3762, figure 1) compare "Gypsy patients" and "Caucasian patients," or, in the same paper, synonymously, "patients of the Gypsy ethnic group" and "patients (of white origin)" (Castella et al., 2011, 3760). Again, other papers (such as Martinez-Cruz et al. 2015, 5, figure 2) compare Y-chromosome and mtDNA lineages in "Roma populations" and "Host populations" (with sub-labels "Bulgaria," "Romania," "Hungary," "Ukraine," "Slovakia," and "Spain"). In forensic papers, the comparison group is most often labeled "Caucasian."

These examples reveal that visualizations in DNA studies of Roma perform the function of devices that fix, bring forth and circulate population labels. Visualizations appear as the final outcome of the research process underpinning the theoretical-conceptual research model, recruitment on the field, sampling, labeling, data processing and interpretation. At the same time, they are *vehicles* that carry population labels from the field (usually small, isolated rural communities or institutionalized populations in health units, maternity wards prisons, and armies) to the laboratory and onward to scientific journals and audiences. Visualizations in DNA studies of Roma contribute to a leap of faith at an ontological level: they stretch argumentative claims from small, unrepresentative samples to entire populations or even continental groups, from familial disease mutations to ethnic disease mutations, and from historical migration routes to narratives of origins and ancestry.

Case Study: "Refining the South Asian Origin of the Romani People"

The image we chose to analyze in this paper comes from a genetic study published in 2017, titled "Refining the South Asian Origin of the Romani people" (Melegh et al. 2017). In the abstract, the subsection "Results" reads:

> According to our analyses, Roma showed significant IBD sharing of 0.132 Mb with Northwest Indian ethnic groups. The most significant IBD sharings included ethnic groups of Punjab, Rajasthan and Gujarat states. However, we found also significant IBD sharing of 0.087 Mb with ethnic groups living in Pakistan, such as Balochi, Brahui, Burusho, Kalash, Makrani, Pashtun and Sindhi. (Melegh et al. 2017, 1)

From the title and the abstract, the reader takes away the message that Romani people have predominantly South Asian ancestry, and specifically share a significant portion of their ancestry with populations in Northwest India and Pakistan. There is no mention whatsoever of any West Eurasian or European ancestry of the DNA donors who have been labeled Roma. In the conclusions, the authors go into detail regarding the South Asian ancestry of their probands, while they devote no more than a half sentence to the "West Eurasian ancestry."

The most important method employed in the paper is a visualizing one: PCA. The geneticist Eran Elhaik has recently provided a painstaking critique of this method, as well as a detailed explanation of how it works (Elhaik, 2022). To represent genetic distances, PCAs are used to plot the genetic data of different populations in a two-dimensional space even though the original representation of the data set is three-dimensional. The metric distance within the diagram is meant to denote the genetic distance between the populations. Each colored dot or symbol in the image represents one individual; each color represents one studied population. The populations that are mapped together in Melegh et al. (2017) are the following: Roma, Europeans, Han Chinese, North Indian groups, and South Indian groups. In this case, "Europeans" means, following the sampling logic of the data used in this study, descendants of people who migrated to Utah in past times, where Mormon communities tried to establish a racialized White community.

Notably, readers might easily feel inclined to read historical patterns and movement over time into these images, or to infer a phylogenetic-tree-like relationship between the groups and samples depicted. However, a PCA plot cannot show such causal relationships. It cannot determine how the displayed relationship—whether proximal or distant—between any two sample clusters has come about historically. This issue is up for interpretation, if not for speculation, and typically, there are many different scenarios one could consider.

The two images displayed below are taken from the same graph: image (A) includes the North Indian groups, while image (B) includes the South Indian groups. Any of the group names used for this PCA could be an interesting object of analysis, but here, we focus on "Roma."

In both images, all Roma individuals—represented by red crosses—are closest to the Europeans from Utah, represented by green *x*'s. Were this a three-dimensional representation, one could perhaps find many red crosses hidden behind the green *x*'s; this is because the green *x*'s, as the second plotted group, were plotted on top of the red crosses. In fact, some red dots shine through the green cloud constituted by the green "Europeans," indicating that there is overlap between the two populations. In the logic of PCAs, the population overlap implies that there is no such thing as two distinguishable populations (Elhaik 2022, 14683). Hence, according to this image, Roma, whose symbols are more spread out on the graph than those of the Europeans—who cluster together very closely, as one would expect from a more isolated population—might be understood as a population with a more varied ancestry than Europeans. However, in light of the considerable overlap, Roma and Europeans cannot be distinguished from one another.

Importantly, while the wording of the abstract and the handling of population labels seems to suggest homogeneity among human groups, the graph shows that the group of Roma is not homogeneous, and instead far more scattered than the claim in the abstract suggests. This is also true for the other populations represented in the image, but few of them show a degree of scattering similar to the "Roma" dots. In addition, a population cluster may also look more or less homogeneous depending on the scale at which it is examined: a local population will look very homogeneous when plotted together with populations that have been sampled elsewhere, far away on the globe, while the very same population might look very heterogeneous when plotted together with other local populations from the same region.

One might also ask what is missing in the plot, or what the white spaces between the populations signify. Some North Indian populations in version A are more tightly plotted than any of the South Indian populations in version B. But in both images, there is a big gap between Roma/Europeans and all other populations. This raises the question of what kind of population samples might be able to close the gap, and whether those populations might be more closely related to Roma than any Indian population. If so, one could ask whether the "out-of-India" scenario makes any sense at all, given that there must be alternative scenarios, or whether the most likely scenario can even be inferred from a graph like this, given the variety of other potential scenarios. For example, the red dots labeled "Roma" could show close proximity to samples used to represent Greece or Turkey, which might fall between the dots labeled Roma and Northwest Indian. However, nobody would use this data to infer that Greeks and Turks are more closely related to Northwest Indians than Roma, as it would not resonate with any cultural narrative. A much more obvious interpretation, namely, that Roma seem to be closely related to South East Europeans and have as little to do with Northwest India as Greeks or Turks, would also not stick as it aligns neither with cultural narratives nor with what the authors intend to show.

Furthermore, if one reads the image in such a way that any other (unplotted) population between Roma and Northwest Indian were also a candidate for being closely related to Northwest Indians, this would seem counterintuitive, given that the historical movement implied by the image is one of westbound migration of "Roma," and not of a back-and-forth-migrational pattern between other populations, whether close or distant.

The uncertainties multiply, and the data set labeled "Roma" begins to appear even fuzzier when one takes the sampling scheme into consideration. The authors' team states that their sample representing Roma had more individual DNA data than usual: in this case, 158 individuals. These 158 were merged from two different sources. First, 27 samples were taken from Moorjani et al. (2013). Second, another dataset with 152 individual data labeled "Roma" was taken from Mendizabal et al. (2012). Both datasets had been sampled in international collaborations. For the smaller data set, their authors state:

> We collected 27 Roma samples belonging to six groups that were sampled from four countries in Europe from Hungary (3 linguistically and culturally separated sub-groups: 7 samples from Olah (Vlah), 4 samples from Beas (Boyash) and 4 samples from Romungro) [sic], 4 samples from Romania, 4 samples from Spain and 4 samples from Slovakia (Slovakian speaking Roma). (Moorjani et al. 2013, 8)

And as for their sampling criteria, the authors state: "Roma individuals self-reported as being descendants of the same tribe for at least three generations" (ibid).

Figure 1: reproduced from: Melegh et al. 2017. "Refining the South Asian origin of the Romani people." BMC Genetics 18(1), 1–13, fig. 2, 6. The original figure caption reads: "Relationship of Roma to European and South Asian Populations. The principal component analysis results were plotted on the two principal components with the highest eigenvalue. One symbol represents one individual. (a) Relationship of Roma to European and North Indian groups. (b) Relationship of Roma to European and South Indian groups. Separating North and South Indian populations have only practical purpose in order to give a better overview. Both graphs are the result of the same PCA."

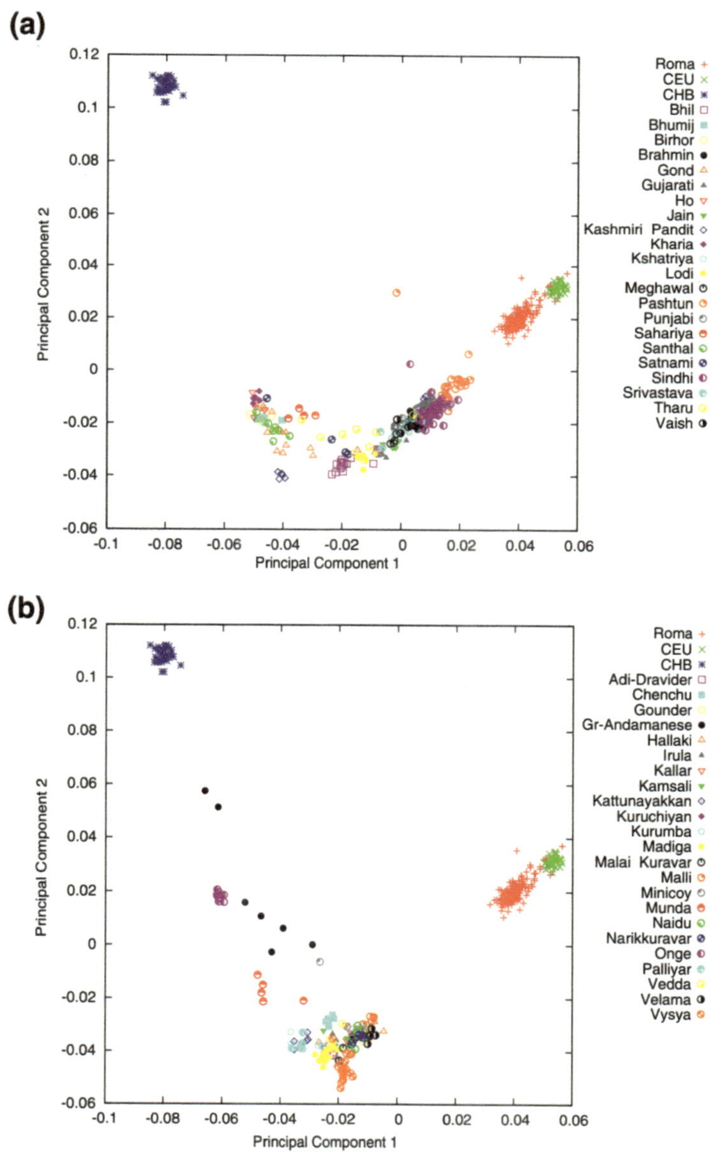

CEU - Utah residents with Northern and Western European ancestry from the CEPH collection.
CHB - Han Chinese in Beijing, China.

The second set of data, used in Mendizabal et al. 2012, is much more complex to grasp. 152 individuals from 13 countries were asked to contribute their DNA data. The largest sample comes from Bulgaria (18 individuals), the smallest from Estonia and Wales (7 individuals each). The authors state that "all individuals included in this study were self-identified as Romani" (Mendizabal et al. 2012, Supplemental Information, 1). They also state that "importantly, the self-identification as Romani is a delicate matter in some European countries due to the social stigma attached to Romani identity; hence additional information obtained in sampling can be scant" (ibid). Nonetheless, some information is given for each of the thirteen data sets; most often, the authors state that the people labeled Romani were native speakers of the dominant national language. Some of the datasets had been published before; for others no such information is provided although the authors had made use of them in other studies where publication information is given.

However, not all the samples gathered in this dataset involved self-identification as Romani. Two sub-datasets come from forensic genetic studies (Irwin et al. 2007, Deligiannidis et al. 2006), and one of them—Irwin—lists coauthors from law enforcement institutions. Parts of the data had been collected more than a decade prior to publication. Hence, informed consent was most probably not provided in those studies—and none of the studies mention any self-identification procedure. Self-identification is questionable for at least part of the data, and identification by others—state authorities, for example—can be assumed (for further sampling issues of said data set, see Lipphardt et al. 2021a).

Altogether, the overall sample of 206 individuals seems patchy and incoherent with regard to the sampling scheme. 54 individuals had to be excluded from the sample "for various reasons." The remaining 152 were analyzed and later transferred to the Melegh et al. (2017) study.

Here, another exclusion took place which affected the data represented in the image. Melegh et al. state that they merged the two datasets, that is, 27 and 152 individuals, for a total of 179, but then reduced it to 158. Hence, they subtracted 21 individual datasets, which is more than 10 percent of the overall data. The rationale for excluding these datasets was the following:

Based on PCA and clustering methods, we removed Roma individuals from the merged Roma dataset, which showed significant admixture with non-Roma Europeans. (Melegh et al. 2017, 2)

If an individual dataset showed "significant admixture with non-Roma Europeans" in the lab analysis, it was to be removed from the overall "Roma" dataset. In the twenty-seven individuals from the first source publication, the authors had paid attention to sampling individuals descended from "the same tribe" for at least three generations. It is likely that in the other study (152 individuals) similar sampling strategies were in place. This is already an attempt at filtering out individuals with mixed ancestry in the field before a DNA sample is donated. The second filtering step happens in the lab, in a first PCA run which is not published and is meant to exclude those who have been included in the DNA donor panel when field samples were collected, despite making tribal descendance the criterion for inclusion. Since this first PCA is not published, one cannot ascertain whether other filtering mechanisms have played a role as well. The authors, however, have not only filtered out the Roma sample in the lab but also the sample of Indian populations. As the paper explains in its methodological section, seven Indian groups were removed from the analysis: Siddis because they "have significant recent African related ancestry"; and six other Indian populations because they "showed complex, mainly East Asian related ancestry in the preliminary analyses" (ibid., 2).

Accordingly, the dataset representing Roma has gone through a two-step filtering process to represent individuals who are as distinct as possible from Europeans. Similarly, the Indian data were filtered to remove individuals from the analysis that presented complex, mixed ancestries. Despite these filtering mechanisms, the graph shows no gap between the Roma and Europeans but considerable overlap.

In the results, the authors state: "The Roma are on this cline between the Europeans and South Asians, but are closer to the European samples. Pakistani groups are the closest South Asian groups to the Roma." (ibid., 5) This is not incorrect, but we have noted a clear gap between any South Asian group and Roma, while we have also noted considerable overlap between Roma and Europeans, even though Roma data had been filtered to make them more distinct from Europeans. Thus, the authors make the result of close relatedness or even identical ancestry between Roma and Europeans sound less clear and strong than it is. In addition, in the results, the authors state, "applying F4 Ratio Estimation on this setup, our results showed that Roma have on average 81.08 +/– 0.53 percent West Eurasian related ancestry." (ibid., 5). In the discussion, they state:

> Population structure and ancestry estimation analyses using PCA and model-based clustering methods placed Roma between Europeans and South Asians. These analyses anticipated that European (or more precisely West Eurasian) ancestry in Roma is significant and its proportion is higher than the proportion of the Indian ancestry. (Ibid., 10)

Throughout the discussion, the authors report that all their methods have yielded a strong European ancestry in Roma. But in the conclusions, they do not mention this quantitative aspect when they summarize their findings:

> In conclusion, the results of our study suggest that the West Eurasian ancestry of Roma originates likely from Central and East Europe, and Northwest India plays an even more important role in the South Asian ancestry of Roma than previous studies suggested. Our results also suggest that besides Northwest Indian populations, Pakistani populations play also an important role of the source of South Asian ancestry of Romani people. These new findings extend the South Asian origin of the Romani people making the Pakistani region a similarly important source of ancestry for the Romani people as the Indian subcontinent. (Ibid., 11f.)

This summary does not demonstrate that, according to their analysis, an overwhelming amount of Roma ancestry is European (or, as the authors specify, "more precisely West Eurasian"). The largest part of the summary is dedicated to South Asian ancestry. Considering how the results are presented in the abstract at the beginning of the paper makes this neglecting even more striking: the overwhelming European ancestry of Roma is not even mentioned there – it's only the Asian ancestry that made it to this summarized result section in the abstract on the front page.

With our case study, we aim to demonstrate how labels, sampling, and visualization are interconnected procedures that stabilize truth claims about the Indian origin of the Roma and their difference from "Europeans." Firstly, the research design operates with assumptions about groupness, and assumes the existence of neatly separable groups that fit the labels chosen by the researchers. Our problematization of these assumptions begins on the methodological level with sampling schemes: datasets assembled from various sources through incoherent sampling schemes cannot adequately represent any group. Had the geneticists wanted to show genetic relatedness in a specific social group, they would have had to ensure that all members of the group under study were included under the same sampling criteria. Instead, they lumped together data, including data from people who did *not* self-identify as Roma, which was collected under heterogeneous sampling schemes, and then curated it to make it fit within the confines of a more homogeneous group. We agree that genetic differences exist and can be visualized; however, we contest that these differences easily and clearly align with meaningful social groupings.

Second, no matter whether the grouping is done in/correctly or un/problematically, the forced tracing back of "Roma" ancestry to India, and the active construction of such an ancestry, is at best biased; at worst, it is racist, even if not deliberately so. The sampling strategy proceeds with the deliberate selection only of those individuals who maximize the hypothesized isolation and Indian origin of the Roma (the criterion, "descendants of the same tribe for at least three generations," serves as a proxy to that hypothetical ancestral group); the small sample sizes used by the researchers exclude "admixed" individuals by sampling design. The third step represents the culmination of circular reasoning: in the laboratory, data are further filtered, and those samples of Roma and Indians which show "admixture" are removed from the analysis. In the fourth step, the two-dimensional PCA graph and the order of data plotting work together to obscure, albeit not quite successfully, the juxtaposition between data of Roma and "Europeans." The PCA plot serves to demonstrate that "Roma" are outsiders to Europe and that they are distinct from those who inhabit "Europe" as "natives." The underlying rationale and narrative of the genetic studies on Roma thus supports exclusionist and racist political agendas that are expanding in EU member states today. While the PCA plot is not sufficiently convincing in showing genetic distances between Roma and "Europeans," the interpretation of the graph, which contradicts what is shown, textually accomplishes the final operation in stabilizing the truth claim that Roma are from India.

Conclusions

Our chapter analyzes the forms of thinking and the scientific practices that generate (visualizations of) genetic distances in DNA studies of Roma. The switch from a "representational idiom" to a "performative idiom" (Pickering 1995) and to the scientific practices, a methodological route that we follow, is a well-established research tradition in studies of science (e.g., Hacking 1983; Barad 2003; Latour 2005; Cetina 1999; Law 2004) including the analysis of scientific visualizations (e.g., Burri 2012; Daston 2014; Dumit and de Laet 2014; Latour 1987; Law 2014; Lynch 1985, 1988, 2014) to which we aim to contribute. In our approach, we follow Bredekamp, Dünkel and Schneider (2015, 2) who argue that technoscientific images are "productive agents and distinctive multi-layered elements of the epistemic process."

Visualizations of different sorts are commonplace in DNA studies of Roma; in more recent publications since 2010, they occupy a central role in the economy of scientific publications. The visualizations we have analyzed are part of a trend towards abstraction in pictorial traditions of human variation, which already began in the second half of the twentieth century due to increased sophistication in statistical and graphical techniques. Within this development towards abstraction, the preferred form is that of the diagrammatic tree (Lipphardt and Sommer 2015; Lipphardt 2015). The latest genre within the visual repertoire for picturing human variation is the PCA scatterplot; for our case study, we have chosen a PCA scatterplot to problematize the performativity of technoscientific images in DNA studies of Roma. Within human population genetics, PCA methods and scatterplots are already contested techniques; Elhaik (2022) calls for a reevaluation of between 32,000 and 216,000 genetic studies based on them due to their biased findings. Regardless of whether the field of human population genetics will heed Elhaik's call, the effects produced by these studies—the naturalization of social differences between human groups cemented by labels and cultural narratives—will surely be long-lasting.

As our case study demonstrates, visualizations of human diversity are artifacts that may misrepresent the populations they purport to depict as they often conceal biased sampling, the arbitrariness of population labelling and the manipulation of data in the laboratory. As we have shown, visualizations are part of simplification work carried out by data filtering in the field and in the laboratory resulting in an exaggeration of differences (or distances) among the represented populations. The role of visualizations is to fix an absent reference – populations substituted by labels—and to arrest a dynamic process of continuous mixing among groups, which are instead presented as delimited entities. Visualizations act as vehicles of typological thinking as they operate at an ontological level with continental clusters of populations, and as they present populations as homogenous, discrete entities (hence "types"). Visual strategies and labels are being used to demonstrate genetic distances: at its core, the diversity is both represented and *generated* by genetic distances.

Our contribution aims to critically address the "beyond race" claim in the life sciences and, to use the words of M'charek (2022, 1), "to stay curious about what race is made to be in practice, how it manifests and what politics it does." Our argument is that the *punctum* (in Barthes's 1984 understanding) of past and present visions shared by scientific and expert practices we have analyzed is *race*, understood as a "relational object" created through scientific practices and situated "beyond fact and fiction" (M'charek 2013). The racialization of Roma in DNA papers we have analyzed occurs in the interplay of sampling, labeling, and visualization. Roma are often sampled under the assumption of a consanguineous, inbreeding "genetic isolate"; further on in the laboratory, only those samples deemed to best represent Indian origins are retained, bolstering assumptions related to Roma populations' "purity" and "admixture." Samples gathered and manipulated on essentialist assumptions are then labeled and grouped mostly under continental labels, except for a few cases when racial labels are used (e.g., "Whites" and "Gypsies"). Finally, the interpretation of the results and the reduction of information in the abstract is biased as it overemphasizes the importance of South Asian ancestry. In this way, visualizations can naturalize racial descriptors and racial differences while concealing the ways in which they have been produced.

Finally, we have been asked to constructively propose a more nuanced way of capturing groupness, or of visualizing genetic relationships with PCA images. As a preliminary proposal, we would respond as follows: one could use a large, heterogeneous dataset in a PCA plot. The dataset could include data from individuals who have self-identified clearly and freely using any population label of their choice. Datasets would not be colored or symbolized using population labels; instead, one would print all dots in gray. Next, one could look for clusters of gray dots. If there were clearly identifiable clusters, one could ask whether the annotating data correlated to this cluster had something in common: perhaps self-identification with the same population label, birthplace, or language? If there were clear correlations, one could speak of a social groupness that is also reflected in genetic data. If not, one would have to concede that the relationship between socially ascribed, self-assigned, and genetic groupness is much more complex than the simplistic groupness assumptions within these genetic studies of Roma.

References

Barad, Karen. 2003. "Posthumanist Performativity: Toward an Understanding of How Matter Comes to Matter." *Signs: Journal of Women in Culture and Society* 28 (3): 801–831.

Barthes, Roland. 1984. *Camera Lucida: Reflections on Photography*. London: Flamingo.

Barthes, Roland. 1980. "Rhetoric of the Image." In *Classics Essays on Photography*, edited by Alan Trachtenberg, 269–285. New Haven, CT: Leete's Island Books.

Ben-Eghan, Chief, Rosie Sun, Jose Sergio Hleape et al. 2020. "Don't Ignore Genetic Data from Minority Populations." *Nature* 585 (7824): 184–186.

Birch, Kean, and Fabien Muniesa, eds. 2020. *Assetization: Turning Things into Assets in Technoscientific Capitalism*. Cambridge, MA: MIT Press.

Birch, Kean. 2019. *Neoliberal Bio-Economies? Co-Constructing Markets and Natures*. Cham: Palgrave Macmillan.

Bourdieu, Pierre, and Luc Boltanski, eds. 1990. *Photography: A Middle-Brow Art.* Cambridge: Polity Press.

Bredekamp, Horst. 2021. *Image Acts: A Systematic Approach to Visual Agency.* Berlin: de Gruyter.

Bredekamp, Horst. 2019. *Darwin's Corals: A New Model of Evolution and the Tradition of Natural History*. Berlin: de Gruyter.

Bredekamp, Horst, Vera Dünkel, and Birgit Schneider, eds. 2015. *The Technical Image: A History of Styles in Scientific Imagery*. Chicago: University of Chicago Press.

Bredekamp, Horst. 2015. "A History of Styles of Technical Imagery: Between Description and Interpretation. A Conversation with Horst Bredekamp." In *The Technical Image*, edited by Horst Bredekamp, Vera Dunkel, and Birgit Schneider. 18–32.

Bredekamp, Horst, Vera Dünkel, and Birgit Schneider, 2015. "Introduction: The Image – A Cultural Technology: A Research Program for a Critical Analysis of Images." In *The Technical Image*, edited by Horst Bredekamp, Vera Dunkel, and Birgit Schneider. 1–8.

Bruhn, Matthias. 2015. "Beyond the Icons of Knowledge: Artistic Styles and the Art History of Scientific Imagery." In *The Technical Image*, edited by Horst Bredekamp, Vera Dunkel, and Birgit Schneider. 36–48.

Burri, Regula Valérie. 2012. "Visual Rationalities: Towards a Sociology of Images." *Current Sociology* 60 (1): 45–60.

Callon, Michel. 1984. "Some Elements of a Sociology of Translation: Domestication of the Scallops and the Fishermen of St. Brieuc Bay." *The Sociological Review* 32 (1 Suppl.): 196–233.

Castella, Maria, Roser Pujol, Elsa Callén et al. 2011. "Origin, Functional Role, and Clinical Impact of Fanconi Anemia FANCA Mutations." *Blood* 117 (14): 3759–69.

Cetina, Karen Knorr. 1999. *Epistemic Cultures: How the Sciences Make Knowledge*. Cambridge, MA: Harvard University Press.

Catelijne Coopmans, Janet Vertesi, Michael E. Lynch, and Steve Woolgar, eds., 2014. In *Representation in Scientific Practice Revisited*, edited by Catelijne Coopmans, Janet Vertesi, Michael E. Lynch, and Steve Woolgar. Cambridge, MA: MIT Press.

Curtis, David, and François Balloux. 2020. "Topical Ethical Issues in the Publication of Human Genetics Research." *Annals of Human Genetics* 84 (4): 313–314.

DeCODE Genetics. 2010. "The New deCODE Map of the Human Genome." Accessed November 20, 2024. www.decode.com/map-of-the-human-genome/

Daston, Lorraine. 2014. "Beyond Representation." In *Representation in Scientific Practice Revisited*, edited by Catelijne Coopmans, Janet Vertesi, Michael E. Lynch and Steve Woolgar. 319–323.

Daston, Lorraine, and Peter Galison. 1992. "The Image of Objectivity." *Representations* 40: 81–128.

Dumit, Joseph, and Marie de Laet. 2014. "Curves to Bodies: The Material Life of Graphs." In *Routledge Handbook of Science, Technology, and Society*, edited by Daniel Lee Kleinman and Kelly Moore, 93–112. London/New York: Routledge.

Elhaik, Eran. 2022. "Principal Component Analyses (PCA)-based Findings in Population Genetic Studies Are Highly Biased and Must Be Reevaluated." *Scientific Reports*, 12 (1): 14683.

Fiore, Mathieu, Xavier Pillois, Paquita Nurden, Alan T Nurden, Frédéric Austerlitz. 2011. "Founder Effect and Estimation of the Age of the French Gypsy Mutation Associated with Glanzmann Thrombasthenia in Manouche Families." *European Journal of Human Genetics* 19 (9): 981–987.

Fleck, Ludwik. 1979. *Genesis and Development of a Scientific Fact*, translated by Frederick Bradley and Thaddeus J. Trenn. Chicago: University of Chicago Press.

Gabrikova, Dana, Martin Mistrik, Jarmila Bernasovska et al. 2013. "Founder Mutations in NDRG1 and HK1 Genes Are Common Causes of Inherited Neuropathies among Roma/Gypsies in Slovakia." *Journal of Applied Genetics* 54 (4): 455–460.

García-Magariños Manuel, Thore Egeland, Ignacio López-de-Ullibarri, Nils L Hjort, Antonio Salas. 2015. "A Parametric Approach to Kinship Hypothesis Testing Using Identity-By-Descent Parameters." Statistical Applications in *Genetics and Molecular Biology* 14 (5): 465–479.

Goffman, Erving. 1979. *Gender Advertisements*. Cambridge, MA: Harvard University Press.

Hacking, Ian. 1983. *Representing and Intervening: Introductory Topics in the Philosophy of Natural Science*. Cambridge: Cambridge University Press.

Hacking, Ian. 1999. "Making Up People." In *Reconstructing Individualism: Autonomy, Individuality, and the Self in Western Thought*, edited by Thomas C. Heller, Morton Sosna, and David Wellbery, 222–236. Stanford: Stanford University Press.

Hindorff, Lucia. A., Vence L Bonham, Lawrence C Brodye et al. 2018. "Prioritizing Diversity in Human Genomics Research." *Nature Reviews Genetics* 19 (3): 175–185.

Latour, Bruno. 2005. *Reassembling the Social: An Introduction to Actor-Network-Theory*. Oxford: Oxford University Press.

Latour, Bruno. 1987. *Science in Action*. Cambridge, MA: Harvard University Press.

Latour, Bruno. 1986. "Visualization and Cognition." *Knowledge and Society: Studies in the Sociology of Culture Past and Present* 6: 1–40.

Law, John. 2014. "Indistinct Perception." In *Representation in Scientific Practice Revisited*, edited by Catelijne Coopmans, Janet Vertesi, Michael E. Lynch and Steve Woolgar. 337–343.

Law, John. 2004. *After Method: Mess in Social Science Research*. London: Routledge.

Lee, Sandra Soo-Jin, Joanna Mountain, Barbara Koenig et al. 2008. "The Ethics of Characterizing Difference: Guiding Principles on Using Racial Categories in Human Genetics." *Genome Biology* 9 (7): 1–4.

Lipphardt, Veronika, Gudrun A. Rappold, and Mihai Surdu. 2023, unpublished. "Ethical standards in forensic genetic research – a critical appraisal of Roma studies."

Lipphardt, Veronika, Gudrun A. Rappold, and Mihai Surdu. 2021. "Representing Vulnerable Populations in Genetic Studies: The Case of Roma." *Science in Context* 34 (1): 69–100.

Lipphardt, Veronika, Mihai Surdu, Nils Ellebrecht, Peter Pfaffelhuber, Matthias Wienroth, and Gudrun A. Rappold. 2021. "Europe's Roma People Are Vulnerable to Poor Practice in Genetics." *Nature* 599: 368–371.

Lipphardt, Veronika, Gudrun A. Rappold, and Mihai Surdu. 2021. "Protect Minorities in Genetic Research." *Science* 373 (6562): 1452–1452.

Lipphardt, Veronika, and Mihai Surdu. 2018, unpublished. Genetic studies on Roma 1921–2018. Internal Report.

Lipphardt, Veronika. 2015. "Traditions and innovations: Visualizations of human variation, c. 1900–38." *History of the Human Sciences* 28 (5): 49–79.

Lipphardt, Veronika, and Marianne Sommer. 2015. "Visibility Matters: Diagrammatic Renderings of Human Evolution and Diversity in Physical, Serological and Molecular Anthropology." *History of the Human Sciences* 28 (5): 3–16.

Lynch, Michael, and Steve Woolgar. 2014. "Preface." *Representation in Scientific Practice Revisited*, edited by Catelijne Coopmans, Janet Vertesi, Michael E. Lynch and Steve Woolgar. VII–IX.

Lynch, Michael, and Steve Woolgar. 1990. "Introduction: Sociological Orientations to Representational Practice in Science." In *Representation in Scientific Practice Revisited*, edited by Catelijne Coopmans, Janet Vertesi, Michael E. Lynch and Steve Woolgar. 1–19.

Lynch, Michael. 2014. "Representation in Formation." In *Representation in Scientific Practice Revisited*, edited by Catelijne Coopmans, Janet Vertesi, Michael E. Lynch and Steve Woolgar. 323–329.

Lynch, Michael. 1985. "Discipline and the Material Form of Images: An Analysis of Scientific Visibility." *Social Studies of Science* 15 (1): 37–66.

Lynch, Michael. 1988. "The Externalized Retina: Selection and Mathematization in the Visual Documentation of Objects in the Life Sciences." *Human Studies* 11: 201–234.

Martínez-Cruz, Begoña, Isabel Mendizabal, Christine Harmant et al. 2016. "Origins, Admixture and Founder Lineages in European Roma." *European Journal of Human Genetics* 24 (6): 937–943.

Mašindová, Ivica, Andrea Šoltýsová, Lukáš Varga et al. 2015. "MARVELD2 (DFNB49) Mutations in the Hearing Impaired Central European Roma Population – Prevalence, Clinical Impact and the Common Origin." *PloS one* 10 (4): e0124232.

M'charek, Amade. 2022. "Race and Sameness: on the Limits of Beyond Race and the Art of Staying with the Trouble." *Comparative Migration Studies* 10 (1): 1–16.

M'charek Amade. 2013. "Beyond Fact or Fiction: On the Materiality of Race in Practice." *Cultural Anthropology* 28 (3): 420–442.

Melegh, Bela I., Zsolt Banfai, Kinga Hadzsiev, and Attila Miseta. 2017. "Refining the South Asian Origin of the Romani people." *BMC Genetics* 18 (1): 1–13.

Mendizabal, Isabel, Oscar Lao, Urko M. Marigorta et al. 2012. "Reconstructing the Population History of European Romani from Genome-wide Data." *Current Biology* 22 (24): 2342–2349.

Milanov, Ivan, Nikolay Topalov, and T. S. Kmetski. 1999. "Prevalence of Multiple Sclerosis in Gypsies and Bulgarians." *Neuroepidemiology* 18 (4): 218–222.

Moorjani, Priya, Nick Patterson, Po-Ru Loh et al. 2013. "Reconstructing Roma History from Genome-wide data." *PloS one* 8 (3): e58633.

Morar, Bharti, Sashka Zhelyazkova, Dimitar N Azmanov et al. 2011. "A Novel GEFS+ Locus on 12p13. 33 in a Large Roma Family." *Epilepsy Research* 97: 198–207.

Parson, Walther, and Lutz Roewer. 2010 "Publication of Population Data of Linearly Inherited DNA Markers in the International Journal of Legal Medicine." *International Journal of Legal Medicine* 124 (5): 505–9.

Pickering, Andrew. 1995. *The Mangle of Practice. Time, Agency and Science*. Chicago: University of Chicago Press.

Ramal, L. M., M. A. López-Nevot, J. M. Sabio et al. 2004. "Systemic Lupus Erythematosus in Southern Spain: a Comparative Clinical and Genetic Study between Caucasian and Gypsy Patients." *Lupus* 13 (12): 934–940.

Roewer, Lutz, M. Krawczak, S. Willuweit et al. 2001. "Online Reference Database of European Y-chromosomal Short Tandem Repeat (STR) Haplotypes." *Forensic Science International* 118 (2–3): 106–113.

Ruan, J., O. Peyruchaud, P. Nurden et al. (1998). "Family Screening for a Glanzmann's Mutation Using PCR-SSCP Thrombasthenia." *Platelets* 9 (2): 129–136.

Schlegel, Nicole, O. Gayet, M. C. Morel-Kopp et al. 1995. "The Molecular Genetic Basis of Glanzmann's Thrombasthenia in a Gypsy Population in France: Identification of a New Mutation on the alpha IIb Gene." *Blood* 86 (3): 977–982.

Sekula, Allan. 1984. *Photography Against the Grain: Essays and Photo Works 1973–1983*. Halifax: The Press of Nova Scotia College of Art and Design.

Sirugo, Giorgio, Scott M. Williams, and Sarah A. Tishkoff. 2019. "The Missing Diversity in Human Genetic Studies." *Cell* 177 (1): 26–31.

Sommer, Marianne. 2015. "Population-genetic Trees, Maps, and Narratives of the great Human Diasporas." *History of the Human Sciences* 28 (5): 108–145.

Sontag, Susan. 1977. *On Photography.* New York: Farrar, Straus and Giroux.

Surdu, Mihai. 2019. "Why the 'Real' Numbers on Roma Are Fictitious: Revisiting Practices of Ethnic Quantification." *Ethnicities* 19 (3): 486–502.

Surdu, Mihai. 2016. *Those Who Count. Expert Practices of Roma Classification*. Budapest: Central European University Press.

Surdu, Mihai, and Martin Kovats. 2015. "Roma Identity as an Expert-Political Construction." *Social Inclusion* 3 (5): 5.

Star, Susan Leigh. 1983. "Simplification in Scientific Work: An Example from Neuroscience Research." *Social Studies of Science* 13 (2): 205–228.

Tufte, Edward R. 2001. *The Visual Display of Quantitative Information*. Cheshire, CT: Graphics Press.

Werner, Gabriele. 2015. 'Discourses about Pictures: Considerations on the Particular Challenges Natural-Scientific Pictures Pose for the Theory of the Picture." In *The Technical Image*, edited by Horst Bredekamp, Vera Dunkel, and Birgit Schneider. 8–13.

Woolgar, Steve. 2014. "Struggles with Representation: Could It Be Otherwise?" In *Representation in Scientific Practice Revisited*, edited by Catelijne Coopmans, Janet Vertesi, Michael E. Lynch and Steve Woolgar. 329–333.

Contextualizing the Representations of Finland as a Population Isolate

Aaro Tupasela and Heta Tarkkala

Abstract *The Finnish population is often represented in the scientific and medical literature as isolated and homogeneous. According to the proponents of this view, these characteristics resulted from bottleneck effect dating back some 4,000 years. Many scientific papers and journals have visually represented this asserted distinctiveness in maps or pictures, as have websites portraying the results of studies arguing for this thesis and depicting the country's population history from its perspective. In this chapter, we discuss these visual representations through three analytical lenses that each provide their own view on the narrative of population homogeneity and isolation. First, we will highlight the specific background against which the understanding of an isolated population emerged as part of wider research project on certain genetically enriched rare diseases understood to be particularly Finnish. Second, we will discuss the way Finns have been portrayed as outliers by discussing visual representations of Finns in the medical and scientific research literature. Third, we will discuss the characteristics of this supposedly isolated population that in fact indicate an inner heterogeneity, more likely pointing toward two (multiple) populations of Finns rather than one, clearly defined, isolated, and homogeneous population. Our analysis reveals how the concept of a population isolate has not been a stable and fixed concept through the years. Instead, the concept has drawn from several different methodological and conceptual approaches to what constitutes a population isolate, resulting in diverging understandings about that very same object of study. We suggest that this "flexibility" in the interpretation of isolation has helped researchers attract interest to Finnish genetics and create bridges to other population groups for purposes of comparison.*

Introduction

The Finnish population has been described as one of the "best-studied" population isolates in the world (Peltonen et al. 1999, 1913). In their article on Finnish population history, for instance, Palo et al. (2009, 1336) note: "As an isolated outlier population on the European genetic landscape, the Finns have attracted a great deal of interest among the geneticists." Drawing on the concept of isolation has served Finnish research in genetics well, but the promises of a population isolate for research has also evolved over the years (see Uusimaa et al. 2022; Kääriäinen et al. 2017).

Population isolates "are subpopulations resulting from the founder effect of a small number of individuals as a consequence of some type of bottleneck" (Arcos-Burgos and Muenke 2002). In addition to Finns, groups such as the Hutterites, old order Amish, and the Paisa community of Antioquia in Columbia are considered to be population isolates. In these populations it has been easier to identify the genetic causes of some diseases and medical conditions, such as asthma and multiple sclerosis. Historically, population isolates were seen as groups that had remained "unmixed" for generations. During the twentieth century, groups of people came to be seen as populations instead of races; the historical weight of eugenics and the race sciences of the late nineteenth and early twentieth centuries also prompted researchers to study human variation and human populations as dynamic rather than fixed categories (Lipphardt 2012, 2013). A focus on isolation was also part of this shift, as it helped establish certain ideas of "originality" to which later developments could be compared (Tupasela and Tamminen 2015; Lipphardt 2013).

In late twentieth-century biomedical research, population isolates were considered efficient in the study of monogenetic diseases (Hatzikotoulas 2014). Today, they are framed as promising in applications that include finding "entry points into the biology of common diseases through low-frequency, high impact variants" (Kurki et al. 2023, 508). Such recent framings serve to highlight the continued relevance of these populations for biomedical science and help identify why isolation sometimes plays such an important role in the characterization of populations. Understanding Finns as a population isolate characterized by "founder effect, genetic drift and isolation" (Peltonen et al. 1999, 1913) has been at the center of Finnish genetics research for decades, because of the small size of the country's total population. Finland's largest public-private genomics project to date, Finngen, is founded precisely on such characteristics (https://www.finngen.fi/en). Over the years, the project has highlighted several of these characteristics, such as the homogeneity of the population, as relevant, important, and particularly promising for genetic and genomic research both nationally and internationally (Helén et al. 2024). These claims, however, are not new. In a 2001 article, Finnish genetic researcher Juha Kere wrote that the Finnish population has "out of proportion for its size … by example shaped research in human disease genetics" (Kere 2001, 103). Even three decades earlier in 1973, pediatric researcher Reijo Norio and colleagues described the Finnish population as "rare flora in rare soil" (Norio et al. 1973) and, as such, possessing with certain unique attributes for biomedical research.

The history of Finns as a population isolate is thus in many ways both grounded in and a result of medical research. In this chapter, we want to contextualize how isolation has figured in different, yet related medical and scientific disciplines in Finland over time. We do not, for example, question the existence of a bottleneck effect as such, or that Finns may be a population isolate. Rather we aim to situate and discuss how this idea of an isolated population has been represented and what gives this population its characteristics. This includes examining how the understanding of a particularly defined population came to serve as a reference allowing Finns to be seen and claimed as unique, and how the inner heterogeneity of the population in Finland accords with the representation of a homogeneous population isolate. We argue that these discussions and scientific studies represent a preoccupation with two important aspects in genetic studies of populations. *First*, where do present-day populations come from? This is to say, what are the origins and "roots" of a people that have come to be known since the establishment of the Finnish nation-state in 1917 as Finns. *Second*, how have Finns, as they have been selected for various comparative genetic studies, been represented visually as relating to "other" populations. We see these two discussions as being fundamental to science and medicine, and more broadly to how the Finnish population is related to other populations in Europe and further afield. We also consider these debates and discussions to be central to the stabilization of Finnishness to a fixed time and place in history, enabling claims of origin and authenticity (Tupasela and Tamminen 2015).

We have previously discussed Finnish population within biomedical research from various viewpoints, such as population branding (Tupasela 2022; 2017), authenticity (Tupasela and Tamminen 2015), and the characteristics by which it is framed as valuable research material for biomedical research and development (Tarkkala 2019; Tarkkala and Tupasela 2018). In this chapter, however, we focus on representations of the Finnish population isolate and show the underlying dynamics that can contribute to the idea of isolated, homogeneous population. We also point out that "population isolate" is not a static concept but has varied over time and in different contexts in Finland.

Our discussion focuses on how the claim of an isolated population is represented visually in Finnish scientific work. Our interest in visualizations relates to how certain theoretical and methodological choices are represented visually, and how they simultaneously encapsulate certain ways of thinking and certain approaches in science, specifically in relation to the idea of the Finnish population isolate. We reiterate that the notion of population isolates might not prove to be valid if the selection criteria or pool of individuals selected for analysis were wider and less selective as it has been in Finland. In the following, we first will introduce visualizations to lay the groundwork for our analytical discussions, which will then draw attention to the way understandings of populations have been constructed and conditioned by the tools used, the availability of materials and methods, and the decisions made by researchers in various phases of their research. To this end, we augment our analysis of visual representations with a consideration of written and published materials from researchers working in this field. We have collected scientific articles on topic published between the years of 1972 and 2023 that discuss the characteristics of the Finnish population, either explicitly or as part of reporting results of certain analyses. We also utilize material from the internet that has been published by researchers and research institutions on their home page and contain information and visualizations about Finnish population history (Tupasela 2022).

Visual Representations in Genetics

In sciences, visualizations are a form of portraying evidence and condensing findings (Coopmans et al. 2014). Visualizations present data, expressing trends and relationships or even space-time interrelationships (Gross and Harmon 2014, 53). Visualizations might "witness what the scientist saw in the field, laboratory, office, or observatory" as well as "reveal the function of equipment" (Gross and Harmon 2014, 53). They can be produced as maps, images, plots, graphs, tables, and so on. In population genetics, visualizations are used for applications such as maps showing how people have moved from Africa to other continents, various kinds of evolutionary trees, or figures showing certain patters, variations, and connections (Sommer 2015; Palsson 2007).

More generally, images in science are usually produced based on certain instruments or imaging procedures, and they encapsulate the reasonings and methodological choices of their own time (Bredekamp et al. 2015, 1–3). As Bredekamp et al. (2015, 1) argue: "The transformation of observations, findings, and insights into images partakes actively in the constitution of knowledge." Images and visualizations can "create persuasive representations" (Burri and Dumit 2008, 297) for readers. In a similar vein, Lipphardt and Sommer (2015, 3) have argued that "images are at the heart of strategies of persuasion." Burri (2012, 54) calls for sociologists to pay attention to images, and while doing so "not focus on images alone but take the *social practices and contexts of image production, interpretation and use* into account" (emphasis in the original). Indeed, visual representations are always created and used in certain contexts, to deliver, strengthen, and stabilize certain understandings and arguments and portray them as stripped of too much complexity (Lynch and Woolgar 1990).

This is also the case in visual representations of the Finnish population and the population history claimed to have caused the isolated gene pool for this specific group of people (see Tupasela 2022). Some of the first notable images found in medical genetics in Finland date to the early 1970s, when doctors such as H. R. Nevanlinna (one of the most prominent doctors and geneticists in Finland to study population structure) published articles seeking to describe the history and development of early settlements in Finland. Similar maps can be found in historical and archaeological accounts of Finnish history. Historical texts and maps on the Treaty of Pähkinäsaari in 1323 were quite common in subsequent discussions on the genetic division of Finns between east and west and in creating a boundary between Swedish and Russian influence. This boundary also evinces striking correlations between Finnish dialects, medical reimbursements, as well as genetic differences (Palo 2020). Medical doctors such as Nevanlinna, however, began to link these forms of history to medical conditions and population genetics. Images of population structure and differences help to create distinctions that are easy to understand. The translation of similarity and difference to the visual realm, rather than offering only numerical figures, helps to convey the message of the researchers. Yet images and visualizations also help to conceal the meaning and methods necessary to create such images, thus producing a certain level of opacity.

Hence in this paper, we are not interested in analyzing the images per se, but in their function as representations of certain scientific settings (methodological, theoretical, etc.) that also contribute to different ways of seeing and understanding the Finnish population isolate.

Understanding Finns as a Population Isolate–How Was This Understanding Formed?

The understanding of the Finnish population as a population isolate emerged as a by-product of an effort to understand certain rare diseases appearing in Finnish families. The studies on what is known as Finnish Disease Heritage (FDH), a group of over thirty monogenetic, rare diseases that have been overrepresented in Finland, formed a basis on which wider understanding of Finnish population history could form (Norio 2000). FDH was first introduced in the Finnish medical literature in 1972 (Perheentupa 1972) and two years later in an English-language publication (Norio et al. 1974). For this endeavor, certain inclusion criteria were applied to delineate which diseases could be included under the umbrella term. The inclusion criteria for the Finnish population in this context included that the conditions had to be overrepresented in the Finnish population, and that a minimum of ten cases have been clinically diagnosed. The study of FDH produced several narratives of specific causes for these rare conditions, which drew on notions of migration and isolation. Figure 1 provides an example of one such map where early migration history is used to explain the isolation and early settlement history of Finns, which has resulted in the emergence of various FDH conditions.

Reijo Norio, a prominent Finnish pediatrician and perhaps the best-known expert in FDH has noted that the criteria for FDH are not particularly strict, in that some have been included in the list since they were not found in the literature or elsewhere in the world but have a relatively high prevalence in Finland. For Norio and his colleagues, the method for studying rare inherited diseases related specifically to studying family pedigrees and inheritance. In many, but by no means all, cases FDH diseases were the outcome of a relatively small population in which there were carriers of recessive genes. In addition, those carriers had to reproduce with another carrier. In this sense, many FDH conditions have been the outcome of a specific gene pool and general isolation of smaller populations within Finland (Norio, 2000: 25).

So-called superisolates or subisolates within Finland provide an example of how some FDH conditions have come about. The people who originated from founder populations of Kuusamo and Salla (Varilo et al. 2000), for instance, are known to be carriers of several recessive genes. One study by Varilo et al. (2000) compares Finnish and Kuusamo-area samples. Kuusamo and its inhabitation history, Varilo et al. (2000) argue, is an example of how long distances have resulted in population pockets where a very small founder population can be identified, and where inbreeding invariably followed. The Kuusamo area, for instance, was inhabited at the end of the seventeenth century and "39 families survived the great famine of 1695–1697" (Varilo et al. 2000, 605). Thus, "all inhabitants of Kuusamo born before 1939 have ancestors from these 39 families and show numerous genealogical connections" (Varilo et al. 2000, 605). Local historical conditions such as this have left a mark in the genetics of the population. As Peltonen has noted:

> One additional feature in the population history of Finland has greatly emphasized the concept of isolation. After the first founders the slow inhabitation process of this relatively large country, broken by hundreds of lakes, resulted in the formation of internal subisolates. Typically, some 20–40 families formed a settlement which could remain

isolated for centuries. Additionally, most eastern and northern parts of the country became inhabited relatively late, the main population expansion beginning as late as the 17th century when people from southern parts of the country moved towards northern and eastern regions. Population records such as taxation documents and church records are well preserved from the 17th century and reveal highly restricted immigration to many subregions of Finland as well as limited population expansion in these subisolates, predicted to result in the regional enrichment of some alleles and in the loss of others. (Peltonen 1997, 553–554)

Figure 1: "Figure 1 A" and "B" as published in Varilo et al. 2003, showing the internal settlement history of Kuusamo and Finland as well as the way the sampling was done based on the birthplaces of grandparents from both sides of the family as inclusion criteria.

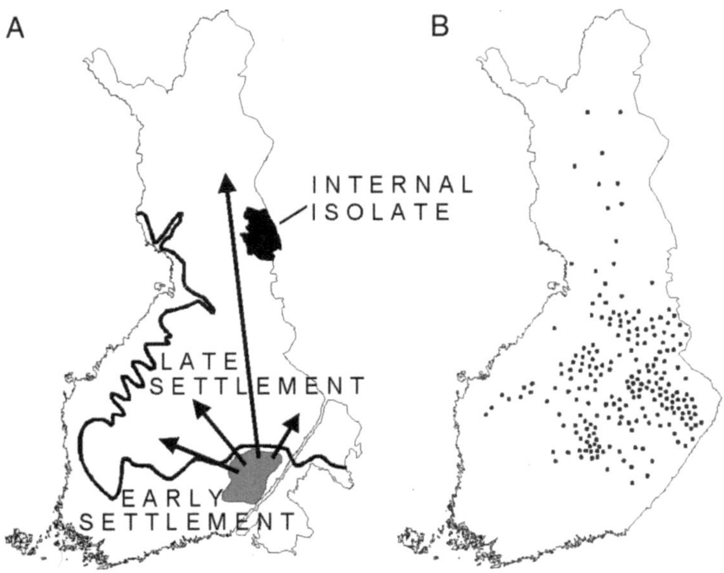

Figure 1 also shows how the different settlement histories of early and late settlements have been identified as key factors for population formation. These have been reproduced in several articles showing borders, areas, and arrows at the side of a map of birthplaces of grandparents specific to the late settlement that formed the essence of what the article identifies as the "Finnish population isolate." The frequent reproduction of these images in other sources has, moreover, constantly reinforced this understanding. Similar maps containing essentially the same information seen in figure 1, as taken from Varilo 2003, have been circulating since at least the 1970s (see Norio et al. 1973).

One result has been the emergence of an understanding of a genetically homogeneous population isolate of Finns. In his book *Suomi-neidon geenit: Tautiperinnön takana juurillemme johtamassa* (The genes of the Finnish fair maiden), Norio emphasizes both national isolation and regional isolation as causes for many of the inherited rare diseases listed in FDH (Norio 2000, 26). The work done on FDH in Finland served specific purposes in both consolidating rare disease research in Finland to provide critical mass when there were relatively few patients per condition. Norio additionally notes that being listed as an FDH condition also afforded publicity and visibility to conditions that needed support and funding from public authorities.

In addition to being referred to as a population isolate, in genetics the Finnish population has also been described as originating from a "founder population" (Lim et al. 2014; Wang et al. 2014; Järvelä et al. 2021). The specific term "founder population" refers to a specific group of people who usually traveled and settled to an isolated area, and whose isolation over several generations led to the enrichment and possible inbreeding of individuals who did not know they were related. As a concept it is more historically grounded and specific than isolated population as such. The same goes for the concept of a "bottlenecked" population (Minton 2023), which has been used to refer to specific events in Finnish history that have had an impact on the country's population structure and size. The notion of a founder population does not necessarily signify that a population has been isolated over the years, but rather that a small number of individuals originally inhabited a specific region.

Selecting Finns for Research

Early studies of rare diseases were always premised on following family pedigrees based on patients and families who had been identified as carriers of specific rare diseases. As mentioned above, the emergence of the concept of Finnish Disease Heritage came to represent a group of diseases that were thought to be overrepresented in the Finnish population. As Reijo Norio has noted, however, it could not be ruled out that such conditions did not exist elsewhere, since not all countries had healthcare systems that reported and studied rare conditions such as those found in Finland (see Tupasela 2016).

In contrast to these studies on rare diseases, other early studies into the Finnish population structure drew heavily on the idea of selecting individuals from rural areas. As H. R. Nevanlinna noted in an article published in 1972: "In order to eliminate the effect of different groups on the national level (and that of individuals not belonging to the original population in the communities), the sampling was made as a weighted one from the rural communities only" (Nevanlinna 1972, 199). The notion of an "original population" is not exactly clear in this context, in that Nevanlinna does not specify in his publication whether he is referring to a founder population or another type of criterion regarding how long someone has been living in a specific area. This selection criteria appears rather problematic given that, only several lines earlier in the same article, he observes: "The Swedish-speaking urban population seems to be of mixed origin" (Nevanlinna 1972, 198). Although the urban population represented a relatively small portion of the whole population, its diversity was significant in that it included Jews, Russians, and Roma. Despite the small sizes of these groups, arguments for who counts as a Finn have played an important role in subsequent selection criteria in Finnish studies. In addition to excluding Sami, early studies thus sought to clearly define which individuals could be considered to represent Finnishness at a genetic level. The exclusion criteria also included women, as well as all individuals who lived north of a diagonal line that went approximately from the northwest of Finland to the southeast (see figure 2). The criteria that were introduced in these studies have played an important role in stabilizing who is included in studies of population genetics in Finland.

Figure 2: Selected populations for study

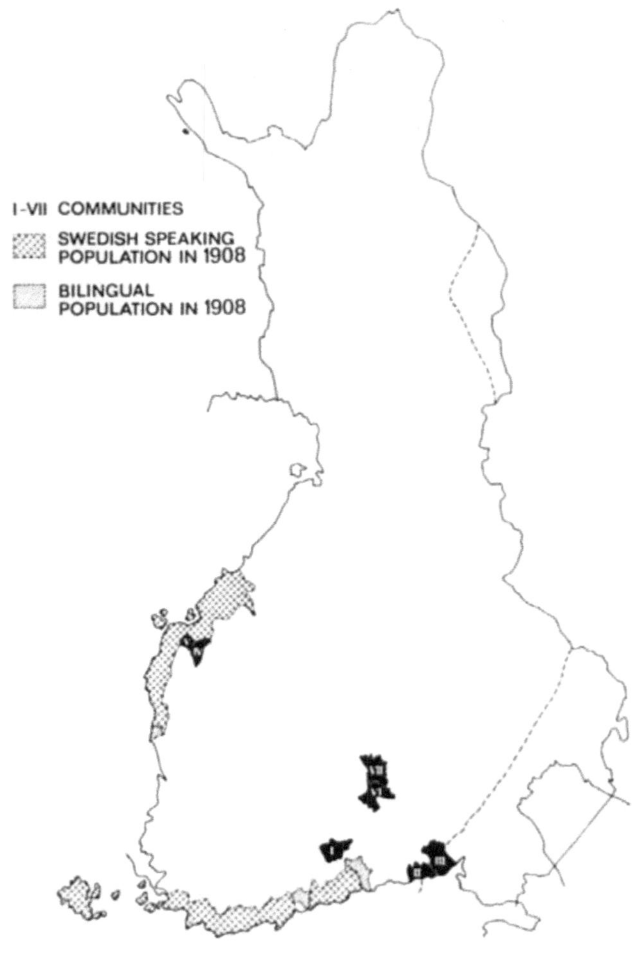

Source: Nevanlinna 1972, 201

Figure 2 provides an example from Nevanlinna's article. The type and form of mapping found in this piece was used quite prominently throughout the end of the last millennium and can be also found in subsequent articles and publications by such prominent doctors as Reijo Norio (2000) and Jaakko Perheentupa (1972). More contemporary use of population genetic images is based on methods such as principal component analysis and often also includes reference populations from other countries (see figure 3 for an example).

Certain images and the information they contain are consistently used in publications on the topic (e.g., figures 1 and 2), which reinforces the persuasiveness of the image and what it represents. But the reuse of images and narratives tends to detach the images from the original methodological and sampling-related discussions and framings. This can be seen, for instance, in several newspaper articles that have sought to popularize the study of Finnish genetics and the notion of Finns as "two people" ("kaksi kansaa") (see Kotkavirta 2019; Puttonen 2017).

These early selection criteria have contributed in part to the notion of Finland as an isolated population. One might go as far as argue that the early selection criteria served to solidify the notion of Finland as a homogeneous population. Interestingly, the early exclusion of the Finnish-Swedish population found along the western coastline of Finland would also provide the basis for later findings relating to genetic differences among the Finnish population.

Connecting Rare Diseases and Population Genetics

The historical narratives of Finland as a population isolate developed through the study of Finnish Disease Heritage were later connected with the broader field of population genetics and molecular genetics in the scientific literature in the mid-1990s. The work of Finnish geneticist Leena Peltonen has especially emphasized and drawn on the connection between the history and development of rare diseases in Finland and the more general population and its impact on national genetic structure and substructure (Helén et al. 2024). But the shift from the study of rare diseases, which focused on families and family pedigrees, to the study of more complex diseases required that connections needed to be made regarding the relevance of isolation to complex diseases and to the identification of genetic loci that played a role in their development. This "translation" from monogenic to complex on the one hand, and rare to common on the other, is an important bridge between the two periods and methods of studying disease in Finland (see Tarkkala 2019).

One such article establishes a connection between the major causes of rare diseases (isolation, bottleneck, and founder population) and "major loci" contributing to complex diseases. The piece focuses on the examples of Aspartylglucosaminuria (AGU) and infantile neuronal ceroid lipofuscinosis (INCL) to make this link. As it notes: "In both of these lysosomal accumulation diseases one major (point) mutation is carried by 98% of diseases alleles in Finland" (Peltonen et al. 1995, 699). The connection between rare disease research and the causes of common diseases is an important step within the Finnish context as it helps to reaffirm and accentuate the narrative of an isolated population. Despite the significant differences in relation to methodology and sampling between rare diseases and common/complex diseases, the story of isolation, bottlenecks, and founder population has played an important role within subsequent population studies in Finland. This linkage is especially significant as, during the past decade, Finnish researchers have had to "backpedal" somewhat on the claims of homogeneity and uniqueness to make Finnish samples and data of more relevant internationally (Tarkkala and Tupasela 2018).

After the mid- to late 1990s, the sequencing of the whole human genome sequence introduced a host of new techniques and methods for studying complex diseases in Finland. An important development during this time concerns the ways in which visual representation also began to develop to include new types of methods and techniques in analysis and representation. A significant feature of this shift was the introduction of comparison to other populations as a means of understanding Finnish uniqueness and difference.

One example of the representation of Finnish isolation in relation to other European populations can be found in a comparative study of European populations by Lao et al. (2008). As seen in figure 3, a comparison of twenty-three European subpopulations generated an image where Finns are presented as distinct from other European populations. Images such as this have played an important role in enforcing and emphasizing the unique and isolated nature of the Finnish population. The image in figure 3 is somewhat misleading in that similarity and difference are represented on two axes. On the y-axis, Finns are quite close to Swedes, Norwegians, Danes, and even the Polish population. On the x-axis, similarity is with Poles and Hungarians. The image is an important milestone in the study of populations, however, due to the way populations are represented visually and in relation to others.

Figure 3: SNP-Based PCA of 2,457 European individuals from twenty-three subpopulations

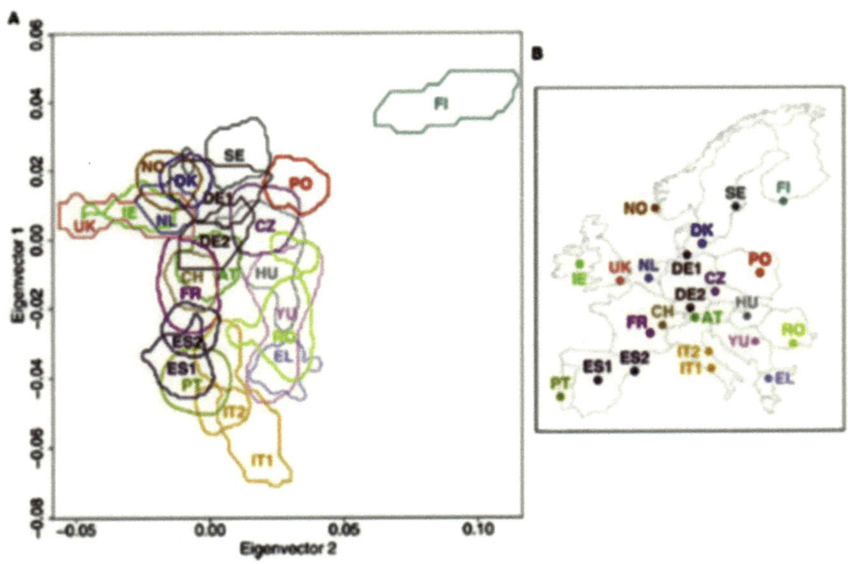

Source: Lao et al. 2008

Figure 4: A statistical summary of genetic data from 1,387 Europeans based on principal component axis one (PC1) and axis two (PC2).

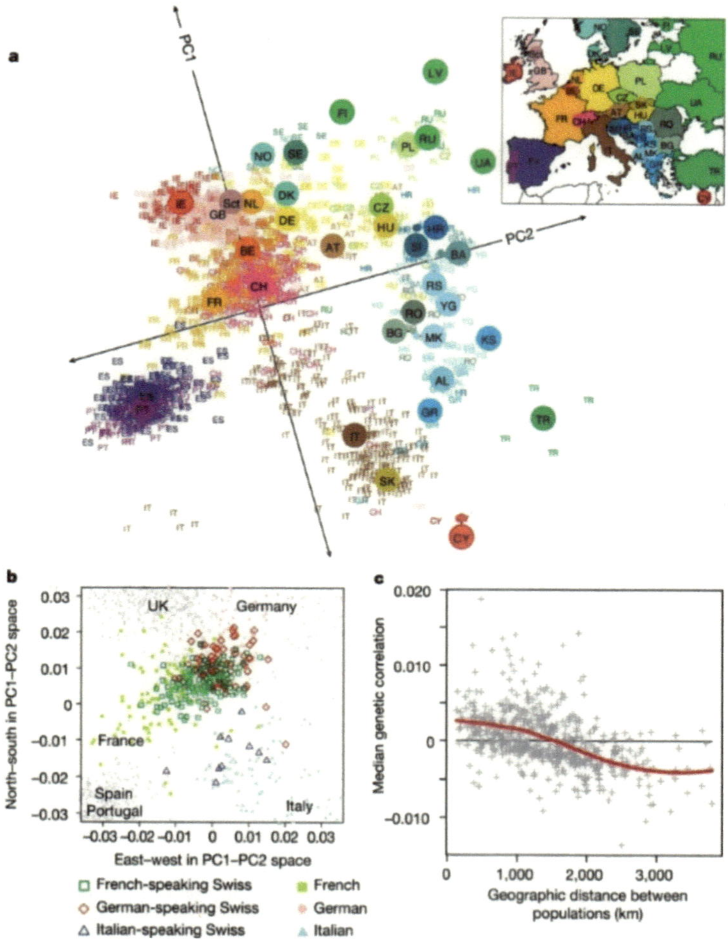

Source: Novembre et al. 2008

This image can be compared to another similar study published by Novembre et al. (2008), which includes more subpopulations. In this study and the accompanying image (figure 4), the Finnish population appears far less isolated with the inclusion of samples from Russia and Latvia. The way the visualization uses dots, rather than circles, also plays an important role in how visualization contributes to how we understand similarity and difference, and thus isolation as well. The role that reference populations have played in the representation of difference and similarity has been similarly influential in constructions of Finns as a population isolate. The inclusion and exclusion of specific reference populations impacts the relative distance that the chosen population representatives are from each other. Therefore, adding a new reference population may either increase or decrease relative distances from other populations, which will either emphasize or de-emphasize the visual representation of the isolation of specific populations. This impact can be seen in comparing the differences between figures 3 and 4. As more populations are added to comparisons, the relative genetic differences between each population becomes less and less distinct.

In genomics and medicine, understanding a certain population as distinct, and thus as different from another population, can play a significant role in perceptions and practice. Each population can thereby be seen to carry exceptional characteristics, which might prove to be important for certain scientific findings, methods, and approaches, as happened in the case of Finnish Disease Heritage. Certain methods, together with certain samplings of Finns, produced a corpus of knowledge of these monogenetic, rare diseases: linkage disequilibrium studies, for instance, were conducted for this purpose (Hästbacka et al. 1992; Varilo et al. 2000).

Each population gains its specific, exclusive characteristics from the comparison to other populations, seen to embody something different. This logic is, at its core, simple. However, when 99.9 percent of human DNA is shared, the effort to draw lines between groups of people is a significant task requiring careful sampling and scientific precision. In medicine, to repeat what was noted above, these differences might be significant, and scientists continuously fine-tune their approaches to be able to detect even rare variants (Martin et al. 2017). Yet because many studies have been done in European populations in the past, there has been a call to diversify the field, as the "lack of ethnic diversity in human genomic studies means that our ability to translate genetic research into clinical practice or public health policy may be dangerously incomplete, or worse, mistaken" (Sirugo et al. 2019, 26). Nevertheless, the question of comparison and of the reference point against which a certain group is compared is always meaningful for settings of genetic research. Potential "noise" in data often needs to be controlled, and in a homogeneous population there is less noise to start with due to high level of similarity—a fact that Finnish biomedical research has leveraged. As we have shown in our previous studies (Tarkkala and Tupasela 2018; Tarkkala 2019), exaggerating the uniqueness of the Finnish population may produce a conclusion that the Finnish population does not provide the necessary means for comparison if it is considered as mere outlier population in which results based on analyses of other populations cannot be reproduced or replicated. However, in European comparisons Finns often seem to stand out (Novembre et al. 2008). The emphasis of Finns as unique and isolated has therefore required a kind of balancing act on the part of Finnish researchers.

The inclusion of Finns as "European" is important for Finnish biomedicine. There has been a concern about whether the Finnish samples could be compared with other samples—that is to say, whether they have been handled and stored in a way that does not add variation in the analysis, and whether findings in the Finnish population can be replicated in others (Tarkkala 2019). European samples have been used in biomedicine to a great extent and have been considered useful; hence it is important to include the following:

> European GWA studies have produced many successes that can be replicated in different sets of individuals from the same European population as that in which the association was originally detected; and associations in one population of European descent

are often replicable in other European populations, sometimes in groups that are quite geographically distant within the European continent. (Rosenberg et al. 2010, 358)

It is worth asking whether the perceived uniqueness of the Finnish population as an isolated population would remain as strong if samples from a broader pool of individuals living in the geographical area of the country had been included from the beginning. The role of visualizations and relevance to European GWA studies cannot be overstated, however, as Finnish genetics has sought to find a place in the global market for genomics research using the strengths that it has developed over decades.

One, Two or More Than Two Populations

Even though Finns have been widely represented as a homogeneous population isolate, the population in Finland is in fact also heterogenous, as several publications have documented (Salmela et al. 2008; Huyghe et al. 2011; Kerminen et al. 2017; Virtaranta-Knowles et al. 1991). Discussions centered on homogeneity and heterogeneity are crucially related to both the reference populations used for comparison and the scale at which population structure is analyzed. And yet these discussions are also related to how isolation is understood and represented. Within the context of what constitutes a population isolate, there is no commonly agreed upon definition or standard against which isolation and homogeneity are compared. Therefore, there are multiple and sometimes even different definitions of what constitutes homogeneity within a population. And since there is no agreed upon definition of what scale of analysis is meaningful for claims of homogeneity or heterogeneity, the term ends up being somewhat fluid and fuzzy. As Kerminen and colleagues note:

Our haplotype-based population assignment using a data set that evenly covers a major part of Finland therefore provides unprecedented information on the fine-scale genetic structure of the Finnish population. In general, the fine-scale structure that we detected is highly geographically clustered with little overlap between the populations. (Kerminen et al. 2017)

This observation points to the way in which scale can play an important role in how a larger population isolate, such as Finns, can further be subdivided into smaller distinct populations depending on the scale of analysis.

Interestingly, the division between people originating from western Finland and people originating from eastern Finland is clear-cut and has even been said to point to two distinct populations. This internal heterogeneity is important to the notion of Finns as a population isolate, as when the heterogeneity is accounted for, many Finns have much closer ties to other populations, such as Swedes, than previously claimed. The dichotomy is also something that Nevanlinna noted in his early studies and recognized as contributing to difference within Finland, and thus as impacting selection criteria for early studies. Another example of how internal heterogeneity can be generated comes from an article by Salmela et al. (2008), which sought to study the population structure among western Finns (most of whom are considered Swedish-speaking). To understand the differences among this population, the authors compared samples from western Finns with those of Swedes. They note:

> Interestingly, in the MDS plots the Finnish-Swedes stood out from the rest of Western Finland *only when* Sweden was included in the analysis, which highlights the importance of ***relevant reference populations*** also when detecting patterns of variation within a country. (Salmela et al. 2008) (emphasis added)

The generation of such differences within the Finnish population, and all populations for that matter, raises two important questions. *First*, what reference populations are relevant in studying the population structure of any given group of people? *Second*, when differences are created through comparison, which differences can we consider to be significant and on what basis is that significance determined? How, for example, might one compare the significance of one so-called isolated population to another? What criteria of quality or quantity signify an important difference or benefit that one population may have compared to another?

A further issue of concern in relation to definitions of isolation is the criteria that are relevant for defining isolation and differentiation from other populations. As the article by Kerminen et al. (2017) shows, the scale of analysis can have significant impact on delineating what constitutes the borders and composition of a specific population cluster. Which individuals are included or excluded from the group and why?

The discussion of similarity and difference within Finland has further implications for international comparisons and studies of populations. What is the role of difference and similarity in these studies, and what significance does it have? What relevance does it have, for example, in identifying disease-causing genes? Furthermore, to what extent should studies include more diverse populations and what impact might this have on generation of noise and the need for larger samples sizes? Or to what extent does the introduction of strict selection criteria mask and hide the identification of other disease-causing genes that may have less relevance for a small population like Finland, and more relevance for larger populations?

Discussion: Changing Representations of a Population Isolate

In this chapter, we have traced the origin and changes of the concept of a population isolate in Finland. We focused first on the late 1960s and early 1970s, when the term was first associated with the idea of Finnish Disease Heritage. Over time, especially after the turn of the millennium, the concept became increasingly linked to molecular genetics and population genetics more broadly. This historical arc demonstrates that the concept has not been scientifically stable. Instead, it has been quite flexible, proving adaptable to various perspectives, methodological approaches, and selection criteria when determining which people to study in Finland. Within the Finnish context, some aspects of isolation have been constructed both methodologically and conceptually.

We have shown how these various concepts of population isolation have been popularized and represented visually, in certain images that have been widely reproduced and circulated in both academia and popular literature. These images are often intended to persuade readers to accept certain claims without fully understanding the underlying methodological and conceptual choices. However, visual representations can also be misleading and confusing, as they may obscure the criteria used to select individuals deemed representative of a given population.

The combination of historical perspective and more recent forms of visualization serves as a powerful tool of persuasion. In the Finnish context, this persuasion has been crucial in the construction of knowledge (Bredekamp et al. 2015), presenting a unified narrative despite the availability of alternative possibilities and explanations (Översti et al. 2019; Översti et al. 2017). As several scholars have noted (Burri and Dumit 2008; Lipphardt and Sommer 2015), images play a significant role in this process by condensing information and simplifying complex issues (Lynch and Woolgar 1990).

Additionally, we have extended the understanding of images of populations isolates by highlighting the role of social practices and interests that underpin the generation of these images. In Finland, different notions of population isolation have often been tied to the criteria and focus of analytical methods and selection criteria. This is evident in early studies by Nevanlinna (1972), as well as more recent research by others who examine the genetic characteristics of Finland's population structure in studies on Y chromosomes (Palo et al. 2009). Our study does not argue that isolation has never been a feature of people living in different parts of Finland. Instead, it suggest that the concept of isolation and its connection to genetic analysis is heavily influenced by social factors, including researchers' decisions on selection methods and which parts of the genome to explore for similarity or difference.

Since population isolates have historically been seen as groups that have remained 'unmixed' for generations, we suggest that the criterion for this determination is highly fluid and susceptible to interpretation. As Lipphardt (2012) has argued isolation or 'unmixing' has usually been based on other factors than genetics first. Therefore, issues related to sampling selection, scale as well as reference populations are not brought forth or raised as possible concerns when images are circulated. Thus, the images seek to convince individuals of a specific truth related to notions of isolation. As said in the beginning, our intention is not to question the validity of genetic drift, founder population, bottleneck effect, or isolated homogeneous population as concepts applicable to Finland or elsewhere. Rather, we have sought to contextualize these notions within a broader perspective on populations and in relation to visual representations that have circulated for decades. By doing so, we demonstrate that the notion of an isolated population is not uniform but has some flexibility, as highlighted by the examples given in this chapter.

References

Arcos-Burgos, Mauricio, and Maximilian Muenke. 2002. "Genetics of Population Isolates." *Clinical Genetics* 61 (4): 233–247.

Bredekamp, Horst, Vera Dünkel, and Birgit Schneider B. 2015. "Introduction: The Image—A Cultural Technology: A Research Program for a Critical Analysis of Images." In *The Technical Image: A History of Styles in Scientific Imagery*, edited by Horst Bredekamp, Vera Dünkel, and Birgit Schneider, 1–7. Chicago, IL: The University of Chicago Press.

Burri, Regula Valérie. 2012. "Visual Rationalities: Towards a Sociology of Images." *Current Sociology* 60 (1): 45–60.

Burri, Regula Valérie, and Joseph Dumit. 2008. "Social Studies of Scientific Imaging and Visualization." In *The Handbook of Science and Technology Studies*, edited by Edward J. Hackett, Olga Amsterdamska, Michael E. Lynch, and Judy Wajcman, 297–317. Cambridge, MA: MIT Press.

Coopmans, Catelijne et al, eds. 2014. *Representation in Scientific Practice Revisited.* Cambridge, MA: MIT Press.

Gross, Alan G., and Joseph E. Harmon. 2014. *Science from Sight to Insight: How Scientists Illustrate Meaning.* Chicago, IL: University of Chicago Press.

Hatzikotoulas, Konstantinos, Arthur Gilly, and Eleftheria Zeggini. 2014 "Using Population Isolates in Genetic Association Studies." *Briefings in Functional Genomics* 13 (5): 371–377. DOI: https://doi.org/10.1093/bfgp/elu022.

Helén, Ilpo, Karoliina Snell, Heta Tarkkala, and Aaro Tupasela. 2024. *Genome Finland: From Rare Diseases to Data Economy*. Helsinki: Helsinki University Press. https://doi.org/10.33134/HUP-24

Hästbacka, Johanna, Albert de la Chapelle, Ilkka Kaitila, et al. 1992. "Linkage Disequilibrium Mapping in Isolated Founder Populations: Diastrophic Dysplasia in Finland." *Nature Genetics* 2 (3): 204–211.

Huyghe, Jeroen R. et al. 2011. "A Genome-Wide Analysis of Population Structure in the Finnish Saami with Implications for Genetic Association Studies." *European Journal of Human Genetics* 19 (3): 347–352.

Järvelä, Irma et al. 2021. "Exome Sequencing Reveals Predominantly de Novo Variants in Disorders with Intellectual Disability (ID) in the Founder Population of Finland." *Human Genetics* 140 (7): 1011–1029. DOI: https://doi.org/10.1007/s00439-021-02268-1.

Kere, Juha. 2001. "Human Population Genetics: Lessons from Finland." *Annual Reviews of Genomics and Human Genetics* 2 (1): 103–128.

Kerminen, Sini et al. 2017. "Fine-Scale Genetic Structure in Finland." *G3: Genes, Genomes, Genetics* 7 (10): 3459–3468.

Kotkavirta, Johannes. 2019. "Yksi kuva paljastaa hätkähdyttävällä tavalla, kuinka Suomessa elää kuin kaksi eri kansaa." Ilta Sanomat, 9 October 2019. Accessed June 27, 2024. https://www.is.fi/kotimaa/art-2000006266502.html.

Kurki, Mitja I. et al. 2023 "FinnGen Provides Genetic Insights from a Well-Phenotyped Isolated Population." *Nature* 613 (7944): 508–518. DOI: https://doi.org/10.1038/s41586-022-05473-8.

Kääriäinen, Helena et al. 2017. "Genetics in an Isolated Population Like Finland: A Different Basis for Genomic Medicine?" *Journal of Community Genetics* 8 (4): 319–326.

Lao, Oscar et al. 2008. "Correlation between genetic and geographic structure in Europe." *Current Biology* 18 (16): 1241–1248.

Lim, Elaine T. et al. 2014. "Distribution and Medical Impact of Loss-of-Function Variants in the Finnish Founder Population." *PLOS Genetics* 10 (7): e1004494.

Lipphardt, Veronika. 2010. "The Jewish Community of Rome: An Isolated Population? Sampling Procedures and Bio-Historical Narratives in Genetic Analysis in the 1950s." *BioSocieties* 5 (3): 306–329. DOI: https://doi.org/10.1057/biosoc.2010.16.

Lipphardt, Veronika. 2012. "Isolates and Crosses in Human Population Genetics; or, A Contextualization of German Race Science." *Current Anthropology* 53 (S5): S69–S82.

Lipphardt, Veronika. 2013. "From 'Races' to 'Isolates' and 'Endogamous Communities': Human Genetics and the Notion of Human Diversity in the 1950s." In *Human Heredity in the Twentieth Century*, edited by Bernd Gausemeier, Staffan Müller-Wille, and Edmund Ramsden, 55–68. London: Pickering & Chatto.

Lipphardt, Veronika, and Marianna Sommer. 2015. "Visibility Matters: Diagrammatic Renderings of Human Evolution and Diversity in Physical, Serological and Molecular Anthropology." *History of the Human Sciences* 28 (5): 3–16.

Lynch, Michael E., and Steve Woolgar, eds. 1990. *Representation in Scientific Practice.* Cambridge, MA: MIT Press.

Martin, Alicia R. et al. 2017. "Human Demographic History Impacts Genetic Risk Prediction across Diverse Populations." *The American Journal of Human Genetics* 100 (4): 635–649.

Minton, Kirsty. 2023. "The FinnGen Study: Disease Insights from a 'Bottlenecked' Population." *Nature Reviews Genetics* 24 (4): 207. DOI: https://doi.org/10.1038/s41576-023-0 0584-y.

Nevanlinna, Harri R. 1972. "The Finnish Population Structure: A Genetic and Genealogical Study." *Hereditas* 71 (2): 195–236.

Norio, Reijo. 2000. *Suomi-neidon geenit: Tautiperinnön takana juurillemme johtamassa*. Keuruu: Otava.

Norio, Reijo, Harri R. Nevanlinna, and Jaakko Perheentupa. 1973. "Hereditary Diseases in Finland; Rare Flora in Rare Soil." *Annals of Clinical Research* 5 (3): 109–141.

Novembre, John, Toby Johnson, Katarzyna Bryc, et al. 2008. "Genes Mirror Geography within Europe." *Nature* 456 (7218): 98–101. DOI: https://doi.org/10.1038/nature07331.

Översti, Sanni et al. 2019. "Human Mitochondrial DNA Lineages in Iron-Age Fennoscandia Suggest Incipient Admixture and Eastern Introduction of Farming-Related Maternal Ancestry." *Scientific Reports* 9 (1): 16883.

Översti, Sanni, et al. 2017. "Identification and Analysis of mtDNA Genomes Attributed to Finns Reveal Long-Stagnant Demographic Trends Obscured in the Total Diversity." *Scientific Reports* 7 (1): 6193.

Palo, Jukka U. 2020. "Mikä Pähkinäsaaren rauhan raja oikeasti oli?" *Duodecim* 136 (24): 2796–27800.

Palo, Jukka U., Ismo Ulmanen, Matti Lukka, Pekka Ellonen and Antti Sajantila. 2009. "Genetic Markers and Population History: Finland Revisited." *European Journal of Human Genetics* 17 (10): 1336–1346.

Pálsson, Gísli. 2007. *Anthropology and the New Genetics*. Cambridge: Cambridge University Press.

Peltonen, Leena. 1997. "Molecular Background of the Finnish Disease Heritage". *Annals of Medicine*, 29(6): 553–556. https://doi.org/10.3109/07853899709007481

Peltonen, Leena, Petra Pekkarinen, and Johanna Aaltonen. 1995. "Messages from an Isolate: Lessons from the Finnish Gene Pool." *Biological Chemistry Hoppe-Seyler* 376 (12): 697–704.

Perheentupa, Jaakko. 1972. "Suomalainen tautiperintö kliinikon ja tutkijan kannalta." *Duodecim* 88 (1): 1–3.

Puttonen, Mikko. 2017. "Suomalaiset jakaantuvat kymmeniin geneettisiin alaryhmiin: Perimän erot noudattelevat murrerajoja, kertoo tutkimus." Helsingin Sanomat, 6 October 2017. Accessed June 27. 2024. https://www.hs.fi/tiede/art-2000005398781.html.

Rosenberg, Noah A., Lucy Huang, Ethan M Jewett, Zachary A Szpiech, Ivana Jankovic and Michael Boehnke. 2010. "Genome-Wide Association Studies in Diverse Populations." *Nature Reviews Genetics* 11 (5): 356–366. DOI: https://doi.org/10.1038/nrg2760.

Salmela, Elina, Tuuli Lappalainen, Ingegerd Fransson, et al. 2008. "Genome-wide analysis of single nucleotide polymorphisms uncovers population structure in Northern Europe." *PloS one* 3.10: e3519.

Sirugo, Giorgio, Scott M. Williams, and Sarah A. Tishkoff. 2019. "The Missing Diversity in Human Genetic Studies." *Cell* 177 (1): 26–31.

Sommer, Marianne. 2015. "Population-Genetic Trees, Maps, and Narratives of the Great Human Diasporas." *History of the Human Sciences* 28 (5): 108–145.

Tarkkala, Heta. 2019. *Reorganizing Biomedical Research: Biobanks as Conditions of Possibility for Personalized Medicine*. Helsinki: University of Helsinki. http://urn.fi/URN:ISBN:978-951-51-3384-7.

Tarkkala, Heta, and Aaro Tupasela. 2018. "Shortcut to Success? Negotiating Genetic Uniqueness in Global Biomedicine." *Social Studies of Science* 48 (5): 740–761.

Tupasela, Aaro. 2022. "The Genetic Imagination: Imaging Populations and the Construction of Nationhood." In *Finnishness, Whiteness and Coloniality*, edited by Josephine Hoegaerts, Tuire Liimatainen, Laura Hekanaho, and Elizabeth Peterson, 19–40. Helsinki: Helsinki University Press.

Tupasela, Aaro. 2017. "Populations as brands in medical research: placing genes on the global genetic atlas." *BioSocieties* 12 (1): 47–65.

Tupasela, Aaro. 2016. "Genetic romanticism—Constructing the corpus in Finnish folklore and rare diseases." *Configurations* 24 (2): 121–143.

Tupasela, Aaro, and Sakari Tamminen. 2015. "Authentic, Original, and Valuable: Stabilizing the Genetic Identity in Non-human and Human Populations in Finland." *Studies in Ethnicity and Nationalism* 15 (3): 411–431.

Uusimaa, Johanna, Johannes Kettunen, Teppo Varilo, et al. 2022 "The Finnish Genetic Heritage in 2022: From Diagnosis to Translational Research." *Disease Models & Mechanisms* 15 (10): dmm049490. DOI: https://doi.org/10.1242/dmm.049490.

Varilo, Teppo, Maris Laan, Iiris Hovatta, Victor Wiebe, Joseph D. Terwilliger, and Leena Peltonen. 2000. "Linkage Disequilibrium in Isolated Populations: Finland and a Young Sub-population of Kuusamo." *European Journal of Human Genetics* 8 (8): 604–612.

Varilo, Teppo, Tiina Pauino, Alex Parker, et al. 2003. "The Interval of Linkage Disequilibrium (LD) Detected with Microsatellite and SNP Markers in Chromosomes of Finnish Populations with Different Histories." *Human Molecular Genetics* 12 (1): 51–59.

Wang, Sophie R., Vineeta Agarwala, Jason Flannick, et al. 2014. "Simulation of Finnish Population History, Guided by Empirical Genetic Data, to Assess Power of Rare-Variant Tests in Finland." *The American Journal of Human Genetics* 94 (5): 710–720.

Virtaranta-Knowles, Kirsti, Pertti Sistonen, and Harri R. Nevanlinna. 1991. "A Population Genetic Study in Finland: Comparison of the Finnish- and Swedish-Speaking Populations." *Human Heredity* 41 (4): 248–264.

GenomeIndia

The Epistemologies and Politics of Stitching together Genetic Variation in Indians

Mahendra Shahare

Abstract *Since colonial times, attempts at racial classification and naming the diversity and potpourri of India's population have remained controversial. However, the postgenomic era has created an impetus within the life sciences to study human biological diversity. The ongoing GenomeIndia project (GIP) is a significant development in this direction. The GIP aims to create a catalog of genetic variations in Indians and construct a reference haplotype structure for Indians. But what epistemological assumptions and concepts inform the GIP? And how do such epistemic conceptualizations influence the discourse on identity politics? By focusing on empirical research practices, I highlight here how the GIP relies on taken for granted epistemologies for characterizing human differences. I posit that while population genetics and medical genomics continue to be guided by the normative logic of genetic nationalism, in the case of India, the logic remains closely tied to the diffuse politics of identity and caste. As such, the GIP's notions of a reference haplotype or catalog of genetic diversity cannot fully dissociate science epistemologies from politics.*

Introduction

In January 2020, the Government of India initiated an ambitious mission-mode project, "GenomeIndia: Cataloguing the Genetic Variation in Indians," with a substantial grant of ₹238 crores (approximately USD 30 million). The GenomeIndia Project (GIP), an ongoing collaborative effort between twenty academic and research institutes[1] in India, aims to identify genetic variation in the Indian population through whole genome sequencing of 10,000 representative individuals, with a vision to facilitate the development of precision

1 IIIT Allahabad (Uttar Pradesh), IBSD Imphal (Manipur), MZU Aizawl (Mizoram), NIBMG Kalyani (West Bengal), ILS Bhubaneswar (Odisha), CSIR-CCMB Hyderabad (Telangana), CDED Hyderabad (Telangana), IIT Madras (Tamil Nadu), RGCB Thiru-vananthapuram (Kerala), CBR Bengaluru (Karnataka), IISc Bengaluru (Karnataka), NIMHANS Bengaluru (Karnataka), NCBS Bengaluru (Karnataka), IISER Pune (Maha-rashtra), GBRC Gandhinagar (Gujrat), AIIMS Jodhpur (Rajasthan), IIT Jodhpur (Raja-sthan), SKIMS Srinagar (Jammu and Kashmir), IGIB Delhi (Delhi), IIT Delhi (Delhi).

healthcare and major diseases diagnostics at affordable costs (PIB 2020a). Given the scale and ambition of the GIP and potential long-term consequences on India's sociopolitical spheres, including medicine and law, as well as group, ethnic, and individual identity, it becomes pertinent to critically analyze definitions, concepts, and assumptions employed by the project in framing and understanding human biological diversity.

In engaging with scientific discourse on classifying human diversity in the life sciences, this volume opens up some key questions, including: How do assumptions about the preexistence of human races shape scientific practices? In what ways do such epistemic conceptualizations influence the discourse on identity politics? What are the differences and continuities between historical and present population naming practices in the life sciences? These concerns also resonate with the GIP. The GIP aims to create a catalog of genetic variations in Indians and construct a reference haplotype structure for Indians. But what epistemological assumptions and concepts inform the GIP? Does the cataloging of genetic variation reproduce problematic social constructs, and if so, how? How is diffuse politics around caste and identity implicated in the practices of GIP? And can the notion of a reference haplotype dissociate science epistemologies from politics? In this chapter, I analyze these questions by examining and situating the GIP in the context of scientific discourse around medical genomics and population genetics in India.

Contemporary social sciences scholarship that closely engages with genomic science offers three relevant strands for problematizing the GIP. First, notwithstanding the GIP's well-rehearsed future imaginaries that have been in circulation since the Human Genome Project (HGP), which promised to pinpoint linkages between common human phenotypic traits with common genomic differences, genomic science is facing a conceptual and epistemic quandary. As Reardon (2017, 2) notes, medical genomics is caught up in the problem of meaning, wherein biologists have access to a large number of complete human genomes but are grappling with the question—what does it mean? While at the time of writing, the GIP had completed its target of sequencing 10,000 human genomes, how this massive information will produce meaningful knowledge for "transformative" medicine remains a question (PIB 2024). Second, the GIP is a "national project" that aims to "create a reference haplotype structure for Indians" (CBR 2023). The project therefore can be viewed as a move to assert "genomic sovereignty" through practices of bionationalism that seek to control shared genomes bounded by nation-state borders (Subramaniam 2019, 165). Third and most importantly, the GIP aims to understand India's genetic makeup, putatively constituted by more than 4,600 population groups inhabiting the country. As various scholars have illustrated, within India the discourse on genetic diversity is imbued with questions of origins and ancestry, deeply entangled with complex caste and tribe identities, and is a site of contestation for the politics of "Hindu nationalism" (Sur and Sur 2008, 210). In this chapter, I broadly use these strands to question the epistemic assumptions underlying the GIP. I further show how these assumptions uncritically reproduce the diffuse politics in India around caste and identity. In this article, I draw on my ongoing ethnographic research and documentary analysis of published resources related to the GIP available in the public domain. In particular, I draw on thematically coded semi-structured interviews of three principal investigators working on the GIP, as well as discussions with researchers and stakeholders at a rare diseases conference held in November 2023 in Bengaluru, India.

The chapter is organized as follows. In section two, I provide a brief sketch of the role of science in constituting race discourse in India and its interpolation with the social categories of caste. In section three, I situate the GIP in relation to similar research efforts in the last two decades. Focusing on the idea of cataloging differences, section four details how the GIP is reproducing the problematic population classification of the past and its imbued caste and identity politics. In the fifth section, I analyze the work of stitching together Indian reference haplotype as a way to nationalize ancestry. In the concluding section, I discuss the entanglements between population genetics and medical genomics and argue that the GIP cannot fully dissociate science epistemologies from politics.

Science, Race, and Caste Discourse

Human biological diversity, when principally understood in terms of ancestral collectives, enables a biological concept of human differences. How the life sciences frame and conceptualize human variation shapes and constrains knowledge production processes. Over the last two decades, with technological capacity for DNA sequencing of entire genomes, genomics has become the key mode of knowledge production for anthropological, biomedical, and genealogical studies of human variation. Nonetheless, genomic science operates on a highly politicized epistemic terrain, where the problematic legacy of racial and eugenic theories emanating from the eighteenth- and nineteenth-century paradigms of anthropology, ethnology, and human genetics continues to linger. Through empirically rich case studies, several scholars have argued that research practices in population genetics, which studies human variation, continue to use concepts, classification, and categorization employed by erstwhile race science (Reardon 2005; Burton 2021; Lipphardt 2014). Since the GIP intends to construct a catalog of genetic variations in Indians, it is important to probe and understand the historical ruptures and continuities that inform the conceptualization of human diversity mobilized by the project. As I further elaborate in the next section, the GIP uses caste as its proper unit of analysis. Therefore, it is crucial to contextualize how colonial practices of ordering Indian populations were informed by various contemporary scientific theories and paradigms and how the social construct of caste anchored and mediated these explorations.

Since colonial times, attempts at racial classification and naming the diversity and potpourri of India's population have remained controversial. But long before the colonial push for dividing and labeling natives into ancestral collectives, India had its own preexisting sociocultural system of stratifying populations, i.e., the caste system. The religiously sanctioned Hindu varna system ranks people into four groups—Brahmins (priests), Kshatriyas (warriors), Vaishyas (traders), and Sudras (peasants and artisans). While tribal people are not part of the caste society, the Dalits, formerly known as "untouchables," are outcastes. In general, each varna acts as a broader caste category that comprises multiple subcastes and accords graded social and economic privileges, with Brahmins enjoying the greatest freedoms and Sudras placed at the lower rung of the four groups. The tribal and Dalit populations constitute the most oppressed lot of the society. These historical discriminatory practices are at the heart of India's caste politics. Notwithstanding regional variation and complexity, the social construct of the *jati*/caste system has firmly entrenched a hereditary, hierarchical, and rank-based way of categorizing individuals and communities. The *jatis* commonly signify occupational groups and guilds, comprise myriad subcastes, are governed by complex kinship norms, and generally observe endogamy. Therefore, predictably, questions regarding the genesis, proliferation, and stability of the institution of caste continue to occupy a central place in scholarship attempting to understand the social stratification of Indian society.

The riddle of caste became more complex and pronounced as scholars in the colonial period sought to index and map caste onto race. Science perforce played a pivotal role in efforts to determine racial affinities and label populations in the colonial period. The "science of man" or anthropology, aided by the production of anthropometric data, was the governing paradigm for colonizers to accumulate knowledge about the natives (Philip 2004, 107). These attempts at racial classification were systematized early on through the incorporation of the category of race in the Census of India (Bhagat 2006). In the late 1880s, British colonial administrators initiated ethnographic surveys of the people of India to measure human differences using anthropometric parameters, viz. the cephalic index or nasal index (Bates 1995). Herbert Risley was the key figure of this colonial project. Using anthropometric measurements, he classified Indian populations into seven racial types, namely—Aryo-Dravidian, Dravidian, Indo-Aryan, Mongolo-Dravidian, Mongoloid, Scytho-Dravidian and the Turko-Iranian (Malhotra and Vasulu 2019). Furthermore, the classification of India's tribal populations into distinct racial categories such as the Negrito, the Proto-Australoid, and the Mongoloid was a product of colonial efforts (ibid.). These theories held that caste status and ranking are associated with racial differences.

In parallel to the comparative ethnology work in the nineteenth century, the work of philologists demonstrated close linguistic linkages between Sanskrit and other European languages. Max Müller, an orientalist philologist, argued that linguistic affinity indicates racial kinship and proposed his two-race theory that classified Indian populations into the Hamites (Aboriginals) and the Japhites (immigrant Aryans) (Sur and Sur 2008). Müller's work contributed to the discourse on Aryan and Dravidian racial identity. The two-race theory had a profound impact on India's academic and political spheres and even today acts as a meta category. For example, Govind Ghurye, a pioneering native scholar of Indian sociology, accepted and advanced the racial categorization of the Indian populations. These debates gave rise to what is referred to as the Aryan invasion/migration theory and created its own long protracted identity politics (see Barbosa in this volume). Nevertheless, this discourse directly or indirectly utilized caste as a marker.

During the interwar period, as Projit Mukharji illustrates in his cogent historical account of the development of sero-anthropology, scientists took up the notion of blood groups as a more objective basis to explain human differences. Through the works of Hirszfeld, Macfarlane, and other scientists, Mukharji notes, "from being a metonym for all of India in 1918," the blood group B became a metonym "for the lower-castes in the mid-1930s" (Mukharji 2014, 165). The transnational serological discussion about race therefore got translated into a debate about caste and its origin. Although some of the foremost scholars, including Dhirendranath Majumdar, Iravati Karve, and L. D. Sanghavi, combined anthropometric and serological approaches to study Indian populations, by the late 1950s the field witnessed a decline (ibid.). Though a small number of studies on blood group systems (viz. ABO, MN, Rh) from India were reported till the 1980s (Papiha 1996), these academic debates remained subdued.

Although the notion of race is now understood as a biologically dubious concept, it acquired a new meaning in the nineteenth century as a tool and method for colonial domination and was an active product of "scientific racism." Nonetheless, after independence, India ended the practice of racial classification, and the Decadal Census of India (1951 onwards) does not enumerate or recognize any racial groups in the country. Notwithstanding various social movements between the 1960s and late 1980s, by and large, the scientific discourse in India around human variation, race, caste, and origins remained muted. At the turn of the millennium, a study titled "Genetic Evidence on the Origins of Indian Caste Populations" (Bamshad et al. 2001) became the moment that revived the race and caste discourse in India, bringing genomic science into conversation with the diffuse politics of identity around Hindu nationalism and caste politics. The authors emphasized that their data analysis suggests "upper castes being more similar to Europeans, whereas lower castes are more similar to Asians" (ibid.). These findings revived the "two-race theory of Aryan migration from Central Asia into India" (Sur and Sur 2008, 206), ascribed linkages between race and caste, and threw open to debate "the basis for the majoritarian, communal identity as forged by Hindu nationalism" (Sur and Sur 2008, 215). Furthermore, citing these findings, when a few Dalit organizations demanded that caste oppression be discussed at the United Nations conference against racism at Durban, the Indian government disapproved the move and sought to control the entanglement of the social and the biological through bionationalism (Subramaniam 2019, 155). Debates on ethnicity, ancestry, and indigeneity of the diverse Indian population (Joseph 2021), which were put on ice after the emergence of democratic India in 1947, were thus reopened by advances in population genetics that promised to answer the question—Who are Indians? These developments came with claims of better scientific evidentiality afforded by the DNA and genomic sciences, and placed sciences back at the core of the politically vexed question of origins (Figueira 2015). As a response to the political anxieties stirred up by entanglements of legacy racial frameworks and scientific explorations of the peopling of India, Harvard geneticist David Reich, along with his Indian collaborators published a very influential paper in 2009, which invented two new population labels—Ancestral North Indians (ANI), and Ancestral South Indians (ASI) (Reich et al. 2009). As Barbosa argues (see this volume), these new categorical denominations not only nationalized ancestry but also sought to conceal questions surrounding the foreignness of ancestry, caste inequalities, and religious nationalism. Nevertheless, most models for racial/ethnic classification of the people of India continue to utilize caste as the unit for denoting differences and similarities. In the following sections, I trace and show how the GIP is not disjointed from such historical notions of human variation.

Situating the GenomeIndia Project (GIP)

In the late 1980s, with the availability of commercial DNA sequencers which could facilitate sequencing of the complete genome, the epistemic frame in life sciences witnessed a shift. It was against this backdrop that the Human Genome Project was launched in

1990. With the promise of discovering the "blueprint of life," transforming medicine, and curing cancer, the HGP became a vehicle and marker of exponential growth in genomics. Nonetheless, as Reardon points out, researchers were soon faced with the question: what does genomic data mean? She posits, "this turn to the question of meaning—the question of the uses, significance, and value of the human genome sequence" as the "postgenomic condition" (Reardon 2017, 2). A recent editorial in the leading Indian newspaper echoes the same question about the GIP:

> When the Human Genome Project published its reference "human genome" in 2003 ... it rang with a "brave-new-world" promise of ... mapping every awry gene to a disease and a future of "personalized medicine." Much of Genome India's sales pitch reflects similar promises. However, the subsequent decades have tempered such expectations ... In other words, genome sequencing only opened up new realms of complexity. (Hindu 2024)

But despite the uncertainty and complexity involved, over the last two decades, India has extended public funding to build genomic infrastructure, and the GIP is the culmination of these efforts. While India lacked resources in the 1990s to partner with big science initiatives such as the HGP, it decided not to participate in the International Haplotype Map Project (HapMap) and instead initiated the Indian Genome Variation (IGV) Project in 2003. The Human Genome Diversity Project (HGDP), precursor to the HapMap project, which intended to survey and catalog human genetic diversity, faced public censure in the early 1990s. The criticism of the HGDP as a "vampire project" practicing "biocolonialism" (Burton 2021, 244) had not only alerted countries in the Global South to how genomic resources might be exploited and capitalized by the powerful Global North but also conveyed the political import of safeguarding nations' shared genomes. Already by November 1997, the Indian Council of Medical Research had issued the "Guidelines for Exchange of Human Biological Material for Biomedical Research Purposes," mandating government permission for shipping outside the country any human tissue samples that were collected in India (Hardy et al. 2008).

The IGV project was implemented through six laboratories of the Council of Scientific and Industrial Research (CSIR) in collaboration with the Indian Statistical Institute Kolkata and the Anthropological Survey of India (AnSI). The IGV project was granted USD 5 million to develop markers for predictive medicine and "used SNP [i.e., single nucleotide polymorphism]- based genotyping of 900 genes from over 1800 individuals across fifty-five subpopulations to underscore the heterogeneity of the Indian population" (Jain et al. 2020). Though the IGV project did not perform whole genome sequencing (WGS), in 2009, India announced that a team of scientists at the CSIR Institute of Genomic and Integrative Biology Delhi had completed the entire human genome sequencing of a healthy Indian male (PIB 2009). As Subramaniam has suggested, India thus sought to assert "genomic sovereignty" through practices of bionationalism "that seeks to secure and control their national genomics" (Subramaniam 2019, 165).

However, the commencement of a large-scale study of human variation utilizing WGS happened only around 2017. The Centre for Brain Research (CBR) Bengaluru, which is also spearheading the GIP, inaugurated the Genome India initiative (GII) in collaboration with the Institute of Bioresources and Sustainable Development (IBSD) Imphal. The GII was a precursor to the GIP but with a focus on a specific region of India (and particularly tribal populations). The GII intended to carry out WGS of 2000 individuals from the northeastern states of India to understand "the genetic disease burden which would help in the development of personalized medicine" (IANS 2017). Soon, after a hiatus of a decade, the CSIR also launched the IndiGen Program in April 2019 with the motto—"Genomics for public health in India." Within a span of six months, the IndiGen consortium announced the completion of whole genome sequencing of 1029 healthy Indians and emphasized "the need for an India centric population genomic initiative" (PIB 2020b). Consequently, in early 2020, the Department of Biotechnology (DBT) launched the GIP.

The ambitious "national project" (CBR 2023) since then has acquired the genetic material from a cohort of Indians to catalog genetic variations. To ensure that genetic variation in Indians is well captured, individuals chosen for genome sequencing were seemingly selected in a way so as to represent the country's diverse population. The GIP collected genetic material from individuals belonging to ninety-nine distinct populations residing in different regions of the country (PIB 2024). The CBR at the Indian Institute of Science, Bengaluru, is leading this ambitious collaborative effort of scientists from twenty research institutes across India. The CBR website lists four major aims for the GIP:

1. Create an exhaustive catalog of genetic variations (common, low frequency, rare, single nucleotide polymorphisms or SNPs and structural variations) in Indians.
2. Create a reference haplotype structure for Indians. This reference panel can be used for imputing missing genetic variation in future GWA [genome-wide association] studies.
3. Design genome wide [*sic*] arrays for research and diagnostics at an affordable cost.
4. Establish a biobank for DNA and plasma collected for future use in research. (CBR 2023)

The GIP thus seems to be an attempt at identifying and characterizing genetic differences within Indian populations and constructing a knowledge base for healthcare and medical purposes. On 27 February 2024, the GIP consortium announced that it has achieved the milestone of whole genome sequencing of 10,000 Indian individuals (PIB 2024). Accordingly, the GIP is another addition to various worldwide genomics resource creation efforts, embracing a postgenomic epistemological framework. However, the cataloging of variations and biobanking of DNA would also be a valuable resource for population genetics studies, including reconstructing human migration history. How such aggregation of genetic material will simultaneously characterize differences amongst the Indian populations and generate a reference haplotype that encodes similarities between Indians needs to be problematized. In the following section, I probe epistemological assumptions and concepts that inform the GIP to open up this debate.

Catalog of Genetic Variation or Differences

The HGP made a strong assertion about human DNA similarity, i.e., the DNA sequences of two randomly selected humans would be 99.9 percent identical. As a corollary, the epistemological base of human genetics and genomics entirely lies in that 0.1 percent fraction. Even so, this fraction translates into differences of about two to three million base pairs, as the human genome comprises more than three billion nucleotide base pairs. Hence medical genomics and population genetics research essentially relies on human (genetic) differences. The degree of genetic variation between two populations then becomes a marker of their differences or similarities. However, as Troy Duster has argued, "[i]t is possible to make arbitrary groupings of populations (geographic, linguistic, self-identified by faith, identified by others by physiognomy, etc.) and still find statistically significant allelic variations between those groupings" (Duster 2005). Nevertheless, the differences population genetics valorizes in practice often are not between any arbitrary groupings of individuals or populations but rather between discrete populations identified as relevant using prevalent norms, viz. race/caste. Consequently, epistemic assumptions that inform scientific theories and practices shape how knowledge claims are contested and legitimated and how knowledge is used and applied in multiple contexts. As Lipphardt has argued, for empirical work, researchers continue to rely on race concepts through their "sampling practices, group labels, narratives and the concept of the isolate" (Lipphardt 2014, 51). In the following paragraphs, I illustrate similar continuities within the GIP.

Since the GIP aims to construct an exhaustive catalog of genetic variations in Indians, it is vital to probe how the notion of variation or difference is operationalized. In practice, human biological diversity is conceptualized not in terms of differences between individuals but between populations. As a consequence, the science of genomics inevitably deifies social constructs (viz. ethnicity, geography, race), which is akin to presuming (pre)existence of human races or caste. The GIP is no exception. In fact, an important justification for the GIP is that the majority of genomic studies are based on individuals of European descent. The CBR website puts the rationale for the GIP as follows:

> The Indian population of 1.3 billion consists of >4,600 population groups, and several thousand of them are endogamous … Thus, the Indian population harbors distinct variations and often many disease-causing mutations are amplified within some of these groups. Therefore, findings from population based or disease based human genetics research from other populations of the world cannot be extrapolated to Indians. (CBR 2023)

Therefore, the non-European population, specifically the Indian population, is a distinctive lens for the GIP. Furthermore, the Indian population itself is conceptualized as an aggregate of more than 4,600 separate populations, which are presumed to significantly differ biologically from each other. So, two-level distinctions are operationalized—one, the Indian population is genetically distinct from the rest of the global human population; and second, multiple Indian population groups exhibit distinct gene pools. The GIP then simultaneously intends to look at the distinctiveness of genetic diversity amongst

Indian population groups while paradoxically constructing sameness across thousands of population groups. But what is the ground for the GIP to assume that there are more than 4,600 population groups in India that need to be cataloged? The GIP has adopted the categorization of the Indian population postulated by the AnSI. Between 1985 and 1992, AnSI undertook an ambitious project, "The People of India" (POI), to generate an anthropological profile of all communities living in independent India (Singh 1993). The POI project primarily utilized caste groups as a marker and basis for recording the socio-cultural diversity of India. It identified 4,635 stratified communities across India (castes, tribes, minorities, etc.), many of which practice endogamy (Joshi, Gadgil, and Patil 1993). If anthropologically defined discrete groups are the criteria, then the imagined biological Indian identity and the sameness thereof is then necessarily a mosaic of small and large population groups rather than a single large homogeneous population.

The GIP in practice is therefore ambiguously reproducing the historical/colonial practices of population classification that reinforce problematic social constructs (viz. caste, tribe). The sampling strategy and methodology applied for the collection of physical DNA offers a way to discern the assumptions. Institutes that form the GIP consortium are spread across India (twelve states and two union territories) and functioned as DNA collection centers for each region. The CBR, as the nodal center, provided guidance to these centers on which population groups (read caste/tribe) in the respective region are to be sampled and the number of samples to be drawn. A principal investigator of the GIP whom I interviewed as a part of my fieldwork added that DNA samples from ninety-nine distinct population groups were collected for the WGS. They also noted that, based on the absolute population size of the concerned community, the number of individual DNA samples collected from each group varied from 160 to 350. In general, a greater number of DNA samples were collected from communities with large absolute population sizes and vice versa.

The catalog of genetic variations in Indians, as envisaged by the GIP, is in practice deeply rooted in specific forms of sociomateriality. In particular, differences are of prime concern for the GIP, and how this delineation was operationalized in relation to human bodies and embedded DNA provides insights into underlying epistemic assumptions. In order to obtain DNA, the GIP collected blood samples of volunteering individuals from the relevant population groups. Since many participating institutes are located in urban areas, whereas most relevant population groups reside in small towns, villages, and remote hamlets, sampling posed logistic challenges. Furthermore, ethical considerations limited nonmedical academic institutes in collecting samples themselves. Therefore, some institutes partnered with doctors and medical institutes, who organized special blood sample collection camps at various places. Approximately 10 ml of blood sample was collected from each volunteer on site. Furthermore, to ensure precision, apart from unrelated individuals, samples were also collected in trios (mother-father-child) from each respective population group. According to a PI, "We collected extra samples because it is extremely difficult to go back to the community and get samples." This practice reflects both economic and social constraints. They further added that, in the process, the GIP in its first phase of the study has collected blood samples of approximately 20,000 individuals belonging to various Indian population groups, i.e., twice the number of samples it aims to process. However, only those samples that were found to be originating from "normal and healthy" individuals were sent for genome sequencing. As the CBR website notes, "The study individuals from diverse socioeconomic backgrounds undergo a plethora of anthropometric tests, blood biochemical analyses informative of complex conditions (such as diabetes, dyslipidemia, kidney function, and liver function)" (CBR 2023). The primary (population group) and secondary (blood screening tests) inclusion criteria point to continuities with colonial practices of population classification. Furthermore, as corroborated by a PI in an interview, anthropometric measurements of each participating individual were also carried out. Thus, as Ray notes, "[e]ven though the experimental system of anthropometry is dead, the experimental reasoning of race science ... continues to animate genomic research" (Ray 2018). Such epistemic practices associated with naming and grouping populations thus give rise to their own contradictions and politics (Sur and Sur 2008).

Nonetheless, the DNA of the members of discrete population groups is central to the GIP's conceptualization of human biological diversity and its catalog of variations. In an interview, a PI explained, though it was through self-reporting and on occasion through local community vetting, blood samples for the GIP were collected only from the individuals in whose family no "cross-marriage" had taken place in the last five generations, i.e., caste/tribe endogamy. Such methodological and epistemic moves are not neutral but intrinsically linked with identity practices and politics, wherein researchers frame the questions as to what constitutes a valid and legitimate sample of DNA. Nonetheless, if the individual was found to be normal and healthy through blood biochemical analyses, the collection center transferred the blood sample into multiple tubes and added a unique GIP tag. The centers used one tube to extract DNA and dispatched two sample tubes with GIP tags to one of the four sequencing centers. The sequencing center utilized these tubes to extract and store DNA and perform WGS. In the process, the GIP has constructed a new material basis and genomic infrastructure for studying human biological diversity. It has done so first through genome sequencing and construction of baseline data for Indians, with the eight petabytes of data generated by the GIP now hosted at the Indian Biological Data Centre Faridabad; second, by establishing DNA biobanks at the collection as well as four sequencing centers; and third, by means of a biobank of blood and plasma samples at the collection centers and the CBR (housing 20,000 samples).

As evident from the discussion above, the GIP sampling is not only based on typological reasoning provided by AnSI but also relies on the assumption that caste/tribe affords a rational unit for the project. Two types of rationales have been employed to justify this historical continuity—1) representing the country's diverse population, and 2) the need to identify disease-causing mutations resulting from prevalent endogamy for medical purposes. It must be noted that India is home to about 17 percent of the world's population, and as the POI project indicated, there is a remarkable resilience of regional identities (Singh 1993). Furthermore, disease-causing mutations could also be detected using regional sampling in combination with an effective disease surveillance network. Nevertheless, the GIP seems to assert that not every difference counts the same. While some differences are valued as significant, others are judged as trivial and to be ignored safely. The ways of ordering and classifying populations employed by the GIP are thus not entirely divorced from historical practices. As a consequence, the GIP would contribute to the naturalization and biological essentialization of caste/tribe constructs (Egorova

2010). In effect, the GIP is essentially an exercise of difference (re)production—not just vis-à-vis global populations but also between people inhabiting India—which relies on taken-for-granted epistemologies and reinforces problematic social constructs. In the following section, I focus on the notion of similarities and the ways in which the GIP intends to stitch together the category of "Indian" through the imaginary of a reference haplotype.

Stitching Together Indian Reference Haplotype

One of the key motivations and assumptions guiding the GIP is that the Indian population has a distinct genetic makeup compared to other populations of the world. Yet, notwithstanding the myriad differences, the project emphasizes unity in diversity through the idea of a single Indian population demarcated by geographical and nation-state boundaries. The ambition of the GIP is to stitch together all such distinct variations and rich genetic diversity of Indian population groups under one single ancestral group. Apropos, one explicitly stated goal of the GIP is to "create a reference haplotype structure for Indians," with the aim to employ the resulting reference panel "for imputing missing genetic variation in future GWA studies" (CBR 2023). I explore below close linkages between concepts of reference haplotype and descent, the production of similarities thereof, and the embedded politics of nationalism.

A haplotype refers to a single or a set of markers (polymorphisms) present along a single chromosome that tend to be conserved or inherited together. Such specific DNA variations at different locations on a single chromosome provide a genetic signature. In practice, such DNA signatures are used to understand disease predispositions of population groups and to trace migration history and evolutionary patterns. International HapMap Project, which constructed a reference panel of 420 haplotypes, established this approach to throw light on human genetic variation (IHC 2003). A reference panel is constituted by genetic sequences of individuals that belong or are attributed to the same ancestral group or collective (read geography, ethnicity, caste, etc.). A reference data or panel made up of individuals sharing genetic ancestry can be used as a baseline for conducting genome-wide association studies to statistically predict variants associated with a specific trait and imputing missing genetic variation. The governing idea here being, "[i]n samples of unrelated individuals, the haplotypes of the individuals over short stretches of sequence will be related to each other by being identical by descent (IBD)" (Marchini and Howie 2010, 500). Thus, notions of similarity such as descent, ancestry, and kinship are closely tied to the creation of a reference haplotype. Furthermore, while the GIP is invested with (re)production of difference amongst Indian population groups, as discussed in the previous section, a reference haplotype paradoxically aims at both the production of similarity within the Indian population and the simultaneous construction of difference between Indian and non-Indian populations. The GIP is thus a Janus-faced enterprise, where scientific knowledge-making is shaped by population labels that both construct and subsume differences, simultaneously invoking different scientific meanings and practices.

Nonetheless, what might it mean to construct a reference haplotype that represents the genetic diversity of the Indian population? The GIP has obtained DNA samples from ninety-nine discrete population groups, which is inadequate to truly represent genetic diversity. The GIP researchers are currently engaged in genotyping and comparing each individual genome with other genomes from the same population groups, as well as comparing genomes of one population group with other population groups. Such comparative analysis of genomic variation within and between sampled Indian population groups will lead to a collection of SNPs and haplotypes, which would be used to construct a reference haplotype structure for Indians. During a field interview, a PI emphasized that the biggest challenge the GIP researchers are facing is that of genome alignment and stitching together genetic information. Preliminary findings released by the GIP, denoting differential prevalence of variations across Indian and other world populations, suggest that 69.02 percent of genetic variations have less than 5 percent differences in prevalence (CBR 2023). On the other hand, approximately 10 percent of genetic variations are unique to the Indian population and not found in other populations (ibid.). But are those variations uniformly observed in all population groups in India? Drawing attention to the GIP data, a PI commented that there is no singular Indian genome but rather multiple genomes, and in particular, northeastern region of the country exhibits the most diverse genetic makeup. Therefore, in practice, multiple contradictions challenge the assumptions of similarity made by the GIP researchers.

Even so, how can we make sense of the politics of stitching together a reference haplotype for Indians? The adjective Indian is an operational trope in this context. By naming the proposed baseline haplotype as Indian, i.e., on a national basis, the GIP is attempting to formulate a form of homogeneous biological identity. Such aggregation is necessary for the making of what Sung terms as "bionation," which engages in the politics of resource-making—be it through people's bodies or information coded in their genes (Sung 2010). The imaginary of Indian reference haplotype then is a way to reinscribe India as a nation on biological terms, distinct from other global populations. This however requires establishing "genomic sovereignty" through the formation and assertion of a genetically unique and homogeneous identity of the Indian population. As Warwick Anderson has suggested, "the clinic and laboratory should be added to those sites where the nation—any nation—may be imagined" (Anderson 2006). At another level, this aggregation of genetic diversity by subsuming differences could also be viewed as a move to create a secular biological identity detached from traditional caste/tribe markers. The reference haplotype might thus occasion a partial collapse of two contemporary population labels, i.e., ANI and ASI. Nevertheless, as Barbosa (see this volume) suggests, it needs to be read as (de)politicization of diffuse caste politics. Furthermore, the GIP is also a means for the state to govern its populations through the promise of science-based healthcare for its citizens, viz., rare diseases (Guardian, Sivasubbu, and Scaria 2019). Yet, as discussed above, Indian reference haplotype draws on the notion of descent. In practice, for mapping disease-causing variations onto an individual's genomic data, ethnic information would play a key role because of the defining assumptions and epistemic conceptualizations of the GIP. As Burton has suggested, these continuities reiterate that "nationalism and human genetics are so thoroughly co-constituted, neither better technology nor better intentions are sufficient ... to produce politically neutral data about ancestry" (Burton

2021, 248). The production of similarities through the trope of a reference haplotype and shifting focus from population genetics to medical genomics thus does not fully dissociate science epistemologies from the politics of nationalism.

Conclusions

The postgenomic era in the new millennium has created an impetus within the life sciences to study human biological diversity using DNA information. But as Burton suggests, "[a]lthough the capabilities of DNA sequencing far exceed older technologies [...] geneticists repeat the process of reifying religious, linguistic, and social differences into biologically detectable ethnic or racial groups" (Burton 2021, 260). In this article, I examined the epistemological assumptions and concepts that inform the GenomeIndia Project and their entanglements with caste and identity politics. The GIP aims to create a catalog of genetic variations in Indians through the whole genome sequencing of 10,000 individuals and construct a reference haplotype structure for Indians. By focusing on empirical research practices employed for sampling and labeling groups, I have highlighted how the GIP relies on taken for granted epistemologies for characterizing difference, which reinforces social constructs of caste/tribe. To be sure, the GIP does promise to remove the personal identifiers of the participating individuals. Although data in its entirety may not be made public, in order to create and maintain a catalog of genetic variations in Indians as well as DNA and blood biobank of samples, caste/tribe labels that trace the provenance of data have been preserved by all the GIP centers. The difficulties in unmooring identities are well reflected in the statement of the DBT secretary, who admitted that "[t]his hidden treasure of genomes also comes with a hidden privacy concern that the Department is striving to address ethically" (Mishra 2024). While there are disagreements amongst researchers on the relationship between the concept of race and caste (Natarajan and Greenough 2009), caste differentiation and identity discourse linked with the question of race and migration continue to remain a politically vexed subject.

Another aim of the GIP is to construct a reference haplotype structure for Indians, with a vision to facilitate the identification of disease-causing mutations. Yet by naming the baseline haplotype as Indian and using a nonbiological criterion in the production of similarities, the GIP seeks to forge a homogeneous biological identity and nationalize ancestry. I suggest that notwithstanding the uncertainty and complexity posited by the postgenomic condition, the GIP must also be understood as a genomic infrastructure and means to assert "genomic sovereignty." Through accumulating a baseline genomic digital database, establishing DNA biobanks, and setting up a blood and plasma biobank, the GIP has constructed a new material basis for studying human biological diversity. Practices of bionationalism steer such aggregation and enable the making of a "bionation." Consequently, human genetics and nationalism are coconstituted. I posit that while population genetics and medical genomics continue to be guided by the normative logic of genetic nationalism, in the case of India, the logic remains closely tied to the diffuse politics of identity and caste. As such, the GIP's notions of a reference haplotype or catalog of genetic diversity cannot fully dissociate science epistemologies from politics.

References

Anderson, Warwick. 2006. The Cultivation of Whiteness: Science, Health, and Racial Destiny in Australia. Durham, NC: Duke University Press.

Bamshad, Michael, Toomas Kivisild, W. Scott Watkins, et al. 2001. "Genetic Evidence on the Origins of Indian Caste Populations." *Genome Research* 11 (6): 994–1004. DOI: https://doi.org/10.1101/gr.173301.

Bates, Crispin. 1995. "Race, Caste and Tribe in Central India: The Early Origins of Anthropometry." In *The Concept of Race in South Asia*, edited by Peter Robb, 219–259. Delhi: Oxford University Press.

Bhagat, Ram B. 2006. "Census and Caste Enumeration: British Legacy and Contemporary Practice in India." *Genus* 62 (2): 119–134.

Burton, Elise K. 2021. *Genetic Crossroads: The Middle East and the Science of Human Heredity.* Stanford, CA: Stanford University Press.

Centre for Brain Research. 2023. GenomeIndia. Accessed March 27, 2024. https://www.cbr.iisc.ac.in/research/flagship-projects/genomeindia.

Duster, Troy. 2005. "Race and Reification in Science." *Science* 307 (5712): 1050–1051. DOI: https://doi.org/10.1126/science.1110303.

Egorova, Yulia. 2010. "Castes of Genes? Representing Human Genetic Diversity in India." *Genomics, Society and Policy* 6 (3): 32–49. DOI: https://doi.org/10.1186/1746-5354-6-3-32.

Figueira, Dorothy M. 2015. *Aryans, Jews, Brahmins: Theorizing Authority through Myths of Identity.* New Delhi: Navayana.

The GUaRDIAN Consortium, Sridhar Sivasubbu, and Vinod Scaria. 2019. "Genomics of rare genetic diseases—experiences from India." *Human Genomics* 13. DOI: https://doi.org/10.1186/s40246-019-0215-5.

Hardy, Billie-Jo, Béatrice Séguin, Peter A. Singer, Mitali Mukerji, Samir K. Brahmachari, and Abdallah S. Daar. 2008. "From Diversity to Delivery: The Case of the Indian Genome Variation Initiative." *Nature Reviews Genetics* 9 (S1): S09–S14. DOI: https://doi.org/10.1038/nrg2440.

The Hindu. 2024. Decoding the Script: On the Genome India Project and Its Sequencing 10,000 Indian Genomes. Accessed March 27, 2024. https://www.thehindu.com/opinion/editorial/decoding-the-script-on-the-genome-india-project-and-its-sequencing-10000-indian-genomes/article67899979.ece.

Indio-Asian News Service. 2017. Genome India Project Launched in Northeast. Accessed March 27, 2024. https://economictimes.indiatimes.com/news/science/genome-india-project-launched-in-northeast/articleshow/61203159.cms?from=mdr.

The International HapMap Consortium. 2003. "The International HapMap Project." *Nature* 426 (6968): 789–796. DOI: https://doi.org/10.1038/nature02168.

Jain, Abhinav, Rahul C. Bhoyar, Kavita Pandhare, et al. 2021. "IndiGenomes: A Comprehensive Resource of Genetic Variants from over 1000 Indian Genomes." *Nucleic Acids Research* 49 (D1): D1225–D1232. DOI: https://doi.org/10.1093/nar/gkaa923.

Joseph, Tony. 2021. *Early Indians: The Story of Our Ancestors and Where We Came From.* New Delhi: Juggernaut.

Joshi, N. V., Madhav Gadgil, and Suresh Patil. 1993. "Exploring Cultural Diversity of the People of India." *Current Science* 64 (1): 10–17. https://www.jstor.org/stable/24095541.

Lipphardt, Veronika. 2014. "'Geographical Distribution Patterns of Various Genes': Genetic Studies of Human Variation after 1945." *Studies in History and Philosophy of Science Part C: Studies in History and Philosophy of Biological and Biomedical Sciences* 47 (A): 50–61. DOI: https://doi.org/10.1016/j.shpsc.2014.05.006.

Malhotra, K. C., and T. S. Vasulu. 2019. "Development of Typological Classification and Its Relationship to Microdifferentiation in Ethnic India." *Journal of Biosciences* 44 (3), article no. 64 DOI: https://doi.org/10.1007/s12038-019-9880-8.

Marchini, Jonathan, and Bryan Howie. 2010. "Genotype Imputation for Genome-Wide Association Studies." *Nature Reviews Genetics* 11 (7): 499–511. DOI: https://doi.org/10.1038/nrg2796.

Mishra, Akanksha. 2024. "India Largest Genetic Lab in the World": What Completion of India Genome Project Means. Accessed March 27, 2024. https://theprint.in/health/india-largest-genetic-lab-in-the-world-what-completion-of-india-genome-project-means/1982036/.

Mukharji, Projit Bihari. 2014. "From Serosocial to Sanguinary Identities: Caste, Transnational Race Science and the Shifting Metonymies of Blood Group B, India c. 1918–1960." *The Indian Economic and Social History Review* 51 (2): 143–176. DOI: https://doi.org/10.1177/0019464614525711.

Natarajan, Balmurli, and Paul Greenough, eds. 2009. *Against Stigma: Studies in Caste, Race and Justice since Durban.* New Delhi: Orient BlackSwan.

Papiha, S. S. 1996. "Genetic Variation in India." *Human Biology* 68 (5): 607–628. https://www.jstor.org/stable/41465511.

Philip, Kavita. 2004. *Civilizing Natures: Race, Resources, and Modernity in Colonial South India.* New Brunswick, NJ: Rutgers University Press.

Press Information Bureau, Government of India. 2009. CSIR Completes First Ever Human Genome Sequencing in India. Accessed March 27, 2024. https://pib.gov.in/newsite/erelcontent.aspx?relid=55470.

Press Information Bureau, Government of India. 2020a. GenomeIndia: Cataloguing the Genetic Variation in Indians' Project. Accessed March 27, 2024. https://pib.gov.in/PressReleseDetailm.aspx?PRID=1605509.

Press Information Bureau, Government of India. 2020b. CSIR IndiGenome Resource of 1029 Indian Genomes Provides a Compendium of Genetic Variants Representing the Contemporary Indian Population. Accessed March 27, 2024. https://pib.gov.in/PressReleaseIframePage.aspx?PRID=1667949.

Press Information Bureau, Government of India. 2024. GenomeIndia. Accessed March 27, 2024. https://www.pib.gov.in/PressReleaseIframePage.aspx?PRID=2009505.

Ray, Subhadeepta. 2018. "Studying Laboratories." In *Towards a New Sociology*, edited by Mahuya Bandyopadhyay and Ritambhara Hebbar, 141–176. New Delhi: Orient Blackswan.

Reardon, Jenny. 2005. *Race to the Finish: Identity and Governance in an Age of Genomics*. Princeton: Princeton University Press.

Reardon, Jenny. 2017. *The Postgenomic Condition: Ethics, Justice, and Knowledge after the Genome*. Chicago: University of Chicago Press.

Reich, David, Kumarasamy Thangaraj, Nick Patterson, Alkes L. Price, and Lalji Singh. 2009. "Reconstructing Indian Population History." *Nature* 461 (7263): 489–494. DOI: https://doi.org/10.1038/nature08365.

Singh, K. S. 1993. "People of India: The Profile of a National Project (1985–92)." *Current Science* 64 (1): 5–10. https://www.jstor.org/stable/24095540.

Subramaniam, Banu. 2019. *Holy Science: The Biopolitics of Hindu Nationalism*. Seattle: University of Washington Press.

Sung, Wen-Ching. 2010. "Chinese DNA: Genomics and Bionation." In *Asian Biotech: Ethics and Communities of Fate*, edited by Aihwa Ong and Nancy N. Chen, 263–292. Durham, NC: Duke University Press.

Sur, Abha, and Samir Sur. 2008. "In Contradiction Lies the Hope: Human Genome and Identity Politics." In *Tactical Biopolitics: Art, Activism, and Technoscience*, edited by Beatriz da Costa and Kavita Philip, 205–218. Cambridge, MA: MIT Press.

Appendix

About the Authors

Isabelle Bartram is a molecular biologist and works at the Institute of Sociology of the University of Freiburg, Germany. She is interested in interdisciplinary perspectives on biomedical research and questions surrounding responsible use and governance of human genetics.

Anna Bredström is a senior associate professor at the Institute for Research on Migration, Ethnicity and Society (REMESO) at Linköping University, Sweden. Her research focuses on bodies, health, and medicine, and she is particularly interested in questions of how race, culture, and ethnicity are constructed in medical research and health practice.

Trudi Buck is an associate professor of biological anthropology at Durham University, England. Her work focuses on the epistemology of biological and forensic anthropological methods for estimating the biological profile of a human skeleton, particularly in relation to pedagogical practices.

Claude-Olivier Doron is a professor in history and philosophy of science at Université Paris Cité. Trained in history, philosophy, and anthropology of health, his work initially focused on the intersecting history of the notions of race and degeneration in the eighteenth and nineteenth centuries. He has also worked extensively on the history and current relations between psychiatry and justice and now works primarily on a long-term political and epistemological history of the various concepts of "race" and, more specifically, on the relationship between liberalism and race in the nineteenth century, as well as on the redefinitions of the notion of race in population genetics and genomics during the twentieth century.

Yulia Egorova is a professor of anthropology at Durham University, England. Her work focuses on the study of minority identities, specifically in relation to constructions of difference in public discourses about race and religion and in the context of science and biotechnology.

Nils Ellebrecht is a sociologist at the University of Freiburg and the University Hospital Freiburg in Germany. His work focuses on the sociology of organization, technology, security, and medicine.

Rafaela Granja is a sociologist and researcher at the Communication and Society Research Centre (CECS), University of Minho, Portugal. Her research focuses on the technological surveillance of criminalized populations at various stages of the criminal justice system, including criminal investigations, electronic monitoring, and imprisonment.

Jaehwan Hyun is an associate professor of history of science and technology at Pusan National University in Busan, South Korea. His research spans a broad spectrum of topics in the East Asian history of science, including the history of human genetics, biological anthropology, and race in South Korea, with a particular emphasis on transnational connections between Japan, South Korea, and the United States.

Veronika Lipphardt is a professor of science and technology studies at the University College Freiburg, Germany. She has worked on the history of the life sciences in the 20th century, particularly on the history of physical anthropology and human population genetics in their political, social, and cultural contexts. Recently, her research has focused on forensic DNA analysis and population genetic studies of vulnerable populations.

Robert Meunier works in the history and epistemology of the modern life sciences. After graduating from the PhD program Foundations of the Life Sciences and their Ethical Consequences at the University of Milan, Italy, and the European School of Molecular Medicine, he conducted research at the Max Planck Institute for the History of Science and the Institute for Cultural Inquiry in Berlin, the Ludwig Maximilian University of Munich, and the London School of Economics and Political Science. Most recently, he has worked at the Institute for History of Medicine and Science Studies (Lübeck, Germany) within the Cluster of Excellence Precision Medicine in Chronic Inflammation.

Ricardo Gomes Moreira is an anthropologist at the University of Lisbon, Portugal. He has worked at the intersection of anthropology and science and technology studies, with a particular emphasis on knowledge production practices and their institutional configurations. Currently, his research focuses on the lab work of population geneticists, especially in terms of methods for creating collections or formulating conceptual frameworks, and in relation to identitarian issues and classificatory systems with sociopolitical impact.

Shai Mulinari is an associate professor and senior lecturer in the Department of Sociology at Lund University, Sweden. He undertakes research on pharmaceutical use and regulation, health and healthcare inequalities, and pharmaceutical industry practices, regulation, and transparency.

Thiago Pinto Barbosa is an anthropologist at Leipzig University in Germany. He is interested in questions of science and power, especially in relation to categorizations of difference (especially race and caste), social inequalities, coloniality, and conflicts.

Tino Plümecke is a sociologist at the University of Freiburg, Germany. His work centers on science and technology studies, theories of discrimination, and critical race studies, with a particular emphasis on the history and contemporary developments in genetics and (post)genomics.

Filipa Queirós is a sociologist and researcher at the Centre for Social Studies (CES), University of Coimbra, Portugal. Her research explores the use of biometric technologies aiming to gather information about the human face and the biogeographic ancestry of suspects for purposes of surveillance and criminal investigation.

Jenny Reardon is a professor of sociology and the founding director of the Science and Justice Research Center at the University of California, Santa Cruz. Her research draws into focus questions about identity, justice, and democracy that are often silently embedded in scientific ideas and practices, particularly in modern genomic research. She has been the recipient of fellowships and awards from, among others, the National Science Foundation, the Max Planck Institute, the Humboldt Foundation, and the Freiburg Institute for Advanced Studies.

Laura Schnieder is a sociologist who works in medical research. Currently, her research interests focus on the potentials and limits of patient participation in interdisciplinary clinical research and healthcare settings.

Mahendra Shahare is an assistant professor of science and technology studies (STS) at the Indian Institute of Technology Bombay, India. His research examines both historical and contemporary sociopolitics of science and technology in India, with emphasis on public health, life sciences, and emerging technoscientific infrastructures and their societal implications.

Nicole Sommer is a doctoral researcher in sociology at Bielefeld University. Her work examines the historical and contemporary dynamics of international human rights discourse, particularly concerning the category of "people of African descent," at the intersection of knowledge, power, and identity politics, with a special interest in processes of categorization and representation in global context.

Mihai Surdu is a visiting researcher in science and technology studies at University College Freiburg. His work examines the scientific and administrative practices that shape and stabilize populations, particularly those that are vulnerable and marginalized. He has written extensively about how state institutions and scientists construct the Roma group.

Heta Tarkkala is an academy research fellow at the University of Helsinki, Finland. She is a sociologist and STS scholar who has studied the making of biomedical knowledge and questions related to uses of health data. Currently, she is focusing on wastewater as health data flow.

Aaro Tupasela works as a senior researcher at the Faculty of Social Science, University of Helsinki. His research interests revolve around medical science and technology studies (STS). For the past fifteen years, he has explored different perspectives surrounding the biomedical collection and use of human tissue collections, as well as health-related data. Most recently, he has studied everyday practices in the development and implementation of AI systems in health care.

Andrea zur Nieden is a sociologist at the University of Freiburg, Germany. Her research operates in the field of science and technology studies resp. sociology of technology, medicine, and science, with a recent focus on human diversity in the new life sciences as part of the research group SoSciBio. She has also worked on the sociology and history of psychiatry.

13866497
9868867